**Progress in Mathematical Physics**
Volume 39

Fabrizio Colombo
Irene Sabadini
Franciscus Sommen
Daniele C. Struppa

# Analysis of Dirac Systems and Computational Algebra

Birkhäuser
Boston • Basel • Berlin

Fabrizio Colombo
Politecnico di Milano
Dipartimento di Matematica
20133 Milano
Italy

Irene Sabadini
Politecnico di Milano
Dipartimento di Matematica
20133 Milano
Italy

Franciscus Sommen
Ghent University
Faculty of Engineering
Department of Mathematical Analysis
9000 Ghent
Belgium

Daniele C. Struppa
George Mason University
Department of Mathematical Sciences
Fairfax, VA 22030
USA

AMS Subject Classifications: Primary: 30G35; Secondary: 16E05, 33N05

**Library of Congress Cataloging-in-Publication Data**
Analysis of Dirac systems and computational algebra / Fabrizio Colombo .. [et al.].
p. cm. – (Progress in mathematical physics ; v. 39)
Includes bibliographical references and index.
ISBN 978-0-8176-4255-6 (acid-free paper)
1. Mathematical physics. 2. Dirac equation. 3. Clifford algebras. 4. Differential equations, Partial. 5. Mathematical analysis. I. Colombo, Fabrizio. II. Progress in mathematical physics ; v. 39.

QC20.A53 2004
530.15'2'57–dc22 2004053657

ISBN 978-0-8176-4255-6 Printed on acid-free paper.



9 8 7 6 5 4 3 2 1 SPIN 10843874

Birkhäuser is a part of *Springer Science+Business Media*

*www.birkhauser.com*

*To Bianca and Clotilde*—FC, IS

*To my sons*—FS

*For Lisa*—DCS

# Contents

# Preface

Clifford algebras have been widely used as a mathematical tool for the description of physical phenomena. The smallest noncommutative Clifford algebra is the algebra $\mathbb{H}$ of quaternions, first introduced by Hamilton, in which it is possible to formulate several physical laws by means of some special first order differential operators. For example, if we write a quaternion $q$ by means of four real coordinates as $q = x_0 + ix_1 + jx_2 + kx_3$, we can introduce the so-called Moisil–Theodorescu operator defined by

$$\mathcal{D} = i\frac{\partial}{\partial x_1} + j\frac{\partial}{\partial x_2} + k\frac{\partial}{\partial x_3}$$

and we can apply it to quaternion valued functions of the form $f = f_0 + if_1 + jf_2 + kf_3$ to get the homogeneous equation $\mathcal{D}f = 0$. From the point of view of physical applications this approach is particularly interesting because if we write $f = f_0 + \mathbf{F}$, with $\mathbf{F} = if_1 + jf_2 + kf_3$, then

$$\mathcal{D}f = \mathcal{D}(f_0 + \mathbf{F}) = -\nabla \cdot \mathbf{F} + \nabla f_0 + \nabla \times \mathbf{F}$$

and hence any $\mathbb{H}$-valued function $f$ in the kernel of $\mathcal{D}$ gives rise to a pair $(f_0, \mathbf{F})$ satisfying the system

$$\begin{cases} \nabla \cdot \mathbf{F} = 0 \\ \nabla f_0 + \nabla \times \mathbf{F} = 0\,. \end{cases}$$

If $f_0 \equiv const$, for example $f_0 \equiv 0$, then we have that irrotational and solenoidal vector fields (i.e., the solutions of the previous system) are purely vector functions regular in the sense of the Moisil–Theodorescu operator $\mathcal{D}$. The book [111] is entirely devoted to showing how quaternions (and quaternion valued functions) are well suited to the study of an array of different physical phenomena. In the same spirit, Imaeda [92] considered functions defined on real quaternions with values in the algebra of complex quaternions, the so-called biquaternions. He then introduced a new operator $D$ whose kernel is the space of the so-called $D$-regular functions. It can be shown that a function $F : \mathbb{H} \to \mathbb{H} \otimes \mathbb{C}$, $F = a_0 + ib_0 + \mathbf{a} + i\mathbf{b}$ is $D$-regular if and only if it satisfies the system

$$\begin{cases} \partial_{x_0} a_0 - \nabla \cdot \mathbf{a} = 0 \\ \partial_{x_0} b_0 - \nabla \cdot \mathbf{b} = 0 \\ \partial_{x_0} \mathbf{a} - \nabla a_0 + \nabla \times \mathbf{b} = 0 \\ \partial_{x_0} \mathbf{b} - \nabla b_0 - \nabla \times \mathbf{a} = 0. \end{cases}$$

This system represents Maxwell's equations if the vectors $\mathbf{a}$, $\mathbf{b}$ are the magnetic and the electric field respectively; $b_0$ is related to the electric density charge and to the electric current density. The scalar $a_0$ is supposed to be constant to avoid the existence of magnetic monopoles. The notion of $D$-regularity is an interesting one because the operator $D$, at least formally, is well known to mathematicians as a simple variation of the Cauchy–Fueter operator. From this point of view, the work of Imaeda shows how a transformation of time into imaginary time can be used to derive the Maxwell equations from the Cauchy–Fueter system. The earlier (though less known) work of Lanczos introduced similar ideas (see his collected works [114]).

In the setting of biquaternions it is also possible to consider other kinds of complexified Fueter-type operators (see [198]) which, as shown in [59], describe spin 1/2 massless fields.

Another fundamental equation in physics is the Dirac equation. As is well known, the classical conservation energy law $E = p^2/2m + V$ and the quantum mechanical operators associated to energy and momentum $i\partial_t$, $-i\nabla$, respectively, give rise to the Schrödinger equation that is first order in time and second order in space. This deduction is not satisfactory, as pointed out by Dirac, since in the theory of relativity, space and time are strictly connected, therefore for the relativistic wave equation we expect first derivatives also with respect to space variables.

Considering the relativistic energy conservation law $E^2 = m^2 + p^2$ (without potential $V$) and replacing the quantum mechanical operators, one obtains a second order equation in space and time whose solutions cannot be interpreted as a probability density. To overcome this difficulty, one has to linearize the equation as $E = \sqrt{m^2 + p^2}$. This was done by Dirac by introducing suitable $4 \times 4$ matrices, called the gamma matrices, and then replacing the quantum mechanical operators obtaining the equation (in covariant form) $(i\gamma^\mu \partial_\mu - m)\psi = 0$ where $\partial_\mu = \partial_{x_\mu}$, for $\mu = 1, 2, 3$, $\partial_0 = -i\partial_{x_0}$, and $\gamma^\mu$ are the gamma matrices.

The Dirac equation is a central one because its four solutions represent the spin $\pm\frac{1}{2}$ for a particle and its anti-particle. The study of systems of quarks deals with particles having isospins. In this case, the Dirac equation does not have enough components to describe this phenomenon, and the first idea which seems useful is to increase the size of the matrices. For example, a way to extend the $(3+1)$ space-time framework to 8 dimensions and to take into account the increasing quantum numbers and internal symmetries assigned to elementary particles, is the use of octonions. The octonions form a non-associative algebra and this seems to prevent the use of a matrix translation of the eight-dimensional Dirac equation, although in [45] we have shown that using a suitable matrix product, defined on complexified octonions, it is possible to write an $8\times 8$ matrix that represents the Dirac equation. This matrix approach allows one to linearize the relativistic energy conservation law $E^2 = m^2+p^2$. This generalized Dirac equation gives rise in a natural way to the Klein–Gordon equation as in the original Dirac equation.

But a more complex situation can be considered if we move to the Clifford algebra setting, and to the study of the so-called Dirac operator in this framework. If a Clifford algebra $\mathbb{C}_n$ has basis $e_0 = 1, e_1, \dots, e_n$ with defining relations $e_ie_j + e_je_i = -2\delta_{ij}$, $i, j = 1, \dots, n$, the Dirac operator is defined as $\partial = \sum_{j=0}^{n} e_j\partial_{x_j}$ (sometimes the sum starts from $j = 1$). In the case $n = 3$ the Dirac operator has a physical meaning (see [35]) since $\mathbb{C}^4$ can be embedded into $\mathbb{C}_3$ in a standard way and we may think of $\mathbb{C}^4$ as the complex Minkowski space-time. Then the complexified Dirac operator coincides with the Weyl equation for massless fields of spin 1/2.

In this book we propose a uniform treatment of these (and related) systems. The point of view we adopt is based on the theory of algebraic analysis. Specifically, we take advantage of the relatively simple form of these systems (they are all systems of linear partial differential equations with constant coefficients) to study the corresponding modules over a suitable ring $R$ of complex polynomials.

One may in fact consider a system of linear differential equations as an $r_1 \times r_0$ matrix of differential operators $P(D)$, and if $\mathcal{S}$ is a space of (generalized) functions, one sees that $P(D)$ defines a natural map

$$P(D) : \mathcal{S}^{r_0} \to \mathcal{S}^{r_1},$$

whose kernel is the object of interest to analysts.

As has already happened in geometry, however, it soon became clear among analysts as well that the best way to treat situations of this kind was to introduce the notion of sheaf, so that we will not consider simply spaces, but rather sheaves of (generalized) functions, to which the entire algebraic theory will be applied. Partial differential operators have the important property of acting on suitable sheaves as sheaf homomorphisms, and so if $\mathcal{S}$ is such a sheaf, we now have that $P(D)$ is a sheaf homomorphism whose kernel (again a sheaf) we customarily denote by $\mathcal{S}^P$.

In what follows we will consider several different sheaves of (generalized) functions. The theory we will describe works for a large class of sheaves, for example the sheaf $\mathcal{A}$ of real analytic functions, $\mathcal{E}$ of infinitely differentiable functions, $\mathcal{O}$ of holomorphic functions, $\mathcal{D}'$ of Schwartz distributions, and finally the sheaf $\mathcal{B}$ of hyperfunctions.

By taking the Fourier transform of $P(D)$ we get the matrix $P = [P_{ij}]$, symbol of the previous system having entries in a polynomial ring $R$. The first important statement in algebraic analysis is the fact that the algebraic object we must be interested in is the $R$-module

$$M = R^{r_0}/P^t R^{r_1}$$

where $P^t$ denotes the transpose of $P$, and therefore we need to understand how to study such a module. The Hilbert Syzygy Theorem guarantees the existence of a finite free resolution of $M$.

**Hilbert Syzygy Theorem.** *There exists an integer $m$ and a finite exact resolution of the finitely generated $R$-module $M$ with free modules as follows:*

$$0 \longrightarrow R^{r_m} \xrightarrow{P^t_{m-1}} R^{r_{m-1}} \longrightarrow \dots \xrightarrow{P^t_1} R^{r_1} \xrightarrow{P^t} R^{r_0} \longrightarrow M \longrightarrow 0.$$

The maps which appear in this resolution are called the syzygies of $M$, and they can be constructed in several different ways so that the importance of the result is not the existence of a resolution, but the fact that one can find a finite resolution, as well as the fact that we have a natural bound on its length given by the number $n$ of variables in the ring $R$. It is however important to remark that such a resolution is not unique.

One can then dualize such a resolution through the use of the Hom functor (essentially we take the duals of the spaces involved, the transpose of the matrices representing the operators, and we reverse the arrows) to obtain

$$0 \longrightarrow R^{r_0} \xrightarrow{P} R^{r_1} \xrightarrow{P_1} \dots \longrightarrow R^{r_{m-1}} \xrightarrow{P_{m-1}} R^{r_m} \longrightarrow 0.$$

The complex we have just obtained is not necessarily exact, so that one can consider its cohomology (the measure of how inexact the complex is) by taking the quotients of kernels and images. The quotient groups

$$\mathrm{Ext}^j(M, R) = \frac{\ker(P_j)}{\mathrm{im}(P_{j-1})}$$

are actually $R$-modules called the Ext-modules of $M$. In those modules and in the previous complex is encoded a lot of information on the system associated to $P(D)$. Note that while the work of mathematicians like Ehrenpreis, Malgrange, Palamodov provided extremely general results about these complexes, the remaining difficulty resided in the fact that it was very difficult to compute needed algebraic objects, such as the Ext-modules, in any case of real interest. The novelty of our treatment lies in our usage of the theory of Gröbner bases to explicitly construct and compute the resolutions and the algebraic objects of interest to us. The possibility of doing so is a relatively recent advance, and we use both theoretical and computational means to gather as much information as possible about these systems.

**Advice for the reader**

The book is addressed to Ph.D students and researchers interested in (hyper)complex analysis, Clifford analysis, systems of partial differential equations with constant coefficients, and mathematical physics.

As indicated in the preface, the book requires knowledge from different fields of mathematics: commutative algebra, Gröbner bases, sheaf theory, cohomology, topological vector spaces, generalized functions (distributions and hyperfunctions). For this reason, we have collected classical basic material in the first chapter with the purpose of providing the reader (student or specialist in a field) with the necessary background that is spread over several books. We do not claim completeness since we have only inserted the material necessary for this book and we have, in general, omitted the proofs. Exceptions are made when, in our opinion, the proofs are particularly significant or when they are illuminating in the chapters to follow. In any case, we provide a list of references for the omitted parts. This book is a pioneering attempt to apply computational and algebraic techniques to hypercomplex analysis, and therefore any specialist in one of the several topics touched upon will find a lot of omitted material. Chapter 2 is the philosophical core of the book. Here, we discuss some aspects of the algebraic analysis for systems of linear partial differential equations with constant coefficients, and we show how to make such theory computationally effective. Chapter 3 is an in-depth analysis of the Cauchy–Fueter system which, historically, was the first to be studied with these methods. Chapter 4 moves to the realm of Clifford analysis, and shows how to apply these ideas to the study of complexes of Dirac operators, as well as of some of their variations. The book comes full circle in Chapter 5, where we go back to the study of some physical systems.

*Acknowledgments.* Our gratitude is expressed to our doctoral student Alberto Damiano, who has painstakingly read several earlier versions of the book, and whose questions and remarks have greatly improved its readability. We also are grateful to our colleagues and friends with whom we have developed this theory over the years, and who have shared with us their insights. A special mention goes to William (Bill) Adams, Carlos Berenstein, Jarolim Bures, Graziano Gentili, Ivan Kupka, Philippe Loustaunau, Domenico Napoletani, Victor Palamodov, Michael Shapiro, Vladimir Soucek, and Peter Van Lancker. Of crucial importance has also been the support and collaboration from the Genova group who has developed the CoCoA software which has been so instrumental for our progress. A special thanks goes to Lorenzo Robbiano, the leader of this group. We are also indebted to our three institutions which have, at different times, supported our work in many ways. We are happy to acknowledge George Mason University, the University of Ghent (in particular Richard Delanghe), FWO-Krediet aan Navorsers 1.5.106.02, and the Department of Mathematics of the Politecnico di Milano. Finally, we wish to thank the unknown reviewers for their helpful and constructive criticism, as well as our publisher who has been extremely patient with us and, in this way, has helped us come up with a better product.

*F. Colombo*
*I. Sabadini*
*F. Sommen*
*D.C. Struppa*
July, 2004

*Analysis of Dirac Systems and Computational Algebra*

# 1
# Background Material

This preliminary chapter provides some necessary background material and the most important tools to be used throughout the book. In order to make the book accessible to readers who are not experts in either the algebraic or analytic aspects, we have given the minimum of exposition of the various needed tools which are usually spread over several books. The chapter is divided into three parts: algebraic tools, analytical tools and elements of hyperfunctions theory. Under the first heading, we have collected the fundamental notions of commutative algebra that underlie both algebraic geometry and algebraic analysis, the basic notions of sheaf theory, and the foundations of the theory of Gröbner bases. Under the analytical heading, we have assembled the fundamental ideas on the space of distributions. The section on hyperfunctions provides the reader with the fundamental notions about these generalized functions. An appendix containing the basic definitions of category theory completes the chapter. Because of the purely instrumental nature of this chapter, we have made no pretense of completeness, and the reader might want to skip this material and refer to it only as the need occurs.

## 1.1 Algebraic tools

### 1.1.1 Commutative algebra

In this section we review basic ideas from commutative algebra with an emphasis on the concept of resolution for a module. Standard references are [11], [49], [50],

[70], [79]. We also will need some fundamental notions from category theory that we have collected in the appendix to this chapter for the convenience of the reader. References for this are, for example, [116], [120].

**Definition 1.1.1.** *Let $R$ be a ring with identity 1. A left $R$-module $M$ over the ring $R$ is a set $M$ with a binary operation $+: M \times M \to M$ and a map $R \times M \to M$, called scalar multiplication, such that*

1. *$M$ is an abelian group with respect to $+$,*
2. *for all $a \in R$ and for all $f, g \in M$, $a(f+g) = af + ag$,*
3. *for all $a, b \in R$ and for all $f \in M$, $(a+b)f = af + bf$,*
4. *for all $a, b \in R$ and for all $f \in M$, $a(bf) = (ab)f$,*
5. *for all $f \in M$, $1f = f$.*

**Definition 1.1.2.** *A submodule of a left $R$-module $M$ is a subset $N$ of $M$ which is closed under addition and scalar multiplication.*

Note that similar definitions can be given in the case of a right $R$-module, in which the scalar multiplication $M \times R \to M$ acts on the right.

If $M$, $N$ are $R$-modules, their direct sum $M \oplus N$ is the set of all ordered pairs $(m, n)$, $m \in M$, $n \in N$, with addition defined componentwise and with the scalar multiplication given by $r(m,n) = (rm, rn)$, $r \in R$. More generally, the direct product $\prod_{i \in I} M_i$ of a family $\{M_i\}_{i \in I}$ of $R$-modules is the set of sequences $\{x_i\}_{i \in I}$, $x_i \in M_i$, with the natural operations; if we suppose that all but finitely many $x_i$ are zero, we have the definition of the direct sum $\oplus_{i \in I} M_i$. It is obvious that the two definitions coincide when the set of indices $I$ is finite.

**Example 1.1.1.** Every abelian group is a $\mathbb{Z}$-module.

**Example 1.1.2.** The additive group of any ring $R$ with identity is trivially a module over itself. More generally, $R^m = R \oplus R \oplus \ldots \oplus R$, $m$ times, is an $R$-module.

**Example 1.1.3.** If $R$ is a (skew) field, any vector space over $R$ is an $R$-module.

**Example 1.1.4.** Let $N$ be a submodule of $M$. The set of equivalence classes of elements of $M$ under the relation $\sim$ defined by $f \sim g$ if and only if $f - g \in N$, is an $R$-module called the quotient of $M$ by $N$ and denoted by $M/N$.

**Remark 1.1.1.** The same set $M$ can be considered as an $R$-module over different rings $R$. In this book the most frequent instance of this situation occurs with the ring of quaternions $\mathbb{H}$ (for its definition see Section 3.1). The ring $\mathbb{H}$ can be considered as module over $\mathbb{Q}$, $\mathbb{R}$, $\mathbb{C}$ and finally $\mathbb{H}$.

The reader may have noticed that in the previous example $\mathbb{H}$ is not only a module but, in particular, a vector space over the field $\mathbb{Q}$, or $\mathbb{R}$, $\mathbb{C}$ and also over the skew field $\mathbb{H}$ itself. A more significant example in which the ring $R$ is not a (skew) field is the following:

**Example 1.1.5.** Let $x_1, \dots, x_n$ be $n$ indeterminates, $\partial_{x_1}, \dots, \partial_{x_n}$ the corresponding partial derivatives and let $A_n(\mathbb{C})$ be the so called Weyl algebra, namely the subring of $\mathbb{C}$-endomorphisms of $R = \mathbb{C}[x_1, \dots, x_n]$ generated by the multiplications by $x_j$ and the partial derivations $\partial_{x_j}$. It is immediate to show that $R$ is a left module over $A_n(\mathbb{C})$: the action of $x_i$ on $R$ is given by the multiplication, while $\partial_{x_i}$ acts by differentiating with respect to $x_i$.

It is important to note that most of the results in this section require commutativity, so from now on we will suppose, if not otherwise stated, that $R$ is a commutative ring with identity and that the modules are (left) $R$-modules.

**Definition 1.1.3.** *Let $M$ be an $R$-module in which a binary operation called (interior) multiplication $\cdot : M \times M \to M$ is defined. We say that $M$ is an $R$-algebra if $M$ is a ring with respect to the multiplication and, for all $a \in R$, and for all $f, g \in M$, $a(f \cdot g) = (af) \cdot g = f \cdot (ag)$.*

**Definition 1.1.4.** *A graded ring is a ring $R$ together with a direct sum decomposition $R = R_0 \oplus R_1 \oplus \dots$ as abelian groups, such that $R_i R_j \subset R_{i+j}$, $i, j \geq 0$. A graded module over a graded ring $R$ is a module $M$ with a decomposition*

$$M = \oplus_{\ell \in \mathbb{Z}} M_\ell,$$

*as abelian groups such that $R_i M_\ell \subset M_{i+\ell}$ for all $i \geq 0$, $\ell \in \mathbb{Z}$.*

**Example 1.1.6.** Let $R$ be the ring $\mathbb{C}[x_1, \dots, x_n]$ of polynomials in $n$ variables over the field $\mathbb{C}$, with the natural graduation induced by the degree, i.e., $R_j$ is the abelian group of homogeneous polynomials of degree $j$. Let $I$ be a homogeneous ideal, i.e., an ideal that can be generated by homogeneous polynomials and let $I_j = I \cap R_j$ be its homogeneous component of degree $j$. Then $I = \oplus_{j=0}^{+\infty} I_j$ and $R_k I_j \subset I_{j+k}$, which shows that $I$ is a graded module over $R$.

As for any algebraic structure, it is important to define the notion of morphism of modules :

**Definition 1.1.5.** *An $R$-module homomorphism $\phi : M \to N$ between two $R$-modules $M$, $N$ is an $R$-linear map between $M$ and $N$, which is also a group homomorphism.*

*If $M$, $N$ are two graded modules over $R$, we will say that $\phi : M \longrightarrow N$ is a graded homomorphism of degree $d$ if $\phi(M_\ell) = N_{\ell+d}$, for all $\ell \in \mathbb{Z}$.*

This definition implies in particular that the collection of modules with their homomorphisms is an example of category (see the appendix to this chapter).

Note that the kernel and the image of a homomorphism $\phi : M \to N$ turn out to be submodules of $M$ and $N$, respectively. Moreover, let us point out that for any nonempty set $N$ of an $R$-module $M$, the set

$$A_N = \{a \in R \mid ax = 0 \text{ for all } x \in N\}$$

is a left ideal in $R$, called the annihilator of $N$. If $N$ consists of a single element $x$, then $A_{\{x\}}$ is called the order ideal of $x$. If $A_{\{x\}}$ is principal, any of its generators is called the order of $x$.

**Definition 1.1.6.** *Let $x \in M$. We say that $x$ is a torsion element if its order ideal $A_{\{x\}}$ is not zero while $x$ is torsion free if its order ideal is zero.*

**Definition 1.1.7.** *If $R$ is a ring without zero divisors, the $R$-module $M$ is called torsion free if all its elements are torsion free.*

We now give the following multiple definition:

**Definition 1.1.8.** *Let $M$ be an $R$-module. A set of elements $F = \{f_i\}_{i \in I}$, $f_i \in M$ such that any element $f \in M$ can be written as $f = a_1 f_1 + \ldots + a_\ell f_\ell$, for some $a_i \in R$, $f_i \in F$ is said a system of generators for $M$.*

*An $R$-module $M$ is said to be finitely generated if there exists a finite set of elements $F = \{f_1, \ldots, f_\ell\}$, $f_i \in M$ such that any element $f \in M$ can be written as $f = a_1 f_1 + \ldots + a_\ell f_\ell$, for some $a_i \in R$.*

*A system of generators $F$ for an $R$-module $M$ is said to be a basis if every element $f \in M$ can be written in a unique way as $f = a_1 f_1 + \ldots + a_\ell f_\ell$, for some $a_i \in R$, $f_i \in F$.*

We will denote by $\langle f_1 \ldots, f_\ell \rangle$ or by $\langle F \rangle$ the submodule of the $R$-module $M$ generated by $F = \{f_1 \ldots, f_\ell\} \subset M$.

**Definition 1.1.9.** *An $R$-module $M$ is said to be free if it has a basis.*

Since it is known that all bases of a finitely generated free module have the same number of elements, the following definition is well posed:

**Definition 1.1.10.** *If $M$ is a finitely generated $R$-module with a basis containing $r$ elements, then $r$ is called the rank of $M$.*

It is important to note that in striking contrast with what happens for vector spaces, not all modules have a basis.

**Example 1.1.7.** Consider the polynomial ring $R = \mathbb{C}[x, y]$ and the $R$-module $M = \mathbb{C}[x, y]/\langle x \rangle$. Then $M$ is generated by $[1]$ as an $R$-module but $[1]$ is not a basis since $x[1] = [x] = 0$.

The definition of free modules can be stated in an alternative way, more suitable in the language of categories:

**Definition 1.1.11.** *Let $M$ be an $R$-module, let $S$ be a set and let $\alpha : S \to M$ be a map of $S$ into $M$. We say that $M$ is a free module on $S$ if for every $R$-module $M'$ and every map $\alpha' : S \to M'$ there is a unique $R$-homomorphism $\mu : M \to M'$ such that $\mu\alpha = \alpha'$.*

Modules with a basis, i.e., free modules, are actually a good example of what, in category theory, is referred to as *universal objects* in the sense of Definition 1.4.7 in the appendix. In fact, define a category whose objects are the pairs

$(M, \alpha)$ where $\alpha : S \to M$ and whose morphisms $(M, \alpha) \to (M', \alpha')$ are $R$-homomorphisms $\mu : M \to M'$ such that $\mu\alpha = \alpha'$. Then $M$ is a free module on $S$ if and only if $(M, \alpha)$ is a universal repelling object in the category.

Other key results on the theory of free modules are the following two propositions:

**Proposition 1.1.1.** *Every free $R$-module is of the form $\oplus_{i \in I} M_i$ where $M_i \cong R$ as $R$-module.*

**Proposition 1.1.2.** *Every $R$-module is an epimorphic image of a free $R$-module.*

**Example 1.1.8.** If the ring $\mathbb{H}$ is considered as $\mathbb{Q}$-module, then $\mathbb{H}$ is a free module with an infinite basis, while if it is considered as $\mathbb{R}$-module, $\mathbb{H}$ is free with a basis of four elements 1, $i$, $j$, $k = ij$; if it is considered as $\mathbb{C}$-module it is free with a basis of two elements $\{1, j\}$ (if a complex number is written as $x + iy$), and finally, as a module over itself $\mathbb{H}$ has a basis of one element $\{1\}$.

Among the possible submodules of $R^s$ there is a special one associated to the relations, also called syzygies, satisfied by a given ordered $s$-tuple of elements in an $R$-module $M$. The etymology of the word syzygy that, in the modern language, is used in astronomy to denote conjunction of planets, comes from the Greek word $\sigma\upsilon\zeta\upsilon\gamma\iota\alpha$ which means "pairing". In mathematics it indicates the solutions to systems of homogeneous linear equations over a ring in the sense of the following definition.

**Definition 1.1.12.** *Let $f = (f_1, \ldots, f_s)$ be an ordered $s$-tuple of elements in an $R$-module $M$. Any $s$-tuple $a = (a_1, \ldots, a_s) \in R^s$ such that*

$$a_1 f_1 + \ldots + a_s f_s = 0$$

*is said to be a syzygy for $(f_1, \ldots, f_s)$.*

The previous equation can also be written as $a \cdot f^t = 0$ ($f^t$ being the transpose of $f$) which explains why the word "pairing" is used for this concept. We leave to the reader the easy exercise to prove the following proposition.

**Proposition 1.1.3.** *The set $Syz(f_1, \ldots, f_s)$ of all syzygies for $f$ is a $R$-submodule of $R^s$ called the first syzygy module of $(f_1, \ldots, f_s)$.*

If we now suppose that $g_1, \ldots, g_\ell$ are generators for $\mathrm{Syz}(f_1, \ldots, f_s)$, we can repeat the reasoning and construct the set of relations $\mathrm{Syz}(g_1, \ldots, g_\ell)$, that we will call the second syzygies. Going ahead with this procedure we will construct a sequence of generators and relations which we will call a resolution of the $R$-module generated by $\{f_1, \ldots, f_s\}$. The concept of resolution is crucial in the theory of $R$-modules and will lead to the celebrated theorem of Hilbert to be introduced after some more definitions.

**Definition 1.1.13.** *Let $M$, $M'$, $M''$ be $R$-modules and let $\alpha : M \to M'$, $\beta : M' \to M''$ be homomorphisms. We say that the pair of homomorphisms*

$$M \xrightarrow{\alpha} M' \xrightarrow{\beta} M''$$

*is exact if* $\ker(\beta) = \mathrm{im}(\alpha)$. *A short exact sequence is a sequence of the type*

$$0 \longrightarrow M \xrightarrow{\alpha} M' \xrightarrow{\beta} M'' \longrightarrow 0$$

*in which any two consecutive maps are exact. (This means in particular that* $\alpha$ *is injective and* $\beta$ *is surjective).*

**Definition 1.1.14.** *A (descending) complex of R-modules is a sequence* $\mathcal{F} = \{F_i\}$ *of modules and maps* $\{\phi_i : F_i \to F_{i-1}\}$ *such that their compositions* $\phi_i\phi_{i+1}$ *are all zero. The quotient module*

$$H_i(\mathcal{F}) = \ker\, \phi_i / \mathrm{im}\, \phi_{i+1}$$

*is called homology of* $\mathcal{F}$ *at* $F_i$.

*Analogously, an (ascending) complex of R-modules is a sequence* $\mathcal{F} = \{F^i\}$ *of modules and maps* $\{\psi_i : F^{i-1} \to F^i\}$ *such that their compositions* $\psi_{i+1}\psi_i$ *are all zero. The quotient module*

$$H^i(\mathcal{F}) = \ker\, \psi_{i+1} / \mathrm{im}\, \psi_i$$

*is called cohomology of* $\mathcal{F}$ *at* $F^i$.

**Definition 1.1.15.** *A free resolution of an R-module M is a complex*

$$\mathcal{F} : \cdots \to F_n \xrightarrow{\phi_n} \cdots \xrightarrow{\phi_2} F_1 \xrightarrow{\phi_1} F_0$$

*of free R-modules such that* $\mathrm{coker}\ (\phi_1) = M$ *and* $\mathcal{F}$ *is exact, i.e.,* $\mathrm{im}\ \phi_{i+1} = \ker\ \phi_i$, *for all* $i \geq 1$. *We shall often write more explicitly a resolution of M as*

$$\cdots \longrightarrow F_n \xrightarrow{\phi_n} \cdots \longrightarrow F_1 \xrightarrow{\phi_1} F_0 \longrightarrow M \to 0.$$

Consider now a graded $R$-module $M$ and its $d$-twist, i.e., the module $M(d)$ isomorphic to $M$ with grading $M(d)_\ell = M_{\ell+d}$. The typical twisted modules we will consider are of the form

$$R(-d_1) \oplus \ldots \oplus R(-d_m)$$

where $R$ is a ring of polynomial with complex coefficients. The standard basis elements have degrees $d_1, \ldots, d_m$; in fact $R(-d_j)_{d_j} = R_0$. Note also that if an $R$-module $M$ is generated by homogeneous elements of degrees $d_1, \ldots, d_m$, we get in a natural way a map

$$R(-d_1) \oplus \ldots \oplus R(-d_m) \longrightarrow M,$$

which by construction has degree zero. In a similar manner, one can construct a graded homomorphism of degree zero:

$$R(-d_1) \oplus \ldots \oplus R(-d_m) \longrightarrow R(-b_1) \oplus \ldots \oplus R(-b_s)$$

by means of an $s \times m$ matrix whose entry at the place $ij$ is a homogeneous polynomial of degree $d_j - b_i$ for all $i$, $j$.

**Definition 1.1.16.** *Let $M$ be a graded $R$-module over a graded ring $R$. A resolution $\mathcal{F}$ of $M$ is called a graded free resolution if the $F_i$ are graded free modules, and the maps are homogeneous maps of degree 0. If for some $n < \infty$ we have $F_{n+1} = 0$, but $F_i \neq 0$ for $0 \leq i \leq n$, then $\mathcal{F}$ is said to be a finite resolution of length $n$.*

Every module has a free resolution and, if $R$ is graded, then every graded module has a graded free resolution. The construction of a free resolution can be carried out as follows: let $M$ be a module with a set $\{f_i\}$ of generators, and let $\{e_i\}$ be a corresponding set of generators for a free module $F_0$. Then the map $\phi_0 : e_i \rightarrow f_i$ shows that $M$ can be represented as an epimorphic image of the free module $F_0$. However to fully describe $M$ we also need to know the kernel of $\phi_0$ which is a submodule of $F_0$ and which can be represented as the epimorphic image of some other free module $F_1$ through a map $\phi_1 : F_1 \rightarrow F_0$ whose image is the kernel of $\phi_0$. This process can now be repeated by representing the kernel of $\phi_1$ as the epimorphic image of a free module $F_2$. By continuing in this way we obtain a (possibly infinite) free resolution for $M$. In some particular cases it is possible to construct a resolution by hand (see Example 2.1.1), although, in general, the construction is not trivial at all, even when it is known that the resolution is finite.

The most important result, in this context, is the classical:

**Theorem 1.1.1. Hilbert Syzygy Theorem.** *If $\mathbb{K}$ is a field and $R = \mathbb{K}[x_1, \ldots, x_n]$, then every finitely generated graded $R$-module has a finite graded free resolution of length $\leq n$, in which all modules are finitely generated free modules.*

Note that in most of our applications the field $\mathbb{K}$ will be either $\mathbb{Q}$ or $\mathbb{R}$ or $\mathbb{C}$. Several proofs are available for this theorem. The first was given by Hilbert in 1890 [87] and uses the classical methods available at that time. A more modern treatment is due to Cartan and Eilenberg and is based on methods from homological algebra [37]. The recent introduction of the theory of Gröbner bases allows also for a constructive proof which can be found for example in [70].

A very important property of a graded resolution of a graded module $M$ is its minimality: the graded resolution

$$\cdots \longrightarrow F_n \xrightarrow{\phi_n} \cdots \longrightarrow F_1 \xrightarrow{\phi_1} F_0 \longrightarrow M \rightarrow 0$$

is called minimal if for any $j > 0$ the nonzero entries of the matrix of $\phi_j$ have positive degree. Minimality ensures uniqueness up to isomorphisms.

**Proposition 1.1.4.** *Any two minimal graded resolutions*

$$\cdots \longrightarrow F_1 \xrightarrow{\phi_1} F_0 \longrightarrow M \longrightarrow 0$$

*and*

$$\cdots \longrightarrow G_1 \xrightarrow{\varphi_1} G_0 \longrightarrow M \longrightarrow 0$$

*of $M$ are isomorphic, i.e., there are graded isomorphisms $\alpha_j : F_j \longrightarrow G_j$ such that $\alpha_{j-1}\phi_j = \varphi_j\alpha_j$, for all $j \geq 1$.*

We now turn our attention to $R$-modules homomorphisms:

**Definition 1.1.17.** *If $M$, $N$ are $R$-modules, we denote by $\mathrm{Hom}_R(M,N)$ the abelian group of all $R$-homomorphisms from $M$ to $N$.*

It is possible to prove that $\mathrm{Hom}_R(M,N)$ is an $R$-module. Moreover, for every module $M$, $\mathrm{Hom}_R(M,-)$ is a covariant left exact functor from the category of $R$-modules to itself in the sense of Definition 1.4.6.

**Remark 1.1.2.** Note that $\mathrm{Hom}_R(M,-)$ is not right exact as it does not preserve surjectivity. It can be verified that $\mathrm{Hom}_R(-,M)$ is a contravariant right exact functor but not a left exact one.

The next notions we need to introduce are necessary to understand in which sense the covariant functor $\mathrm{Hom}_R(M,-)$ and the contravariant functor $\mathrm{Hom}_R(-,M)$ fail to be right (resp. left) exact. To do that, we need to define the concept of derived functor and we begin by defining the notions of injective and projective modules and resolutions. Projective modules are a class of $R$-modules that are almost as simple as free modules. They are defined as follows:

**Definition 1.1.18.** *An $R$-module $P$ is said to be projective if for every surjective homomorphism of $R$-modules $\alpha : M \to N$ and every homomorphism of $R$-modules $\beta : P \to N$ there is an $R$-homomorphism $\gamma : P \to M$ such that $\beta = \alpha\gamma$.*

**Example 1.1.9.** The simplest examples of projective $\mathbb{Z}$-modules are the free modules $\mathbb{Z}^n$.

A useful characterization of the projective modules is the following.

**Proposition 1.1.5.** *An $R$-module is projective if and only if it is a direct summand of a free $R$-module.*

**Remark 1.1.3.** All free modules are projective by the previous theorem; however not all projective modules are free (see for example [70], Appendix A3).

Another class of modules, not as natural as free or projective modules since they are often not finitely generated, is that of injective modules.

**Definition 1.1.19.** *An $R$-module $Q$ is said to be injective if for every injective homomorphism of $R$-modules $\alpha : M \to N$ and every homomorphism of $R$-modules $\beta : M \to Q$ there is an $R$-homomorphism $\gamma : N \to Q$ such that $\beta = \gamma\alpha$.*

Once again, this is a universal property in the sense of the appendix to this chapter.

**Example 1.1.10.** The abelian group $\mathbb{Z}_p$, as a $\mathbb{Z}$-module, is injective if and only if $p$ is prime.

**Example 1.1.11.** Let $S$ be a commutative ring with identity, $R$ an $S$-algebra, and let $M$ be an injective $S$-module. Then $\mathrm{Hom}_S(R, M)$ is an injective $R$-module.

**Definition 1.1.20.** *Let $M$ be an $R$-module. A sequence of $R$-modules*

$$\mathcal{I} : I_0 \stackrel{\phi_1}{\to} I_1 \to I_2 \to \ldots \to I_n \to \ldots$$

*is said to be an injective resolution of $M$ if all the $I_i$ are injective modules,* $\ker \phi_1 = M$*, and $\mathcal{I}$ is exact except at $I_0$ . An exact sequence*

$$\mathcal{P} : \ldots P_n \to \ldots \to P_2 \to P_1 \stackrel{\psi_1}{\to} P_0$$

*is said to be a projective resolution of $M$ if all the $P_i$ are projective modules,* $\mathrm{coker}\,(\psi_1) = M$ *and $\mathcal{P}$ is exact except at $P_0$. We define the projective dimension $pd(M)$ of $M$ as the minimum length of all projective resolutions of $M$. If $M$ has no projective resolutions of finite length, we set $pd(M) = \infty$.*

**Remark 1.1.4.** Using a slight abuse of notation, we will often write an injective resolution of $M$ as follows:

$$0 \to M \to I_0 \stackrel{\phi_1}{\to} I_1 \to I_2 \to \ldots \to I_n \to \ldots$$

and a projective resolution of $M$ as

$$\ldots P_n \to \ldots \to P_2 \to P_1 \stackrel{\psi_1}{\to} P_0 \to M \to 0.$$

Since projective and injective resolutions play a central role in the construction of some fundamental sequences in algebraic analysis, we show how such resolutions can be constructed. This amounts to showing how to write an injective resolution for an $R$-module $M$, since projective resolutions are constructed using the same procedure used for free resolutions by means of modules $F_i$ that are merely projective. For this purpose, we recall the following proposition (see, for example [70]):

**Proposition 1.1.6.** *For every $R$-module $M$ there is an injective homomorphism $i : M \to Q$ with $Q$ injective.*

It is now clear how to use Proposition 1.1.6 to construct (step-by-step) an injective resolution for any $R$-module $M$: starting from $M$, we construct a monomorphism $M \to I_0$, then we find a monomorphism embedding the cokernel $I_0/M$ in an injective module $I_1$, and continuing in this way we get the desired injective resolution

$$0 \longrightarrow M \longrightarrow I_0 \longrightarrow I_1 \longrightarrow \ldots .$$

We are now ready to define the notion of derived functor.

**Definition 1.1.21.** *Let $M$ be an $R$-module and let $F$ be an additive covariant left exact functor with values in the category of abelian groups. Let $\mathcal{I}$ be an injective resolution for $M$*

$$\mathcal{I} : 0 \to M \to I_0 \to I_1 \to I_2 \to \dots$$

*and consider the sequence*

$$F(\mathcal{I}) : \ 0 \to F(M) \to F(I_0) \to F(I_1) \to F(I_2) \to \dots .$$

*The (right) $n$-derived functor $R^n F$ is defined as the $n$-th cohomology of the complex $F(\mathcal{I})$. It is well defined since it does not depend on the injective resolution of $M$.*

**Definition 1.1.22.** *Let $M$ be an $R$-module and let $F$ an additive controvariant right exact functor with values in the category of abelian groups. Let $\mathcal{P}$ be a projective resolution for $M$*

$$\mathcal{P} :\to \dots \to P_2 \to P_1 \to P_0 \to M \to 0$$

*and consider the sequence*

$$F(\mathcal{P}) : \ \dots \longleftarrow F(P_2) \longleftarrow F(P_1) \longleftarrow F(P_0) \longleftarrow F(M) \longleftarrow 0.$$

*The (left) $n$-derived functor $R^n F$ is defined as the $n$-th cohomology of the complex $F(\mathcal{P})$. It is well defined since it does not depend on the projective resolution of $M$.*

The following example of derived functors is probably the most important one for the algebraic considerations which we will make throughout this book.

**Example 1.1.12.** Let $N$, $M$ be $R$-modules and let

$$\mathcal{N} : \ \dots \to N_2 \to N_1 \to N_0 \to N \to 0$$

be a projective resolution for $N$. Then we apply the $\mathrm{Hom}(-, M)$ functor, which is contravariant and right exact to obtain

$$\mathrm{Hom}(\mathcal{N}, M) : \ \dots \longleftarrow \dots \longleftarrow \mathrm{Hom}(N_2, M) \longleftarrow \mathrm{Hom}(N_1, M) \longleftarrow$$

$$\longleftarrow \mathrm{Hom}(N_0, M) \longleftarrow \mathrm{Hom}(N, M) \longleftarrow 0.$$

Let us consider the $n$-derived functor, i.e., the $n$-th cohomology of this complex $H^n\mathrm{Hom}(\mathcal{N}, M)$.

We now write an injective resolution for $M$

$$\mathcal{M} : \ 0 \to M \to M_0 \to M_1 \to M_2 \to \dots$$

and we apply the $\mathrm{Hom}(N, -)$ functor, which is covariant and left exact, to obtain

$$\mathrm{Hom}(N, \mathcal{M}) : \ 0 \to \mathrm{Hom}(N, M) \to \mathrm{Hom}(N, M_0) \to$$

$$\to \mathrm{Hom}(N, M_1) \to \mathrm{Hom}(N, M_2) \to \dots .$$

Let us consider the $n$-th cohomology group of this last sequence $H^n\mathrm{Hom}(N, \mathcal{M})$. It is possible to show (see [79]) that the groups $H^n\mathrm{Hom}(\mathcal{N}, M)$ and $H^n\mathrm{Hom}(N, \mathcal{M})$ are canonically isomorphic.

This motivates the following definition:

**Definition 1.1.23.** *Let $M$ and $N$ be $R$-modules. We define*

$$\mathrm{Ext}^n(N, M) = H^n\mathrm{Hom}(\mathcal{N}, M) = H^n\mathrm{Hom}(N, \mathcal{M}).$$

A completely analogous process can be followed for functors that are covariant and right exact (in this case we consider a projective resolution and then compute the homology of the resulting complex) or for functors that are contravariant and left exact (in this case the resolution is injective and we compute again the homology). The situation is summarized by the following table:

| | Right exact | Left exact |
|---|---|---|
| Covariant | Projective resolution<br>Homology | Injective resolution<br>Cohomology |
| Contravariant | Projective resolution<br>Cohomology | Injective resolution<br>Homology |

The examples we have provided concerning the derived functor of Hom obviously occupy one of the two diagonals in the table above. Another important example occurs when one considers the tensor product functor $M \otimes -$ or $- \otimes M$. Both functors are covariant, see [79], and right exact and therefore their derived functors are obtained by computing the homology of a suitable resolution. In this case the resulting functors are denoted by $\mathrm{Tor}_n(M, -)$ and $\mathrm{Tor}_n(-, M)$.

## 1.1.2 Gröbner bases: a quick introduction

The constructions described in the previous section are of fundamental importance in algebra as well as in algebraic geometry and algebraic analysis. Although they are well understood from a theoretical point of view, it was, until recently, quite difficult to perform such constructions in a concrete way. It became necessary to devise a new approach which would allow concrete computations. The fundamental tool in this direction is a relatively new concept in computational commutative algebra, the notion of Gröbner basis.

Gröbner bases were introduced by Buchberger in his doctoral thesis in 1965, and the reader is referred to the volume [31] for a historical description of the beginnings of the theory, as well as for an up-to-date analysis of their power and applications. Incidentally, [31] contains a readable introduction to the theme of this book by one of us [205].

The central idea of Gröbner bases is to find suitable bases for ideals in the ring of polynomials which will make explicit computations simpler (in particular those computations discussed in the previous section). The theory of Gröbner bases for the ring of polynomials is very well understood, and one can find ways to construct various objects that are relevant for us. We will show that the special form of the operators we study in this book allows in fact sufficiently simple computations, and therefore we will be able to prove very specific and unexpected properties of the syzygies of the systems we are interested in. Since it is our goal to make this book a self-contained reference, we will include in this section all the basic definitions on Gröbner bases, and we will provide sufficient examples to allow the reader to follow the details which will become necessary when we will apply these techniques to the case of Dirac operators and some of their variations. The reader interested in deepening the understanding of this beautiful subject should refer to the monographs [4], [50] or [113].

We start by introducing the notion of order on the set of monomials in several variables, i.e., in the ring $\mathbb{K}[x_1, \ldots, x_n]$, where $\mathbb{K}$ is any field. We will denote a monomial (also called term or power product) $x_1^{\alpha_1} \ldots x_n^{\alpha_n} \in \mathbb{K}[x_1, \ldots, x_n]$ by $X^\alpha$, where $\alpha = (\alpha_1, \ldots, \alpha_n)$ denotes the vector containing the ordered, nonnegative exponents (in this case we will write $\alpha \in \mathbb{N}^n$, where $\mathbb{N}$ denotes the set of nonnegative integers $\{0, 1, 2, 3, \ldots\}$). The sum of two vectors in $\mathbb{N}^n$ is defined componentwise and the total degree of the given monomial $X^\alpha$ is denoted by $|\alpha| = \alpha_1 + \ldots + \alpha_n$. We need to define an order $>$ on the set of monomials such that if the monomial $X^\alpha$ divides the monomial $X^\beta$, then $X^\beta \geq X^\alpha$. Moreover we expect a total order, i.e., that for every $\alpha$, $\beta$, one of the following holds:

$$X^\alpha < X^\beta, \qquad X^\alpha = X^\beta, \qquad X^\alpha > X^\beta.$$

Finally, in order to have remainders of divisions of degree strictly smaller than the degree of the divisor, we need a well-ordering, i.e., an order such that there is no infinite descending chain $X^\alpha > X^\beta > X^\gamma > \ldots$. The following definition summarizes all those properties.

**Definition 1.1.24.** *A monomial order on $\mathbb{K}[x_1, \ldots, x_n]$ is a binary relation $>$ on the set of monomials $X^\alpha$, $\alpha \in \mathbb{N}^n$, satisfying:*

*(i)* $>$ *is a total order on monomials in* $\mathbb{K}[x_1, \ldots, x_n]$*;*

*(ii) if* $X^\alpha > X^\beta$ *and* $\gamma \in \mathbb{N}^n$*, then* $X^{\alpha+\gamma} > X^{\beta+\gamma}$*;*

*(iii)* $>$ *is a well-ordering on monomials, i.e., every nonempty subset of monomials in* $\mathbb{K}[x_1, \ldots, x_n]$ *has a smallest element under* $>$.

In the following, we will denote the set of monomials in $\mathbb{K}[x_1, \ldots, x_n]$ by the symbol $\mathbb{M}_n$ and we will consider the following order on the variables:

$$x_1 > x_2 > \ldots > x_n.$$

We now provide some examples of orders in $\mathbb{M}_n$.

**Definition 1.1.25.** (Lexicographical Order). *Let $X^\alpha$, $X^\beta \in \mathbb{M}_n$. We write $X^\alpha >_{\text{lex}} X^\beta$ if the first elements $\alpha_i$, $\beta_i$ from the left which are different satisfy $\alpha_i > \beta_i$.*

**Definition 1.1.26.** (Degree Lexicographical Order). *Let $X^\alpha$, $X^\beta \in \mathbb{M}_n$. We write $X^\alpha >_{\text{deglex}} X^\beta$ if*

$$|\alpha| > |\beta| \quad \text{or} \quad |\alpha| = |\beta| \text{ and } X^\alpha >_{\text{lex}} X^\beta.$$

**Definition 1.1.27.** (Degree Reverse Lexicographical Order). *Let $X^\alpha$, $X^\beta \in \mathbb{M}_n$. We write $X^\alpha >_{\text{degrevlex}} X^\beta$ if $|\alpha| > |\beta|$ or $|\alpha| = |\beta|$ and the first elements $\alpha_i$ and $\beta_i$ from the right which are different satisfy $\alpha_i < \beta_i$.*

**Remark 1.1.5.** In $\mathbb{K}[x_1] = \mathbb{K}[x]$ the only monomial order is the one given by the degree, i.e.,

$$\ldots > x^n > x^{n-1} > \ldots > x^2 > x > 1.$$

**Example 1.1.13.** Let us consider the monomials $x_1x_2^2x_3^2$, $x_1x_2^3$. Then $x_1x_2^2x_3^2 <_{\text{lex}} x_1x_2^3$ while $x_1x_2^2x_3^2$ is greater than $x_1x_2^3$ with respect to deglex and degrevlex because it has a higher degree. Let us consider $x_1^2x_2x_3^2$ and $x_1x_2^3x_3$. Then $x_1^2x_2x_3^2 >_{\text{deglex}} x_1x_2^3x_3$ while $x_1^2x_2x_3^2 <_{\text{degrevlex}} x_1x_2^3x_3$.

**Definition 1.1.28.** *Let $f = \sum_\gamma a_\gamma X^\gamma$ be a nonzero polynomial in $\mathbb{K}[x_1, \ldots, x_n]$. Let $>$ be a monomial order such that $X^\alpha > X^\beta > \ldots$ so that we can write $f = a_\alpha X^\alpha + a_\beta X^\beta + \ldots$. We define*

*(i) the leading coefficient of $f$ as*

$$\text{lc}(f) = a_\alpha \in \mathbb{K};$$

*(ii) the leading monomial of $f$ as*

$$\text{lm}(f) = X^\alpha$$

*(with coefficient 1);*

*(iii) the leading term of $f$ as*

$$\text{lt}(f) = \text{lc}(f)\text{lm}(f) = a_\alpha X^\alpha.$$

*We define* $\text{lc}(0) = \text{lm}(0) = \text{lt}(0) = 0$.

**Remark 1.1.6.** Note that lc, lm, lt are multiplicative, i.e., $\text{lc}(fg) = \text{lc}(f)\text{lc}(g)$, $\text{lm}(fg) = \text{lm}(f)\text{lm}(g)$, $\text{lt}(fg) = \text{lt}(f)\text{lt}(g)$.

**Example 1.1.14.** Let us set $x > y > z$ and let us consider the polynomial $f = 3xy^3z + 2x^2yz^2 + 4x^3y$. We have (with respect to the order indicated on the left):

| | | | |
|---|---|---|---|
| lex | $\mathrm{lc}(f) = 4$ | $\mathrm{lm}(f) = x^3y$ | $\mathrm{lt}(f) = 4x^3y$ |
| deglex | $\mathrm{lc}(f) = 2$ | $\mathrm{lm}(f) = x^2yz^2$ | $\mathrm{lt}(f) = 2x^2yz^2$ |
| degrevlex | $\mathrm{lc}(f) = 3$ | $\mathrm{lm}(f) = xy^3z$ | $\mathrm{lt}(f) = 3xy^3z.$ |

A choice of a monomial order on $\mathbb{K}[x_1, \ldots, x_n]$ allows a division algorithm.

**Proposition 1.1.7.** (Division algorithm). *Let $>$ be a monomial order in $\mathbb{K}[x_1, \ldots, x_n]$ and let $(f_1, \ldots, f_s)$ be an ordered s-tuple of elements in $\mathbb{K}[x_1, \ldots, x_n]$. Then every $f \in \mathbb{K}[x_1, \ldots, x_n]$ can be written in the form*

$$f = a_1f_1 + \ldots + a_sf_s + r, \qquad a_i,\ r \in \mathbb{K}[x_1, \ldots, x_n],$$

*where either $r = 0$ or no power product appearing in $r$ is divisible by any one of the $lm(f_i)$, $i = 1, \ldots, s$. The polynomial $r$ is said to be a remainder.*

The proof of the proposition consists in the following algorithm for finding $a_i$ and $r$:

**Division Algorithm**

**Input**: $f, f_1, \ldots, f_s \in \mathbb{K}[x_1, \ldots, x_n]$ with $f_i \neq 0$ for $0 \leq i \leq s$
**Output**: $a_1, \ldots, a_s, r$ such that $f = \sum_{i=1}^{s} a_i f_i + r$, no power product appearing in $r$ is divisible by any one of the $\mathrm{lm}(f_i)$, $i = 1, \ldots, s$, and $\max_{1 \leq i \leq s}(\mathrm{lm}(a_i)\mathrm{lm}(f_i), \mathrm{lm}(r)) = \mathrm{lm}(f)$
**Initialization**: $a_1 := 0, \ldots, a_s := 0$, $r := 0$, $h := f$
**While** $h \neq 0$ **do**
  **If** there exists $i$ such that $\mathrm{lm}(f_i)$ divides $\mathrm{lm}(h)$
    **then**
    choose $i$ least such that $\mathrm{lm}(f_i)$ divides $\mathrm{lm}(h)$
    $a_i := a_i + \dfrac{\mathrm{lt}(h)}{\mathrm{lt}(f_i)}$
    $h := h - \dfrac{\mathrm{lt}(h)}{\mathrm{lt}(f_i)} f_i$
    **else**
    $r := r + \mathrm{lt}(h)$
    $h := h - \mathrm{lt}(h)$.

**Remark 1.1.7.** Note that it can happen that a different ordering of the $f_i$ can produce different quotients $a_i$ and a different remainder $r$.

We now give some more definitions that will be useful in what follows; we obviously suppose, from now on, that an order on $\mathbb{M}_n$ has been fixed.

**Definition 1.1.29.** *Given $f$, $g$, $h \in \mathbb{K}[x_1, \ldots, x_n]$, $g \neq 0$ we will say that $f$ reduces to $h$ modulo $g$ in one step if and only if $lm(g)$ divides a nonzero term $X^\alpha$ appearing in $f$ and*

$$h = f - \frac{X^\alpha}{\mathrm{lt}(g)} g.$$

*We will write*

$$f \xrightarrow{g} h.$$

*Given* $f$, $h$, $F = \{g_1, \dots, g_s\}$, $f, h, g_i$ *polynomials in* $\mathbb{K}[x_1, \dots, x_n]$, *we say that* $f$ *reduces to* $h$ *modulo* $F$ *if and only if there exists a sequence of indices* $i_1, \dots, i_r \in \{1, \dots, s\}$ *and polynomials* $h_1, \dots, h_{r-1}$ *such that*

$$f \xrightarrow{g_{i_1}} h_1 \xrightarrow{g_{i_2}} h_2 \xrightarrow{g_{i_3}} \dots h_{r-1} \xrightarrow{g_{i_r}} h.$$

*In this case we will write* $f \xrightarrow{F}_+ h$.

**Definition 1.1.30.** *A polynomial* $f$ *is said to be reduced with respect to* $F = \{g_1, \dots, g_s\}$ *if it cannot be reduced modulo* $F$.

**Definition 1.1.31.** *Let* $I \subseteq \mathbb{K}[x_1, \dots, x_n]$ *be an ideal. A finite set of nonzero polynomials* $G = \{g_1, \dots, g_s\} \subset I$ *is said to be a Gröbner basis for* $I$ *if and only if for any* $f \in I$, $f \neq 0$ *there exists* $j \in \{1, \dots, s\}$ *such that* $lm(g_j)$ *divides* $lm(f)$.

We can characterize Gröbner bases by giving some equivalent conditions (which also explain the importance of Gröbner bases from a computational point of view). To this end we need the following definition.

**Definition 1.1.32.** *Let* $S$ *be a subset of* $\mathbb{K}[x_1, \dots, x_n]$. *The leading term ideal of* $S$ *is defined to be the ideal generated by the leading terms of polynomials in* $S$:

$$\mathrm{Lt}(S) := \langle \mathrm{lt}(s) \mid s \in S \rangle.$$

**Theorem 1.1.2.** *Let* $I$ *be an ideal in* $\mathbb{K}[x_1, \dots, x_n]$, $I \neq 0$ *and let* $G = \{g_1, \dots, g_s\} \subset I$. *The following conditions are equivalent:*

*(i)* $G$ *is a Gröbner basis for* $I$;

*(ii)* $f \in I$ *if and only if* $f \xrightarrow{G}_+ 0$;

*(iii)* $f \in I$ *if and only if* $f = \sum_{i=1}^s h_i g_i$ *with* $\mathrm{lm}(f) = \max_{1 \le i \le s}(\mathrm{lm}(h_i)\mathrm{lm}(g_i))$;

*(iv)* $\mathrm{Lt}(G) = \mathrm{Lt}(I)$.

As a consequence we have the following:

**Corollary 1.1.1.** *1. If* $G = \{g_1, \dots, g_s\}$ *is a Gröbner basis for* $I$ *then* $I = \langle g_1, \dots, g_s \rangle$.

*2. Every nonzero ideal in* $\mathbb{K}[x_1, \dots, x_n]$ *has a Gröbner basis.*

It is important to illustrate an algorithm, due to Buchberger, that takes as input a set of generators for an ideal $I$ and gives as output a Gröbner basis $G$ for it. First, we introduce the notion of $S$-polynomial which is useful to cancel the leading terms of any two given polynomials:

**Definition 1.1.33.** *Let $f, g \in \mathbb{K}[x_1, \ldots, x_n]$, with $f \neq 0$ and $g \neq 0$. Set* $\mathrm{lm}(f) = X^\alpha$, $\mathrm{lm}(g) = X^\beta$ *and*

$$X^\gamma = \mathrm{lcm}(X^\alpha, X^\beta)$$

*(lcm denotes the least common multiple). The $S$-polynomial of $f$ and $g$, denoted by $S(f,g)$, is the polynomial defined by*

$$S(f,g) := \frac{X^\gamma}{\mathrm{lt}(f)} f - \frac{X^\gamma}{\mathrm{lt}(g)} g.$$

**Example 1.1.15.** Let us consider the two polynomials $f = 2x^2y^3 + x^3y^2 - x^2yz$, $g = 4xyz^2 + 2x^2y^3$ in $\mathbb{Q}[x, y, z]$ with $x > y > z$ and the order deglex. Then

$$\begin{aligned} S(f,g) &= \frac{x^3y^3}{x^3y^2}(2x^2y^3 + x^3y^2 - x^2yz) - \frac{x^3y^3}{2x^2y^3}(4xyz^2 + 2x^2y^3) \\ &= 2x^2y^4 - x^2y^2z - 2x^2yz^2. \end{aligned}$$

We now recall the following fundamental result to detect if a subset of an ideal $I$ is a Gröbner basis.

**Theorem 1.1.3.** (Buchberger's Criterion). *A finite set $G = \{g_1, \ldots, g_n\} \subset I$ of nonzero polynomials is a Gröbner basis of $I$ if and only if $I = \langle g_1, \ldots, g_n \rangle$ and*

$$S(g_i, g_j) \xrightarrow{G}_+ 0,$$

*for all pairs $i \neq j$.*

The criterion gives a procedure to write a Gröbner basis for an ideal $I$. The idea is to reduce all the $S$-polynomials of any two generators of $I$ and if a remainder is not zero, then it must be added to the generating set. The procedure is summarized in the following algorithm (see [4]).

**Buchberger's Algorithm**

**Input:** $F = \{f_1, \ldots, f_t\} \subseteq \mathbb{K}[x_1, \ldots, x_n]$ with $f_i \neq 0$, for all $i = 1, \ldots s$

**Output:** $G = \{g_1, \ldots, g_s\}$ a Gröbner basis for $I = \langle f_1, \ldots, f_t \rangle$

**Initialization:** $G := F$, $\mathcal{G} := \{\{f_i, f_j\} |\ f_i \neq f_j,\ f_i, f_j \in G\}$

**While** $\mathcal{G} \neq \emptyset$ **do**

Choose any $\{f, g\} \in \mathcal{G}$
$\mathcal{G} := \mathcal{G} - \{\{f, g\}\}$
$S(f,g) \xrightarrow{G}_+ h$, $h$ reduced with respect to $G$
**If** $h \neq 0$

**then**
$\mathcal{G} := \mathcal{G} \cup \{\{u, h\} |$ for any $u \in G\}$
$G := G \cup \{h\}$.

Note that the Gröbner basis computed as in Buchberger's Algorithm need not be unique. The uniqueness is obtained by adding suitable conditions on the polynomials in the Gröbner basis

**Definition 1.1.34.** *A reduced Gröbner basis for an ideal $I \subseteq \mathbb{K}[x_1, \ldots, x_n]$ is a Gröbner basis $G = \{g_1, \ldots, g_s\}$ for $I$ such that $\mathrm{lc}(g_i) = 1$, for all $i$, and $g_i$ is reduced with respect to $G \setminus \{g_i\}$, i.e., for all distinct $g_i$, $g_j \in G$ no monomial appearing in $g_i$ is a multiple of $\mathrm{lt}(g_j)$.*

**Theorem 1.1.4.** (Buchberger). *Given a monomial order on $\mathbb{K}[x_1, \ldots, x_n]$, each ideal $I$ has a unique reduced Gröbner basis.*

We now further generalize this theory to the case of modules: following [4] we will introduce the theory of Gröbner bases for submodules of the free module $R^m$, $R = \mathbb{K}[x_1, \ldots, x_n]$ and we will generalize to $R$-modules the theory we developed in the ring of polynomials. An element $\mathbf{f}$ in $R^m$ will be considered as a column vector

$$\mathbf{f} = \begin{bmatrix} f_1 \\ f_2 \\ \vdots \\ f_m \end{bmatrix} = f_1 \begin{bmatrix} 1 \\ 0 \\ \vdots \\ 0 \end{bmatrix} + f_2 \begin{bmatrix} 0 \\ 1 \\ \vdots \\ 0 \end{bmatrix} + f_m \begin{bmatrix} 0 \\ 0 \\ \vdots \\ 1 \end{bmatrix} = f_1\mathbf{e}_1 + f_2\mathbf{e}_2 + \ldots f_m\mathbf{e}_m$$

where $f_i \in R$ and $\{\mathbf{e}_i\}_{i=1,\ldots,m}$ is the standard basis of $R^m$. In short, we will write

$$\mathbf{f} = \sum_{i=1}^{m} f_i\mathbf{e}_i.$$

**Definition 1.1.35.** *We will call monomial an element of the form $X^\alpha\mathbf{e}_i$ while $cX^\alpha\mathbf{e}_i$ will be called a term and $c$ its coefficient. If $\mathbf{m} = X^\alpha\mathbf{e}_i$ and $\mathbf{n} = X^\beta\mathbf{e}_j$ are monomials, we say that $\mathbf{m}$ divides $\mathbf{n}$ if and only if $i = j$ and $X^\alpha$ divides $X^\beta$. The quotient $\dfrac{\mathbf{m}}{\mathbf{n}}$ is defined as $\dfrac{X^\beta}{X^\alpha}$.*

**Example 1.1.16.** In $R^4$ the monomial $\mathbf{n} = (0, 0, x_1^2x_2, 0)$ divides the monomial $\mathbf{m} = (0, 0, x_1^4x_2, 0)$ and

$$\frac{\mathbf{m}}{\mathbf{n}} = \frac{(0, 0, x_1^4x_2, 0)}{(0, 0, x_1^2x_2, 0)} = x_1^2.$$

In the set of monomials in $R^m$ one can introduce more than one notion of total order, depending not only on the power products $X^\alpha$ but also on the position in the vector. There are two important ways to take into account the position in the vector: one is called "TOP" ("term over position"), in which the order of terms is more important than their position; the other is "POT" ("position over terms"), in which the position is more important than the order of the terms. In what follows, we will use the order $\mathbf{e}_1 > \mathbf{e}_2 > \ldots > \mathbf{e}_m$. When different orders are used, we will write them explicitly.

**Definition 1.1.36.** *Let $>$ be any monomial order on $R$. Given the two monomials $\mathbf{m} = X^\alpha \mathbf{e}_i$ and $\mathbf{n} = X^\beta \mathbf{e}_j$, we say that $\mathbf{m} > \mathbf{n}$ in the TOP order if and only if $X^\alpha > X^\beta$ or $X^\alpha = X^\beta$ and $i < j$. We say that $\mathbf{m} > \mathbf{n}$ in the POT order if and only if $i < j$ or $i = j$ and $X^\alpha > X^\beta$.*

Given an order $>$ on the monomials in $R^m$, one can generalize the notions given in the Definition 1.1.28.

**Definition 1.1.37.** *Let $\mathbf{f} = \sum_\alpha a_\alpha \mathbf{m}_\alpha$ be a nonzero element in $R^m$ and let $>$ be a given order so that $\mathbf{m}_\alpha > \mathbf{m}_\beta > \ldots$. We define:*

*(i) the leading coefficient of $\mathbf{f}$ as*

$$lc(\mathbf{f}) = a_\alpha \in \mathbb{K};$$

*(ii) the leading monomial of $\mathbf{f}$ as*

$$lm(\mathbf{f}) = \mathbf{m}_\alpha;$$

*(iii) the leading term of $\mathbf{f}$ as*

$$lt(\mathbf{f}) = lc(\mathbf{f}) lm(\mathbf{f}) = a_\alpha \mathbf{m}_\alpha.$$

**Definition 1.1.38.** *Let $S$ be any subset of $R^m$. The leading term module of $S$ is defined as*

$$\mathrm{Lt}(S) = \langle \mathrm{lt}(\mathbf{s}) \mid \mathbf{s} \in S \rangle.$$

*We define $lc(\mathbf{0})=0$, $lm(\mathbf{0})=lt(\mathbf{0})=\mathbf{0}$.*

**Definition 1.1.39.** *Let $\mathbf{m} = X^\alpha \mathbf{e}_i$ and $\mathbf{n} = X^\beta \mathbf{e}_j$. If $i = j$, the least common multiple $lcm(\mathbf{m}, \mathbf{n})$ of $\mathbf{m}$ and $\mathbf{n}$ is $lcm(X^\alpha, X^\beta)\mathbf{e}_i$. If $i \neq j$ we define $lcm(\mathbf{m}, \mathbf{n}) = \mathbf{0}$.*

All those definitions allow the generalization to $R^m$ of many of the results we have stated for the ring $R$. Let us start with the proposition leading to the Division Algorithm in $R$.

**Proposition 1.1.8.** *Let $(\mathbf{f}_1, \ldots, \mathbf{f}_s)$ be an ordered $s$-tuple of nonzero vectors in $R^m$ and let $\mathbf{f}$ be an element in $R^m$. Then there are polynomials $a_1, \ldots, a_s \in R$ and $\mathbf{r} \in R^m$ such that $\mathbf{f}$ can be written as*

$$\mathbf{f} = a_1 \mathbf{f}_1 + \ldots + a_s \mathbf{f}_s + \mathbf{r},$$

*where $lt(a_i \mathbf{f}_i) \leq lt(\mathbf{f})$ for all $i$ and either $\mathbf{r} = \mathbf{0}$ or $\mathbf{r}$ is a linear combination of monomials, none of which is divisible by any one of $lm(\mathbf{f}_i)$, $i = 1, \ldots, s$. The element $\mathbf{r}$ is said to be a remainder.*

The Division Algorithm that proves the proposition is, with suitable changes, nothing but the same algorithm already given for polynomials in the ring $\mathbb{K}[x_1, \dots, x_n]$. Also the notion of reduction of elements in the module $R^m$ follows, with suitable changes, Definition 1.1.29. For the sake of completeness, we will mention the definition of Gröbner basis of a submodule of $R^m$. Here and in what follows we suppose that a monomial order $>$ has been fixed in $R^m$.

**Definition 1.1.40.** *Let $M$ be a submodule of $R^m$. A set of nonzero elements $G = \{\mathbf{g}_1, \dots, \mathbf{g}_s\} \subset M$ is called a Gröbner basis for $M$ if and only if for any nonzero $\mathbf{f} \in M$, there exists $j \in \{1, \dots, s\}$ such that $lm(\mathbf{g}_j)$ divides $lm(\mathbf{f})$.*

**Remark 1.1.8.** The results corresponding to Theorem 1.1.2 and Corollary 1.1.1 follow immediately.

To arrive to the Buchberger Algorithm for modules, we need to define what is the $S$-polynomial of two elements in $R^m$.

**Definition 1.1.41.** *Let $\mathbf{f}$, $\mathbf{g} \in R^m$ and let $\mathbf{m} = \mathrm{lcm}(\mathrm{lt}(\mathbf{f}), \mathrm{lt}(\mathbf{g}))$. The $S$-polynomial $S(\mathbf{f}, \mathbf{g})$ of $\mathbf{f}$ and $\mathbf{g}$ is defined as*

$$S(\mathbf{f}, \mathbf{g}) = \frac{\mathbf{m}}{\mathrm{lt}(\mathbf{f})}\mathbf{f} - \frac{\mathbf{m}}{\mathrm{lt}(\mathbf{g})}\mathbf{g}.$$

**Example 1.1.17.** Let us consider $\mathbf{f} = (5x^2, x^2y + y)$ and $\mathbf{g} = (y + y^2, xy^2)$ in $R^2$ with the deglex order, $x > y$ and $\mathbf{e}_1 < \mathbf{e}_2$. We compute the $S$-polynomial:

$$S(\mathbf{f}, \mathbf{g}) = \frac{(0, x^2y^2)}{(0, x^2y)}(5x^2, x^2y + y) - \frac{(0, x^2y^2)}{(0, xy^2)}(y + y^2, xy^2)$$

$$= y(5x^2, x^2y + y) - x(y + y^2, xy^2) = (5x^2y - xy - xy^2, y^2).$$

To compute a Gröbner basis of a module we still have Buchberger's Criterion.

**Theorem 1.1.5.** (Buchberger's Criterion). *A finite set $G = \{\mathbf{g}_1, \dots, \mathbf{g}_s\}$ is a Gröbner basis for the submodule $M \subset R^m$ generated by $G$ if and only if*

$$S(\mathbf{g}_i, \mathbf{g}_j) \xrightarrow{G}_+ \mathbf{0},$$

*for all pairs $i \neq j$.*

**Remark 1.1.9.** This theorem leads to an algorithm to compute a Gröbner basis for a submodule of $R^m$ similar to Buchberger's Algorithm already given for polynomial rings. We insert it here for the sake of completeness since it will be used in the chapters to follow.

**Buchberger's Algorithm**

> **Input**: $F = \{\mathbf{f}_1, \dots, \mathbf{f}_t\} \subseteq R^m$ with $\mathbf{f}_i \neq 0$
>
> **Output**: $G = \{\mathbf{g}_1, \dots, \mathbf{g}_s\}$ a Gröbner basis for $\langle \mathbf{f}_1, \dots, \mathbf{f}_t \rangle$
>
> **Initialization**: $G := F$, $\mathcal{G} := \{\{\mathbf{f}_i, \mathbf{f}_j\} | \ \mathbf{f}_i \neq \mathbf{f}_j \in G\}$
>
> **While** $\mathcal{G} \neq \emptyset$ **do**

> Choose any $\{\mathbf{f}, \mathbf{g}\} \in \mathcal{G}$
> $\mathcal{G} := \mathcal{G} - \{\{\mathbf{f}, \mathbf{g}\}\}$
> $S(\mathbf{f}, \mathbf{g}) \xrightarrow{G}_+ \mathbf{h}$, $\mathbf{h}$ reduced with respect to $G$
> **If** $\mathbf{h} \neq 0$
> **then**
> $\mathcal{G} := \mathcal{G} \cup \{\{\mathbf{u}, \mathbf{h}\} | \text{for any } \mathbf{u} \in G\}$
> $G := G \cup \{\mathbf{h}\}$.

Let us consider now a particular submodule of $R^m$: the submodule of the first syzygies of an ideal $I$ generated by polynomials $f_1, \dots, f_r$, (see Proposition 1.1.3). This submodule is finitely generated, as any submodule of $R^s$, and so it is possible to compute its generators. We start with the case in which we know a Gröbner basis $G = \{g_1, \dots, g_s\}$ for the ideal $I$. We recall that the $S$-polynomial of two polynomials $g_i$, $g_j$ can be written as

$$S(g_i, g_j) := \frac{X^{\gamma_{ij}}}{\operatorname{lt}(g_i)} g_i - \frac{X^{\gamma_{ij}}}{\operatorname{lt}(g_j)} g_j,$$

so, by the Division Algorithm, it is possible to write the $S$-polynomial as

$$S(g_i, g_j) = h_{ij1} g_1 + \ldots + h_{ijs} g_s$$

where the remainder is 0, $h_{ijk} \in R$, and $\operatorname{lt}(h_{ijk} g_k) \leq \operatorname{lt}(S(g_i, g_j))$. We denote by $\mathbf{h}_{ij}$ the column vector $\mathbf{h}_{ij} = (h_{ij1}, \dots, h_{ijs})^t$. Finally, we set

$$\mathbf{s}_{ij} = \frac{X^{\gamma_{ij}}}{\operatorname{lt}(g_i)} \mathbf{e}_i - \frac{X^{\gamma_{ij}}}{\operatorname{lt}(g_j)} \mathbf{e}_j - \mathbf{h}_{ij}.$$

With these notations, we can prove the next proposition, showing how to compute the syzygy module of the elements $g_1, \dots, g_s$, which we suppose monic for simplicity, in a Gröbner basis for $I$ (see [4], [50]).

**Proposition 1.1.9.** *The R-module $M = Syz(g_1, \dots, g_s)$ is generated by $\{\mathbf{s}_{ij} \mid 1 \leq i, j \leq s\}$.*

*Proof.* Suppose that $\Sigma = \{\mathbf{s}_{ij} \mid 1 \leq i, j \leq s\}$ is not a generating set for $M$; then there exists an $s$-tuple $(a_1, \dots, a_s) \in M$ such that $(a_1, \dots, a_s) \notin \langle \Sigma \rangle$. We will consider $(a_1, \dots, a_s)$ with $X = \max_j(\operatorname{lm}(a_j)\operatorname{lm}(g_j))$ least and the set

$$\sigma = \{j \mid 1 \leq j \leq s,\ \operatorname{lm}(a_j)\operatorname{lm}(g_j) = X\}.$$

Let us define a new $s$-tuple $(a'_1, \dots, a'_s)$ defined by $a'_j = a_j$ if $j \notin \sigma$ and $a'_j = a_j - \operatorname{lt}(a_j)$ if $j \in \sigma$. For $j \in \sigma$, let us write $\operatorname{lt}(a_j) = c_j \operatorname{lm}(a_j)$. We have

$$\sum_{j \in \sigma} c_j \operatorname{lm}(a_j) \operatorname{lm}(g_j) = 0$$

since $(a_1, \dots, a_s) \in M = \mathrm{Syz}(g_1, \dots, g_s)$, so that

$$\sum_{j\in\sigma} c_j \mathrm{lm}(a_j)\mathbf{e}_j \in \mathrm{Syz}(\mathrm{lm}(g_j),\ \ j \in \sigma).$$

It can be shown, see for example Proposition 3.2.3 in [4], that

$$\sum_{j\in\sigma} c_j \mathrm{lm}(a_j)\mathbf{e}_j = \sum_{i<j,\ i,j\in\sigma} d_{ij}\left(\frac{X^{\gamma_{ij}}}{\mathrm{lt}(g_i)}\mathbf{e}_i - \frac{X^{\gamma_{ij}}}{\mathrm{lt}(g_j)}\mathbf{e}_j\right)$$

where $d_{ij}$ are suitable elements in the ring $R$ that can be chosen to be a constant multiple of $X/X^{\gamma_{ij}}$ since the left-hand side of the previous equality consists of homogeneous terms and $X = \mathrm{lm}(a_j)\mathrm{lm}(g_j)$. Therefore we get

$$\begin{aligned}(a_1, \dots, a_s) &= \sum_{j\in\sigma} c_j \mathrm{lm}(a_j)\mathbf{e}_j + (a'_1, \dots, a'_s) \\ &= \sum_{i<j,\ i,j\in\sigma} d_{ij}\left(\frac{X^{\gamma_{ij}}}{\mathrm{lt}(g_i)}\mathbf{e}_i - \frac{X^{\gamma_{ij}}}{\mathrm{lt}(g_j)}\mathbf{e}_j\right) + (a'_1, \dots, a'_s) \\ &= \sum_{i<j,\ i,j\in\sigma} d_{ij}\mathbf{s}_{ij} + (a'_1, \dots, a'_s) + \sum_{i<j,\ i,j\in\sigma} d_{ij}(h_{ij1}, \dots, h_{ijs}).\end{aligned}$$

Let us set

$$(u_1, \dots, u_s) = (a'_1, \dots, a'_s) + \sum_{i<j,\ i,j\in\sigma} d_{ij}(h_{ij1}, \dots, h_{ijs}).$$

Then $(u_1, \dots, u_s)$ belongs to $M$ but it does not belong to $\langle\Sigma\rangle$. Let us now compute

$$\mathrm{lm}(u_\ell)\mathrm{lm}(g_\ell)$$

for any index $\ell$, $1 \le \ell \le s$:

$$\begin{aligned}\mathrm{lm}(u_\ell)\mathrm{lm}(g_\ell) &= \mathrm{lm}\left(a'_\ell + \sum_{i<j,\ i,j\in\sigma} d_{ij}h_{ij\ell}\right)\mathrm{lm}(g_\ell) \\ &\le \max\left(\mathrm{lm}(a'_\ell), \max_{i<j,\ i,j\in\sigma}(\mathrm{lm}(d_{ij})\mathrm{lm}(h_{ij\ell}))\right)\mathrm{lm}(g_\ell).\end{aligned}$$

The definition of $a'_\ell$ implies $\mathrm{lm}(a'_\ell)\mathrm{lm}(g_\ell) < X$; moreover $d_{ij}$ is a multiple of $X/X^{\gamma_{ij}}$, hence we get

$$\mathrm{lm}(d_{ij})\mathrm{lm}(h_{ij\ell})\mathrm{lm}(g_\ell) = \frac{X}{X^{\gamma_{ij}}}\mathrm{lm}(h_{ij\ell})\mathrm{lm}(g_\ell) \le \frac{X}{X^{\gamma_{ij}}}\mathrm{lm}(S(g_i, g_j)) < X.$$

The above computations show that $\mathrm{lm}(u_\ell)\mathrm{lm}(g_\ell) < X$ for any $\ell$, thus providing a contradiction to the fact that $X$ is least. Then $\Sigma$ is a generating set for $M$. □

Now that we have computed $\mathrm{Syz}(g_1, \ldots, g_s)$, we come back to the problem of determining $\mathrm{Syz}(f_1, \ldots, f_r)$ where $f_1, \ldots, f_r$ do not form necessarily a Gröbner basis for the ideal $I$ they generate. We may compute a Gröbner basis for $I$ and we may suppose that all the polynomials $g_i$ are monic. Set $F = [f_1 \ \ldots \ f_r]$ and $G = [g_1 \ \ldots \ g_s]$. We claim that there exist a $s \times r$ matrix $S$ and a $r \times s$ matrix $T$ with entries in $R$ such that $F = GS$ and $G = FT$. Using the previous proposition, we can compute a set of generators $\{\mathbf{s}_i\}$ for $\mathrm{Syz}(g_1, \ldots, g_s)$ so that we obtain

$$G\mathbf{s}_i = 0 = FT\mathbf{s}_i = F(T\mathbf{s}_i)$$

and $\langle T\mathbf{s}_i \rangle \subseteq \mathrm{Syz}(F)$. Moreover we have

$$F(I - TS) = F - FTS = F - GS = 0$$

so that also the columns $\mathbf{c}_1, \ldots, \mathbf{c}_r$ of the matrix $I - TS$ belong to $\mathrm{Syz}(F)$. We can now prove the following result.

**Theorem 1.1.6.** *The submodule $\mathrm{Syz}(f_1, \ldots, f_r)$ of $R^t$ is given by*

$$\langle T\mathbf{s}_1, \ldots, T\mathbf{s}_t, \mathbf{c}_1, \ldots, \mathbf{c}_r \rangle.$$

*Proof.* Suppose that $\mathbf{s} = (s_1, \ldots, s_r) \in \mathrm{Syz}(f_1, \ldots, f_r)$. Then $F\mathbf{s} = 0 = GS\mathbf{s}$, so that $S\mathbf{s} \in \mathrm{Syz}(g_1, \ldots, g_s)$ that is generated by $\mathbf{s}_1, \ldots, \mathbf{s}_t$ and so we can write $S\mathbf{s} = \sum_{j=1}^{t} h_j \mathbf{s}_j$, $h_j \in R$. With this notation we can rewrite $TS\mathbf{s}$ as $TS\mathbf{s} = \sum_{j=1}^{t} h_j (T\mathbf{s}_j)$ and we get

$$\mathbf{s} = \mathbf{s} - TS\mathbf{s} + TS\mathbf{s} = (I - TS)\mathbf{s} + \sum_{j=1}^{t} h_j (T\mathbf{s}_j) = \sum_{j=1}^{r} s_s \mathbf{c}_j + \sum_{j=1}^{t} h_j (T\mathbf{s}_j).$$

This means that $\mathrm{Syz}(f_1, \ldots, f_r) \subseteq \langle T\mathbf{s}_1, \ldots, T\mathbf{s}_t, \mathbf{c}_1, \ldots, \mathbf{c}_r \rangle$ and since the reverse inclusion has already been proved, we get the statement. □

Also in this case, it is possible to generalize the statement, with suitable modifications in notation, to the case of submodules $M \subset R^m$: if $M = \langle \mathbf{f}_1, \ldots, \mathbf{f}_r \rangle$ and a Gröbner basis for $M$ is $G = \{\mathbf{g}_1, \ldots, \mathbf{g}_s\}$, then we can write the $S$-polynomial using the division algorithm as $S(\mathbf{g}_i, \mathbf{g}_j) = \sum_{k=1}^{s} a_{ijk} \mathbf{g}_k$. We put $\mathbf{a}_{ij} \in R^s$ equal to

$$\mathbf{a}_{ij} = a_{ij1}\mathbf{e}_1 + \ldots + a_{ijs}\mathbf{e}_s$$

and we define

$$\mathbf{s}_{ij} = \frac{\mathbf{m}_{ij}}{\mathrm{lt}(\mathbf{g}_i)}\mathbf{e}_i - \frac{\mathbf{m}_{ij}}{\mathrm{lt}(\mathbf{g}_j)}\mathbf{e}_j - \mathbf{a}_{ij}.$$

We have the following theorem.

**Theorem 1.1.7.** *The syzygy module $\mathrm{Syz}(\mathbf{g}_1, \ldots, \mathbf{g}_s)$ can be generated by the vectors $\mathbf{s}_{ij}$.*

As above, we can compute the syzygy module for $M = \langle \mathbf{f}_1, \ldots, \mathbf{f}_r \rangle$ once we know a Gröbner basis $\{\mathbf{g}_1, \ldots, \mathbf{g}_s\}$ for $M$. Let us construct the $m \times r$ matrix $F = (\mathbf{f}_1, \ldots, \mathbf{f}_r)$ and the $m \times s$ matrix $G = (\mathbf{g}_1, \ldots, \mathbf{g}_s)$. The columns of $F$ and $G$ both generate $M$, so there exist a $r \times s$ matrix $T$ and a $s \times r$ matrix $S$ such that $G = FT$ and $F = GS$. The matrix $S$ can be computed using the division algorithm expressing each $\mathbf{f}_i$ in terms of the $\mathbf{g}_\ell$, while $T$ is obtained as a byproduct of the computation of the Gröbner basis, taking into account reductions arising from Buchberger's Algorithm.

**Theorem 1.1.8.** *Let $M$, $F$, $G$, $S$, $T$ as above and let $\{\mathbf{s}_i \mid 1 \leq i \leq r\}$ be the basis for Syz$(\mathbf{g}_1, \ldots, \mathbf{g}_s)$ constructed above. Let $\mathbf{C}_1, \ldots, \mathbf{C}_r$ be the columns of the $r \times r$ matrix $I - TS$. Then*

$$\mathrm{Syz}(\mathbf{f}_1, \ldots, \mathbf{f}_r) = \langle T\mathbf{s}_1, \ldots, T\mathbf{s}_t, \mathbf{C}_1, \ldots, \mathbf{C}_r \rangle.$$

### 1.1.3 Sheaf theory

Similarly to what happened for algebraic geometry in the forties, the notion of sheaf was instrumental in ushering in a new phase in algebraic analysis. Though this was obviously a gradual process, a "manifesto" for this new approach was provided by Ehrenpreis in [64]. In this section we provide enough material on sheaves to allow the reader a full understanding of the key ideas in algebraic analysis.

**Definition 1.1.42.** *Let $X$ be a topological space. A presheaf $\mathcal{F}$ (of abelian groups) on $X$ is an assignment $\{U, \mathcal{F}(U)\}$ of an abelian group for every open set $U$ in $X$ together with a family of group homomorphisms*

$$\rho_V^U : \mathcal{F}(U) \to \mathcal{F}(V),$$

*for each inclusion of open set $V \subset U$, such that*

1. *$\rho_U^U = 1_{\mathcal{F}(U)}$ for every open set $U$ in $X$;*

2. *$\rho_W^V \circ \rho_V^U = \rho_W^U$ for every $W \subset V \subset U$ open sets in $X$.*

The maps $\rho$ are called restrictions since this is what they are in most natural examples of presheaves. Accordingly, we will often write $f_{|V}$ instead of $\rho_V^U f$, for $f \in \mathcal{F}(U)$ and $V \subset U$.

**Example 1.1.18.** Let $\mathbb{R}$ be the real line with the natural topology. We define $B(U)$ to be the group of real-valued bounded functions on an open set $U$. It is immediate to verify that this assignment, together with the usual restrictions of functions, is a presheaf on $\mathbb{R}$. If we define $\mathcal{E}(U)$ to be the group of infinitely differentiable real (or complex) valued functions on an open set $U$, we still obtain a presheaf.

**Remark 1.1.10.** The notion of presheaf can be given using the language of categories: the family of all open sets of the topological space $X$ are the objects of a category whose morphisms are the inclusions of sets. A presheaf is a contravariant functor from this category to the one of abelian groups.

**Definition 1.1.43.** *A presheaf $\mathcal{F}$ on a topological space $X$ is said to be a sheaf if for every open set $U$ in $X$ and every open covering $\{U_i\}_{i\in I}$ of $U$ the two following conditions are satisfied:*

1. *If $f, g \in \mathcal{F}(U)$ are such that $f_{|U_i} = g_{|U_i}$ for every $i \in I$, then $f = g$;*
2. *If there are elements $f_i \in \mathcal{F}(U_i)$, $i \in I$ such that*

$$f_{i_{|U_i \cap U_j}} = f_{j_{|U_i \cap U_j}}, \qquad \text{for all} \quad i, j \in I \text{ such that } U_i \cap U_j \neq \emptyset,$$

*then there exists $f \in \mathcal{F}(U)$ such that $f_{|U_i} = f_i$ for all $i \in I$.*

**Remark 1.1.11.** A simple way to rephrase this definition consists in saying that a presheaf is a sheaf when the notions of "equality" and of "membership" are localized (this corresponds exactly to points 1 and 2 in Definition 1.1.43).

**Remark 1.1.12.** Sheaves on a topological space $X$ are a category.

**Remark 1.1.13.** It is possible to introduce the notion of sheaf of rings and other algebraic structures. In particular, if $\mathcal{R}$ is a sheaf of commutative rings, $\mathcal{F}$ is a sheaf of $\mathcal{R}$-modules if $\mathcal{R}(U)$ is a commutative ring, $\mathcal{F}(U)$ is a module over $\mathcal{R}(U)$ for any open set $U \subset X$ and the restriction map $\rho_V^U : \mathcal{R}(U) \to \mathcal{R}(V)$ is a ring homomorphism for each $V \subset U$. In general, we will restrict our attention to the case of abelian groups, unless otherwise stated.

**Example 1.1.19.** The presheaf $B$ of the previous example is not a sheaf since property 2 clearly fails to hold. For a function to be bounded, it is not sufficient to be locally bounded. On the other hand, it is easy to verify that $\mathcal{E}$ is a sheaf since both continuity and differentiability are local properties (compare with Definition 1.2.18).

**Definition 1.1.44.** *For a point $x \in X$ the abelian group*

$$\mathcal{F}_x = \varinjlim_{U \ni x} \mathcal{F}(U)$$

*is called the stalk of the (pre)sheaf $\mathcal{F}$ at $x$. Given $f \in \mathcal{F}(U)$ and $x \in U$, the image of $f$ in $\mathcal{F}_x$ is called a germ of $f$ at $x$ and is denoted by $f_x$.*

**Definition 1.1.45.** *Let $\mathcal{F}$ be a (pre)sheaf on the topological space $X$ and let $S$ be a subset of $X$. Let $f$ be a function assigning to each point $x \in S$ an element $f(x) \in \mathcal{F}_x$ such that for each $y \in S$ there is a neighborhood $U$ of $y$ and $s \in \mathcal{F}(U)$ such that $f(x) = s_x$ for every $x \in U \cap S$. We call $f$ a section of $\mathcal{F}$ over $S$.*

The group of sections of $\mathcal{F}$ over $S$ is denoted by $\Gamma(S, \mathcal{F})$. It is possible to show that, given a presheaf $\mathcal{F}$, the assignment $U \to \Gamma(U, \mathcal{F})$ is a presheaf that turns out to be a sheaf called the sheaf of sections. When $\mathcal{F}$ is a sheaf, it can be identified with its sheaf of sections so, for each open set $U$, we may also identify $\Gamma(U, \mathcal{F})$ and $\mathcal{F}(U)$. Note that the notation $\Gamma(U, \mathcal{F})$ makes sense for any subset $U$ of $X$.

**Definition 1.1.46.** *The support of a section $s \in \mathcal{F}(U)$, denoted by* supp $s$, *is the complement in $U$ of the largest open subset of $U$ on which $s$ is zero.*

By definition, the support of a section is a closed set in $X$. If $S$ is any subset of $X$ we will denote by $\Gamma_S(X, \mathcal{F})$ the group of sections in $\Gamma(X, \mathcal{F})$ whose support is contained in $S$. When $X$ is a locally compact topological space, we will also consider the group $\Gamma_c(X, \mathcal{F})$ of sections having compact support.

Throughout this book, we will continuously use some very important sheaves which we want to introduce right away.

On $\mathbb{R}^n$ we can define:

- the sheaf $\mathcal{E}$ of infinitely differentiable functions;
- the sheaf $\mathcal{D}'$ of Schwartz distributions (i.e., the sheaf whose groups of sections are the dual of the spaces of compactly supported infinitely differentiable functions);
- the constant sheaf $G$ where $G = \mathbb{Z}$, $\mathbb{R}$, $\mathbb{C}$, such that $G(U)$ consists of locally constant $G$-valued functions on $U$;
- the sheaf $\mathcal{A}$ of real analytic functions;
- the sheaf $\mathcal{B}$ of hyperfunctions.

On $\mathbb{C}^n$ we define the sheaf $\mathcal{O}$ of holomorphic functions.

There is an equivalent way to define sheaves on a topological space which we mention for sake of completeness.

**Definition 1.1.47.** *A sheaf is a triple $(\mathcal{S}, X, \pi)$ where $\mathcal{S}$, $X$ are topological spaces, and $\pi : \mathcal{S} \to X$ is a continuous, surjective, local homeomorphism such that $\pi^{-1}(x) = \mathcal{S}_x$ is a topological group with respect to the topology induced by $\mathcal{S}$.*

With this definition of sheaf, a section $s$ on an open set $U \subset X$ is any continuous map $s : U \to \mathcal{S}$ such that $\pi \circ s : U \to U$ is the identity map. The set of sections over $U$ is a group denoted by $\Gamma(U, \mathcal{S})$.

**Remark 1.1.14.** It is possible to show [74] that the assignment $U \to \Gamma(U, \mathcal{S})$ generates a sheaf according to the first given definition. Analogously, one can show that $\mathcal{S}_x$ is the stalk in the sense of the first definition of sheaf. The two definitions of sheaves are fully equivalent, and one can canonically move between one and the other.

Homomorphisms of sheaves are defined as follows:

**Definition 1.1.48.** *Given two sheaves $\mathcal{F}$, $\mathcal{G}$ on the same topological space $X$, a sheaf homomorphism $f: \mathcal{F} \to \mathcal{G}$ is a collection $f = \{f_U\}$ of group homomorphisms $f_U: \mathcal{F}(U) \to \mathcal{G}(U)$ such that for every inclusion $V \subset U$ the following diagram commutes:*

$$\begin{array}{ccc} \mathcal{F}(U) & \xrightarrow{f_U} & \mathcal{G}(U) \\ \rho\downarrow & & \downarrow\rho' \\ \mathcal{F}(V) & \xrightarrow{f_V} & \mathcal{G}(V). \end{array}$$

*If the maps $f_U$ are isomorphisms for all $U$, then $f$ is called an isomorphism.*

From the definition of homomorphisms of sheaves we also get the notion of subsheaf.

**Definition 1.1.49.** *Given a sheaf $\mathcal{F}$ on a topological space $X$, a subsheaf $\mathcal{G}$ on $X$ is an assignment $\{U, \mathcal{G}(U)\}$ with $\mathcal{G}(U)$ subgroups of $\mathcal{F}(U)$ such that the inclusions $\{i_U: \mathcal{G}(U) \hookrightarrow \mathcal{F}(U)\}$ give a sheaf homomorphism.*

**Example 1.1.20.** It is possible to prove the following chain of sheaves inclusions (on $\mathbb{R}^n$):

$$\mathcal{A} \subset \mathcal{E} \subset \mathcal{D}' \subset \mathcal{B}.$$

**Definition 1.1.50.** *Given two sheaves $\mathcal{F}$, $\mathcal{G}$ on the same topological space $X$, $\mathcal{G} \subset \mathcal{F}$, we define the quotient sheaf $\mathcal{F}/\mathcal{G}$ as the sheaf associated to the presheaf $U \to \mathcal{F}(U)/\mathcal{G}(U)$.*

We will use the notation $\mathcal{F}_{|U}$ to denote the restriction of sheaf of $\mathcal{F}$ to $U \subseteq X$: it is the sheaf associated to the presheaf

$$U \cap V \to \varinjlim_{\Omega}\{\mathcal{F}(\Omega) \ : \ \Omega \text{ is an open set in } X \text{ containing } U \cap V\}.$$

Every sheaf homomorphism $\mathcal{F} \longrightarrow \mathcal{G}$ induces a homomorphism $\mathcal{F}_{|U} \longrightarrow \mathcal{G}_{|U}$. Moreover, we have a functor

$$U \longrightarrow \mathrm{Hom}(\mathcal{F}_{|U}, \mathcal{G}_{|U})$$

which defines a presheaf and whose associated sheaf is denoted by

$$\mathcal{H}\mathrm{om}(\mathcal{F}, \mathcal{G}).$$

Note that if $\mathcal{R}$ is a sheaf of rings on $X$ and $\mathcal{F}$, $\mathcal{G}$ are sheaves of $\mathcal{R}$-modules, then one can also define the sheaf (see [79], [101])

$$\mathcal{H}\mathrm{om}_{\mathcal{R}}(\mathcal{F}, \mathcal{G}).$$

Given a sheaf homomorphism $\varphi: \mathcal{F} \to \mathcal{G}$, for every $x \in X$ there is an induced map homomorphism between the stalks $\varphi_x: \mathcal{F}_x \to \mathcal{G}_x$.

**Definition 1.1.51.** *A sequence of sheaves homomorphisms*

$$\mathcal{F} \xrightarrow{\varphi} \mathcal{G} \xrightarrow{\psi} \mathcal{H}$$

*is said to be exact if for every $x \in X$ the sequence*

$$\mathcal{F}_x \xrightarrow{\varphi_x} \mathcal{G}_x \xrightarrow{\psi_x} \mathcal{H}_x$$

*on the stalks is exact, i.e.,* $\ker \psi_x = \operatorname{im} \varphi_x$. *The homomorphism $\varphi : \mathcal{F} \to \mathcal{G}$ is a monomorphism (resp. epimorphism) if $0 \to \mathcal{F} \xrightarrow{\varphi_x} \mathcal{G}$ (resp. $\mathcal{F} \xrightarrow{\varphi_x} \mathcal{G} \to 0$) is exact.*

**Definition 1.1.52.** *Let $\mathcal{F}$ be a sheaf on a topological space $X$. A resolution of $\mathcal{F}$ is an exact sequence of sheaves*

$$0 \to \mathcal{F} \to \mathcal{F}^0 \to \mathcal{F}^1 \to \dots .$$

*The smallest integer $n$ such that $\mathcal{F}^i = 0$ for $i > n$ is called the length of the resolution.*

Resolutions are a classical topic in sheaf theory and constitute a suitable tool to translate some very well-known analytical properties of sheaves. Conversely, it is necessary to find resolutions with suitable properties to compute algebraic invariants of a sheaf, such as its cohomology groups that are the next tool to be defined.

We have seen that exactness for maps between sheaves coincides with exactness at the stalk level, but it does not imply exactness at the section level. In fact, given the short exact sequence

$$0 \to \mathcal{F} \xrightarrow{\varphi} \mathcal{G} \xrightarrow{\psi} \mathcal{H} \to 0$$

one can show that, for any open set $U$, the sequence

$$0 \to \Gamma(U, \mathcal{F}) \xrightarrow{\varphi_U} \Gamma(U, \mathcal{G}) \xrightarrow{\psi_U} \Gamma(U, \mathcal{H})$$

is exact but, in general, the map $\psi_U$ is not surjective. In other words, the functor $\Gamma(U, -)$ on the category of sheaves is left exact (see Definition 1.4.6). Also the functors $\Gamma_c(U, -)$ and $\Gamma_S(U, -)$, where $S$ is any subset of $X$, are left exact. In order to complete the sequence in an exact way, one needs to introduce the notion of cohomology with coefficients in a sheaf via the right derived functors of the above left exact functors:

**Definition 1.1.53.** *Let $\mathcal{F}$ be a sheaf of $\mathcal{R}$-modules on a topological space $X$ and let $S$ be any subset of $X$. We define:*

1. *the sheaf cohomology groups of $\mathcal{F}$ on $S$ as*

$$H^n(S, \mathcal{F}) := R^n\Gamma(S, \mathcal{F}), \quad n \geq 0.$$

2. *the sheaf cohomology groups of* $\mathcal{F}$ *on* $X$ *with support in* $S$ *as*

$$H^n_S(X,\mathcal{F}) := R^n\Gamma_S(X,\mathcal{F}), \quad n \geq 0.$$

3. *the sheaf cohomology groups of* $\mathcal{F}$ *on* $X$ *with compact support as*

$$H^n_c(X,\mathcal{F}) := R^n\Gamma_c(X,\mathcal{F}), \quad n \geq 0.$$

The cohomology groups of a sheaf $\mathcal{F}$ can be computed if one knows any resolution of $\mathcal{F}$ by sheaves that are "acyclic" for the functor $\Gamma$ (or $\Gamma_S$ or $\Gamma_c$), where acyclic means that if

$$0 \to \mathcal{F} \to \mathcal{F}^0 \to \mathcal{F}^1 \to \dots$$

is a resolution of $\mathcal{F}$ then, for every $i$, $R^n\Gamma(U,\mathcal{F}^i) = 0$ for $n > 0$. In fact, the cohomology of this resolution is isomorphic to the cohomology with coefficients in $\mathcal{F}$. This result follows from a general fact. Consider a covariant left exact functor $F : \mathcal{C} \to \mathcal{C}'$ where $\mathcal{C}$, $\mathcal{C}'$ are two abelian categories and $\mathcal{C}$ has enough injectives (hypothesis that is satisfied by sheaves of $\mathcal{R}$-modules). Let

$$\mathcal{I} : A \longrightarrow I_0 \longrightarrow I_1 \longrightarrow \dots$$

be an injective resolution of $A$ in $\mathcal{C}$ by $F$-acyclic objects, i.e., for every $i$ it is $R^nF(I_i) = 0$ for $n > 0$, then $R^nF(A)$ is isomorphic to $H^n(F(\mathcal{I}))$ for any $n \geq 0$ (see also e.g., [79]).

Among the classes of sheaves that are acyclic for the functor $\Gamma$ there are some remarkable ones.

**Definition 1.1.54.** *A sheaf* $\mathcal{F}$ *on* $X$ *is said to be flabby if for every open set* $U \subset X$ *the restriction map* $\mathcal{F}(X) \to \mathcal{F}(U)$ *is surjective.*

Then in a flabby sheaf $\mathcal{F}$, for any section $s$ on $U$ there exist $f \in \mathcal{F}(X)$ such that $f_{|U} = s$.

**Theorem 1.1.9.** *In the category of sheaves of* $\mathcal{R}$*-modules, flabby sheaves are acyclic.*

**Definition 1.1.55.** *A sheaf* $\mathcal{F}$ *on* $X$ *is said to be soft if the restriction map* $\mathcal{F}(X) \to \mathcal{F}(S)$ *is surjective for every closed set* $S \subset X$.

**Definition 1.1.56.** *A locally finite covering of the open set* $U$ *is a family of countable open sets* $U_k$ $(k \in \mathbb{N})$ *such that*

- $U = \bigcup_{k\geq 1} U_k$*, where* $U_k$ *are bounded, compact subsets of* $U$ *such that* $\overline{U}_k$ *are contained in* $U$*;*
- *every compact set* $K$ *in* $U$ *intersects only a finite number of* $\{U_k\}$.

We define partition of the unity for the sheaf $\mathcal{F}$, subordinate to any locally finite covering $\{U_\alpha\}$ of $X$, a family $h_\alpha \in \mathrm{Hom}(\mathcal{F},\mathcal{F})$ such that $\mathrm{supp}(h_\alpha) \subset U_\alpha$ and $\sum_\alpha h_\alpha = 1$.

**Definition 1.1.57.** *A sheaf is called fine if it is has a partition of the unity subordinate to any locally finite open covering of $X$.*

**Remark 1.1.15.** Let $X$ be a paracompact topological space, i.e., for any open covering there is a locally finite refinement (this assumption is certainly satisfied by all the Euclidean spaces we are interested in). Then a flabby sheaf of $\mathcal{R}$-modules is also a soft sheaf. Fine sheaves are soft and soft sheaves are acyclic.

**Example 1.1.21.** The sheaf of infinitely differentiable functions on $\mathbb{R}^n$ is a fine sheaf and any sheaf of modules on it is fine.

Let us discuss some examples of resolutions.

- *De Rham resolution.* Let $\mathbb{C}_{\mathbb{R}^n}$ be the constant sheaf on $\mathbb{R}^n$, let $\mathcal{E}^{(k)}$ be the sheaf of $k$-differential forms with $C^\infty$ coefficients and values in $\mathbb{C}$. Let $d$ be the exterior differential operator acting on $\mathcal{E}^{(k)}$ which, in local coordinates, can be written as

$$d(\phi_{i_1\dots i_k} dx_{i_1} \wedge \dots \wedge dx_{i_k}) = \sum_i \frac{\partial \phi_{i_1\dots i_k}}{\partial x_i} dx_i \wedge dx_{i_1} \wedge \dots \wedge dx_{i_k}.$$

We have the following fine resolution of the constant sheaf $\mathbb{C}$:

$$0 \longrightarrow \mathbb{C} \longrightarrow \mathcal{E}^{(0)} \xrightarrow{d} \mathcal{E}^{(1)} \xrightarrow{d} \dots \xrightarrow{d} \mathcal{E}^{(n)} \longrightarrow 0.$$

The sequence allows the computation of the cohomology with coefficients in $\mathbb{C}$. Analogous sequences can be written by replacing the sheaf $\mathcal{E}^{(k)}$ by $\mathcal{A}^{(k)}$ and $\mathcal{O}^{(k)}$ where the coefficients of the $k$-differential forms are real analytic and holomorphic, respectively.

- *Dolbeault resolution.* If one considers the differential forms of type $(p, q)$, i.e., elements of the form

$$\omega = \sum \phi_{i_1\dots i_p j_1 \dots j_q} dz_{i_1} \wedge \dots \wedge dz_{i_p} \wedge d\bar{z}_{j_1} \wedge \dots \wedge d\bar{z}_{j_q},$$

and if one denotes the space of $(p,q)$-forms with $C^\infty$ coefficients on an open set $U \subseteq \mathbb{C}^n \cong \mathbb{R}^{2n}$ by $\mathcal{E}^{p,q}(U)$, one has the following sequence called the Dolbeault sequence

$$0 \longrightarrow \mathcal{O} \longrightarrow \mathcal{E}^{p,0} \xrightarrow{\bar\partial} \mathcal{E}^{p,1} \xrightarrow{\bar\partial} \dots \xrightarrow{\bar\partial} \mathcal{E}^{p,n} \longrightarrow 0,$$

in which the maps are the usual exterior differentiations (see [110] for the details)

$$\bar\partial\omega = \sum_i \frac{\partial \phi_{i_1\dots i_p j_1\dots j_q}}{\partial \bar{z}_i} d\bar{z}_i \wedge dz_{i_1} \wedge \dots \wedge dz_{i_p} \wedge d\bar{z}_{j_1} \wedge \dots \wedge d\bar{z}_{j_q}.$$

The Dolbeault sequence provides a fine resolution of the sheaf $\mathcal{O}$, so it can be used to compute its cohomology groups.

It is possible to show, see [79], that every sheaf has a flabby resolution that can be constructed in an explicit way and, as we have already observed, we can compute the cohomology through flabby resolutions. Another possible way to compute the cohomology groups of a sheaf, more geometrical, is the use of so-called Čech cohomology. This is a rather complex construction and we provide here only the fundamental ideas to familiarize the reader with this topic which will be useful in the next chapter for several results.

Let $\mathcal{U} = \{U_i\}_{i\in I}$ be an open covering of $X$ and let $\mathcal{F}$ be a sheaf on $X$. A $q$-cochain is an element

$$f = \{f_{i_0 i_1 \ldots i_q}\} \in \prod_{(i_0 \ldots i_q)\in I^{q+1}} \mathcal{F}(U_{i_0} \cap U_{i_1} \cap \ldots \cap U_{i_q})$$

for each nonempty intersection $U_{i_0} \cap \ldots \cap U_{i_q}$. The set of $q$-cochains is denoted by $C^q(\mathcal{U}, \mathcal{F})$ and it is a group under formal addition. We can define a homomorphism (the so-called coboundary operator)

$$\delta^q : C^q(\mathcal{U}, \mathcal{F}) \to C^{q+1}(\mathcal{U}, \mathcal{F})$$

by means of the restrictions as follows:

$$(\delta^q f)_{i_0 i_1 \ldots i_{q+1}} = \sum (-1)^{i_j} \rho^{U_{i_0} \cap \ldots \widehat{U_{i_j}} \ldots \cap U_{i_{q+1}}}_{U_{i_0} \cap \ldots \cap U_{i_j} \cap \ldots \cap U_{i_{q+1}}} f_{i_0 \ldots \widehat{i_j} \ldots i_{q+1}}$$

where, as usual, the hat symbol denotes the omission of the corresponding object. For example, if $f \in C^0(\mathcal{U}, \mathcal{F})$,

$$(\delta^0 f)_{UV} = f_{|V} - f_{|U}.$$

A $q$-cochain is called a cocycle if $\delta^q f = 0$, while it is called a coboundary if there exists a $(q-1)$-cochain $g$ such that $f = \delta^{q-1} g$. Since one can verify that $\delta^{q+1} \circ \delta^q = 0$, the pair $\{C^q, \delta^q\}$ forms a complex whose cohomology groups $\check{H}^q(\mathcal{U}, \mathcal{F})$ are called cohomology groups of the covering $\mathcal{U}$ with coefficients in $\mathcal{F}$. To define the cohomology groups $\check{H}^q(X, \mathcal{F})$ of the space $X$, it is necessary to consider all the possible coverings of $X$ and take an inductive limit. Let $\mathcal{V} = \{V_\beta\}$ be a refinement of the covering $\mathcal{U}$, i.e., a covering for which there exists a map $\mu : \mathcal{V} \longrightarrow \mathcal{U}$ such that for every $V_\beta \in \mathcal{V}$, there exists $U_{\beta'} \in \mathcal{U}$ such that $V_\beta \subseteq U_{\beta'}$. Then there is an induced map

$$\mu : C^q(\mathcal{U}, \mathcal{F}) \longrightarrow C^q(\mathcal{V}, \mathcal{F})$$

and an induced homomorphism

$$\mu^* : \check{H}^q(\mathcal{U}, \mathcal{F}) \longrightarrow \check{H}^q(\mathcal{V}, \mathcal{F}).$$

It is possible to show that two different refining maps $\mu$, $\nu$ give the same monomorphisms at the cohomology level, i.e., $\mu^* = \nu^*$. Since the set of coverings of the topological space $X$ is partially ordered with respect to the notion of refinement, and since the homomorphisms $\mu^*$ acting at the level of cohomology are transitive, it is possible to introduce the inductive limit

$$\check{H}^q(X,\mathcal{F}) = \varinjlim_{\mathcal{U}} \check{H}^q(\mathcal{U},\mathcal{F})$$

called the $q$-th Čech cohomology group of $X$ with coefficients in $\mathcal{F}$. The reader who is interested in making our argument precise may refer to e.g., [30]. Obviously, it is almost impossible to work with the definition of cohomology via the direct limit, so it is better to look for conditions under which the cohomology can be computed using a chosen covering of $X$. This can be done for example under the hypothesis of the following theorem, due to Leray.

**Theorem 1.1.10.** (Leray). *If the covering $\mathcal{U}$ of the topological space $X$ is such that*

$$\check{H}^n(U_{i_1} \cap \ldots \cap U_{i_p}, \mathcal{F}) = 0, \qquad n > 0, \text{ for any } i_1, \ldots, i_p,$$

*then $\check{H}^q(X,\mathcal{F}) = \check{H}^q(\mathcal{U},\mathcal{F})$, for all $q \geq 0$.*

The Čech cohomology allows one to compute the cohomology, in the sense of Definition 1.1.53, by virtue of the following theorem which holds under the assumption that the topological space $X$ is paracompact.

**Theorem 1.1.11.** *If $X$ is a paracompact topological space, then*

$$\check{H}^q(X,\mathcal{F}) \cong H^q(X,\mathcal{F}), \qquad \forall\, q \geq 0.$$

For the applications we are interested in, we need some more tools to compute the cohomology groups with support in a given set $A$, also called relative cohomology groups of the pair $(X, X \setminus A)$. Let us consider a topological space $X$ and let $A \subset X$ be a locally closed set, i.e., a set that can be written as the intersection of an open and a closed set in $X$. Let us denote by $\mathcal{U} = \{U_i\}_{i\in I}$ an open covering of $X$ and by $\mathcal{U}' = \{U_i\}_{i\in J}$, $J \subset I$ a subcovering of $\mathcal{U}$ that is a covering of $X \setminus A$. We will say that $\mathcal{U}$, $\mathcal{U}'$ is a relative covering of the pair $(X, X \setminus A)$. We can introduce the set of relative $q$-cochains $C^q(\mathcal{U},\mathcal{U}',\mathcal{F})$ that is the subset of $C^q(\mathcal{U},\mathcal{F})$ containing the elements of the type

$$f = \{f_{i_0 i_1 \ldots i_q}\} \in \prod_{(i_0 \ldots i_q)\in I^{q+1}} \mathcal{F}(U_{i_0} \cap U_{i_1} \cap \ldots \cap U_{i_q}),$$

$$f_{i_0 i_1 \ldots i_q} = 0 \ \text{ if } U_{i_0}, \ldots, U_{i_q} \in \mathcal{U}'.$$

A coboundary operator $\delta^q$ is defined in the natural way

$$\delta^q : C^q(\mathcal{U},\mathcal{U}',\mathcal{F}) \longrightarrow C^{q+1}(\mathcal{U},\mathcal{U}',\mathcal{F}).$$

Then the pair

$$\{C^q(\mathcal{U},\mathcal{U}',\mathcal{F}), \delta^q\}$$

forms a complex whose cohomology $H^q(\mathcal{U},\mathcal{U}',\mathcal{F})$ is called the $q$-th cohomology group of the relative covering $\mathcal{U},\mathcal{U}'$ with coefficients in $\mathcal{F}$. Now, following a technique similar to the one already used before, one has to define refinements of the covering $(\mathcal{U},\mathcal{U}')$ and show that the cohomologies $H^q(\mathcal{U},\mathcal{U}',\mathcal{F})$ form an inductive family with respect to the refinements. Taking the inductive limit we get a notion of cohomology that does not depend on the covering. The procedure in the relative case is more involved than in the absolute case, but it can be carried out, for example, if both $X$ and $X \setminus A$ are both paracompact spaces. The construction can be repeated in the case $A$ is either a closed or an open set. What we obtain is the following:

$$H^q_A(X,\mathcal{F}) = \varinjlim_{\mathcal{U},\mathcal{U}'} H^q(\mathcal{U},\mathcal{U}',\mathcal{F}),$$

i.e., the relative Čech cohomology of the pair $(X, X \setminus A)$. When $X = A$ we get the (absolute) Čech cohomology $H^q(X,\mathcal{F})$. (Note that in the literature one can also find the notation $H^q(X, X \setminus A, \mathcal{F})$ instead of $H^q_A(X,\mathcal{F})$).

Although this construction may appear exceedingly abstract, it is possible, in many concrete cases, to actually compute the cohomology groups. When the hypotheses of Leray's theorem are satisfied, it is possible, also in the relative case, to compute the cohomology using a suitable covering. We will see some of these computations in the chapter to follow.

The sheaf cohomology satisfies some fundamental properties (usually called the cohomology axioms) that are listed in the following theorem.

**Theorem 1.1.12.** *Let $X$ be a topological space, $S \subset X$ a locally closed set and $A \subset S$ a closed subset of $S$. Let $\mathcal{F}$ be a sheaf on $X$. Then we have*

1. (Excision Theorem). *For any open set $U$ containing $S$ it is $H^p_S(X,\mathcal{F}) = H^p_S(U,\mathcal{F}_{|U})$.*

2. $H^0_S(X,\mathcal{F}) = \Gamma_S(X,\mathcal{F})$.

3. *If $\mathcal{F}$ is a flabby sheaf, then $H^p_S(X,\mathcal{F}) = 0$, $p \geq 1$.*

4. (Fundamental long exact sequence). *Let us consider*

$$0 \to H^0_A(X,\mathcal{F}) \to H^0_S(X,\mathcal{F}) \to H^0_{S\setminus A}(X,\mathcal{F}) \to$$

$$\to H^1_A(X,\mathcal{F}) \to H^1_S(X,\mathcal{F}) \to H^1_{S\setminus A}(X,\mathcal{F}) \to \dots .$$

*If $S = U$ is an open set, then the sequence becomes*

$$0 \to H^0_A(U,\mathcal{F}) \to H^0(U,\mathcal{F}) \to H^0(U \setminus A,\mathcal{F}) \to \tag{1.1}$$

$$\to H^1_A(U,\mathcal{F}) \to H^1(U,\mathcal{F}) \to H^1(U \setminus A,\mathcal{F}) \to \dots .$$

5. *If*

$$0 \to \mathcal{F} \to \mathcal{G} \to \mathcal{S} \to 0$$

*is a short exact sequence of sheaves, then the induced cohomology sequence*

$$0 \to H^0_S(X, \mathcal{F}) \to H^0_S(X, \mathcal{G}) \to H^0_S(X, \mathcal{S}) \to$$

$$\to H^1_S(X, \mathcal{F}) \to H^1_S(X, \mathcal{G}) \to H^1_S(X, \mathcal{S}) \to \dots$$

*is exact.*

**Remark 1.1.16.** The fundamental long exact sequence (1.1) holds also when $A$ is an open set under the additional hypothesis that the topological space $X$ is paracompact.

We conclude this section introducing a notion that will be crucial in the definition of hyperfunctions.

**Definition 1.1.58.** *Let $X$ be a topological space, $S$ a locally closed set in $X$ and $\mathcal{F}$ a sheaf on $X$. The $n$-th derived sheaf of $\mathcal{F}$ is the sheaf, denoted by $\mathcal{H}^n_S(\mathcal{F})$, associated to the presheaf*

$$U \to H^n_{S \cap U}(U, \mathcal{F}).$$

The following result assures that the previous assignment gives a sheaf (see [98]).

**Theorem 1.1.13.** *The presheaf*

$$U \to H^n_{S \cap U}(U, \mathcal{F}) \tag{1.2}$$

*is a sheaf when $n = 0$. If $n \geq 1$ and $\mathcal{H}^j_S(\mathcal{F}) = 0$ for $0 \leq j \leq n-1$, then the presheaf (1.2) is a sheaf and $\mathcal{H}^n_S(\mathcal{F})(U) = H^n_{S \cap U}(U, \mathcal{F})$.*

From the cohomology axioms we have the following result.

**Theorem 1.1.14.** *If*

$$0 \to \mathcal{F} \to \mathcal{G} \to \mathcal{S} \to 0$$

*is a short exact sequence of sheaves, then the sequence*

$$0 \to \mathcal{H}^0_S(\mathcal{F}) \to \mathcal{H}^0_S(\mathcal{G}) \to \mathcal{H}^0_S(\mathcal{S}) \to$$

$$\to \mathcal{H}^1_S(\mathcal{F}) \to \mathcal{H}^1_S(\mathcal{G}) \to \mathcal{H}^1_S(\mathcal{S}) \to \dots$$

*is exact.*

## 1.2 Analytical tools

### 1.2.1 Topological linear spaces

Sets of mathematical objects such as functions, measures, distributions, operators and many others can be seen as vector spaces over the real or the complex field. However, since limit processes in analysis play a fundamental role, we are led to consider vector spaces with topologies that are naturally defined. Such topologies can be introduced in the simplest way by norms or, for a larger class of objects, by seminorms. In this section we recall some of the fundamental definitions on topological linear spaces. For the details of most of the proofs of the results stated in the following we refer to [115]. We denote by $\mathbb{K}$ either $\mathbb{R}$ or $\mathbb{C}$, while $X$ will denote a linear space over the field $\mathbb{K}$. The vector addition is denoted by $+$, and the scalar multiplication is indicated by $\cdot$ ; this symbol is often omitted when no confusion arises.

**Definition 1.2.1.** *A linear space $X$ over $\mathbb{K}$ is called a topological linear space over $\mathbb{K}$ if there exists a topology $\mathcal{T}$ on $X$ such that the vector addition and the scalar multiplication are continuous.*

Instead of giving the topology $\mathcal{T}$ by assigning the family of open sets in $X$, it is sometimes useful to introduce a basis for the topology $\mathcal{T}$, i.e., a subset $\mathcal{T}' \subset \mathcal{T}$ such that every open set in $\mathcal{T}$ is a union of elements in $\mathcal{T}'$. A collection $\tau$ of open sets containing a given point $x \in X$ is said a local basis at $x$ if every open set containing $x$ contains an element of $\tau$. As we will show below, it is possible to assign a topology on $X$ when in it is defined a norm or, more generally, a family of seminorms.

**Definition 1.2.2.** *Let $X$ be a linear space over $\mathbb{K}$. A map $p : X \to \mathbb{R}^+ \cup \{0\}$ is said to be a seminorm on $X$ if*

- $p(u+v) \leq p(u) + p(v)$, *for any* $u, v \in X$,
- $p(\mu \cdot u) \leq |\mu| \cdot p(u)$, *for any* $u \in X$, $\mu \in \mathbb{K}$.

*If $p(u) = 0$ implies $u = 0$ then $p$ is called a norm on $X$ and we will denote $p(u)$ by $\|u\|_X$ or simply by $\|u\|$.*

**Definition 1.2.3.** *Let $X$ be a linear space over $\mathbb{K}$. We say that $X$ is a seminormed linear space if there exists a family of seminorms $\Phi = \{p_\gamma\}_{\gamma \in A}$, where $A$ is a set of indices, such that if $p_\gamma(u) = 0$ for all $\gamma \in A$ then $u = 0$.*

*If $X$ is a seminormed linear space and $\Phi = \{p\}$, i.e., $\Phi$ contains only one element, then $X$ is called a normed linear space and $p(u)$ is called the norm of $u$.*

There may be many families of seminorms under which $X$ is a seminormed linear space, so we will use the notation $(X, \Phi)$, to indicate the family of seminorms in use. In the case of a normed space $X$, we will use the simpler notation

$(X, \|\cdot\|)$ instead of $(X, \Phi)$. Obviously, when $X$ is a normed linear space, the norm is the natural candidate to introduce a topology on $X$. In fact, a normed space $(X, \|\cdot\|)$ is also a metric space in which the distance $d(x, y)$ is given by $d(x, y) = \|x - y\|$. We can assign a topology in $X$ by giving a local basis at $x$, for any $x \in X$, which, for example, is the collection of open balls $B(x, r) = \{y : d(x, y) < r\}$ with center at $x \in X$ and radius $r$. A set $U$ will be called an open set if for any $x \in X$ there exists $\varepsilon > 0$ such that $B(x, \varepsilon) \subseteq U$. If $X$ is a seminormed space, it is still possible to introduce a topology through a family of seminorms $\Phi$ following a similar idea. To this purpose we need some more notations:

**Definition 1.2.4.** *Let $(X, \Phi)$ be a seminormed linear space over $\mathbb{K}$. For every $u \in X$, $\varepsilon > 0$, $n \in \mathbb{N}$, and for any set $\{p_1, \dots, p_n\}$ of seminorms in $\Phi$, we define*

$$\mathcal{U}_{p_1,\dots,p_n}(u, \varepsilon) = \{v \in X : p_j(v - u) < \varepsilon,\ j = 1, \dots, n\}.$$

*Moreover, we set*

$$\mathcal{U}_\Phi(u) = \{\ \mathcal{U}_{p_1,\dots,p_n}(u, \varepsilon) : \quad \varepsilon > 0,\ n \in \mathbb{N},\ p_1, \dots, p_n \in \Phi\ \}$$

*and*

$$\mathcal{U}_\Phi = \bigcup_{u \in X} \mathcal{U}_\Phi(u).$$

The next proposition states that we can introduce a topology in a seminormed space using the family of sets $\mathcal{U}_\Phi$ as a basis. Moreover, $X$ turns out to be a Hausdorff space because of the crucial property of seminormed linear space that not all the seminorms in $\Phi$ vanish at any nonzero vector in $X$.

**Proposition 1.2.1.** *Let $(X, \Phi)$ be a seminormed linear space. Then $\mathcal{U}_\Phi$ is a basis for a Hausdorff topology.*

The topology of the seminormed linear space $(X, \Phi)$ generated by $\mathcal{U}_\Phi$ will be denoted by $\mathcal{T}_\Phi$, while $X$ with this topology will be denoted by $(X, \mathcal{T}_\Phi)$. We wish now to provide some examples of spaces of functions with the topology induced by a norm or a family of seminorms.

A multi-index $\alpha = (\alpha_1, \dots, \alpha_n)$ is an ordered set of integers $\alpha_j$ for $j = 1, \dots, n$, and the sum $|\alpha| = \alpha_1 + \dots + \alpha_n$ is called its length. If we set $\dfrac{\partial}{\partial x_j} = \partial_{x_j}$, we use the symbol $D$ to define the vector of first order differential operators

$$D = (\partial_{x_1}, \dots, \partial_{x_n}),$$

while the partial derivatives of order $\alpha$ of a (generalized) function $\phi$ will be denoted by

$$D^\alpha \phi(x) = \frac{\partial^{|\alpha|}}{\partial_{x_1}^{\alpha_1} \dots \partial_{x_n}^{\alpha_n}} \phi(x).$$

Let $\Omega \subseteq \mathbb{R}^n$ be an open set. We define $\mathcal{C}^k(\Omega)$, for $k \in \mathbb{N}$, to be the set of all continuous functions $\phi$ which are continuous with all their derivatives $D^\alpha \phi$ for $|\alpha| \leq k$. For $k = 0$, the space of continuous functions $\mathcal{C}^0(\Omega)$ will be denoted by $\mathcal{C}(\Omega)$. The set of functions $\phi \in \mathcal{C}^k(\Omega)$ which admit a continuous extension to the closure $\overline{\Omega}$ of $\Omega$, together with all their derivatives $D^\alpha \phi$ for $|\alpha| \leq k$, will be indicated by $\mathcal{C}^k(\overline{\Omega})$. We denote by $L^q(\Omega)$, $q \geq 1$ the usual Lebesgue spaces of measurable functions $\phi$ such that $|\phi|^q$ is integrable, and with $L^q_{\text{loc}}(\Omega)$ the set of measurable functions $\phi$ such that $|\phi|^q$ is integrable on every compact set of $\Omega$.

**Example 1.2.1.** Denote by $\mathcal{C}^\infty([0,1])$ the space of all the functions which are continuous with all their derivatives. With the family of seminorms $p_n(u) := \|D^n u\|_\infty$, $(\mathcal{C}^\infty([0,1]), \{p_n\})$ becomes a seminormed linear space.

**Example 1.2.2.** Important examples of spaces of functions with their natural norms are

- $\mathcal{C}^k(\Omega)$, for $k \in \mathbb{N}$, with the norm

$$\|\phi\|_{\mathcal{C}^k(\Omega)} = \sup_{x \in \Omega, |\alpha| \leq k} |D^\alpha \phi(x)|,$$

- $L^q(\Omega)$, for $q \geq 1$, with the norm

$$\|\phi\|_{L^q(\Omega)} = \left( \int_\Omega |\phi(x)|^q \, dx \right)^{1/q}.$$

For the reader acquainted with complete spaces, we point out that those two spaces are Banach spaces (see below for their definition).

We are now interested in looking for the conditions under which if $(X, \mathcal{T})$ is a topological linear space, then there exists a family of seminorms $\Phi$ on $X$ such that $(X, \Phi)$ is a seminormed linear space and $\mathcal{T} = \mathcal{T}_\Phi$. This is not possible in general as the following example shows.

**Example 1.2.3.** The spaces $L^q([0,1])$, with $0 < q < 1$, are neither normed linear spaces, since $\||u| + |v|\|_{L^q([0,1])} \geq \|u\|_{L^q([0,1])} + \|v\|_{L^q([0,1])}$ for any $u$, $v \in L^q([0,1])$ (see [1]), nor seminormed linear spaces, but they are topological linear spaces if one chooses the topology generated by the neighborhoods $\mathcal{U}(u, \varepsilon) = \{v \in L^q([0,1]) : \|u - v\|_{L^q([0,1])} < \varepsilon\}$. It is common to use the notation $\|\cdot\|_{L^q([0,1])}$ introduced in Example 1.2.2 even if, for $0 < q < 1$, $\|\cdot\|_{L^q([0,1])}$ represent only a positive functional.

The fundamental point of this problem lies in the nature of the open neighborhoods. In fact we have the following important proposition which clarifies the nature of the local basis for seminormed linear spaces.

**Proposition 1.2.2.** *Let $(X, \Phi)$ be a seminormed linear space over $\mathbb{K}$. Then $\mathcal{T}_\Phi$ contains a local basis at each point in $X$ that consists of convex open sets.*

Since the existence of such convex open sets is the main point, we are led to give the definition of locally convex topological linear space.

**Definition 1.2.5.** *A topological linear space $X$ over $\mathbb{K}$ is called a locally convex topological linear space if each point $u \in X$ has a local basis consisting of convex open sets.*

We can state the fundamental result of this theory: seminormed linear spaces and locally convex topological linear spaces are the same object.

**Theorem 1.2.1.** *Let $X$ be a topological linear space over $\mathbb{K}$. Then $X$ is a locally convex topological linear space over $\mathbb{K}$ if and only if there exists a family of seminorms $\Phi$ on $X$ such that $(X, \Phi)$ is a seminormed linear space over $\mathbb{K}$ and the topology $\mathcal{T}$ coincides with the seminorm topology $\mathcal{T}_\Phi$, defined by $\mathcal{U}_\Phi$.*

Among locally convex spaces, there is a class that is particular important: the so-called Fréchet spaces. From a historical point of view, the theory of Fréchet spaces and of their inductive limits (so-called LF spaces) are of great importance. As described in [119], during the early 40s Schwartz investigated how much of Banach duality theory could be extended to Fréchet spaces. This work, apparently abstract, culminated in a justly celebrated paper [57] on duality theory in FS and DFS spaces (we will use these notions in Chapter 2), and opened the way for one of the most impressive constructions of 20th century mathematics: the theory of distributions which is the topic of next section.

For our purposes, it is necessary to characterize the notion of limit in terms of seminorms for which we need a preliminary definition, generalizing the notion of sequence:

**Definition 1.2.6.** *Let $A$ be a directed set, i.e. a set with an order relation $\prec$ such that if $\alpha$ and $\beta \in A$ there exists $\gamma \in A$ with $\alpha \prec \gamma$ and $\beta \prec \gamma$. A net $\{u_\gamma\}$ is a map from $A$ into a topological space $X$.*

Note that the set of indices $A$ can be in particular $\mathbb{R}$ or $\mathbb{N}$, and in this last case a net is simply a sequence. Using nets, one can give the definition of complete space without having a metric space (compare with Example 1.2.2) but simply a topological space:

**Definition 1.2.7.** *Let $(X, \mathcal{T})$ be a topological linear space over $\mathbb{K}$. Then $X$ is said to be complete if, for every Cauchy net $u_\gamma$ in $X$, there exists $u \in X$ such that $u_\gamma \to u$ in the topology $\mathcal{T}$.*

In the case of seminormed linear spaces, the convergence of nets may be characterized in terms of seminorms:

**Theorem 1.2.2.** *Let $(X, \Phi)$ be a seminormed linear space over $\mathbb{K}$, let $\{u_\gamma\}$ be a net and $u \in X$. Then the net $\{u_\gamma\}$ converges to $u$ in $(X, \mathcal{T}_\Phi)$ if and only if for every $p \in \Phi$ we have $\lim_\gamma p(u_\gamma - u) = 0$.*

We can now give the definition of Fréchet space:

**Definition 1.2.8.** *Let $(X,\mathcal{T})$ be a topological linear space over $\mathbb{K}$. If $\mathcal{T}$ is metrizable (i.e., there exists a metric $\rho$ on $X$ which induces the topology $\mathcal{T}$) and $(X,\mathcal{T})$ is complete, then $X$ is called the Fréchet space. If $X$ is a normed complete linear space, then it is called a Banach space.*

The space $(\mathcal{C}^\infty([0,1]),\{p_n\})$, where $p_n(u) := \|D^n u\|_\infty$, is an example of Fréchet space which it is not a Banach space.

The characterization of metrizable locally convex linear spaces is given by the following result:

**Theorem 1.2.3.** *Let $(X,\mathcal{T})$ be a locally convex linear space over $\mathbb{K}$. Then $\mathcal{T}$ is metrizable if and only if there exists a countable family $\Phi$ of seminorms on $X$ such that $(X,\Phi)$ is a seminormed linear space.*

At this point it is necessary to introduce the concept of inductive and projective limit that is defined, in the general context of categories, in the appendix of this chapter. In the case of locally convex spaces it can be formulated as explained below after some preliminary definitions. In what follows, $A$ will always denote a directed set.

**Definition 1.2.9.** *Let $\{X_\alpha\}_{\alpha\in A}$ be a family of linear spaces. For any $\alpha,\beta$ with $\alpha \prec \beta$ let $\rho^\alpha_\beta : X_\alpha \to X_\beta$ (resp. $\rho^\beta_\alpha : X_\beta \to X_\alpha$) be a linear map. We say that $\{X_\alpha,\rho^\alpha_\beta\}$, $\alpha,\beta\in A$ (resp. $\{X_\alpha,\rho^\beta_\alpha\}$, $\alpha,\beta\in A$) is an increasing (resp. decreasing) family of linear spaces if $\rho^\alpha_\alpha$ is the identity map and for $\alpha\prec\beta\prec\gamma$ it is $\rho^\beta_\gamma\cdot\rho^\alpha_\beta=\rho^\alpha_\gamma$ (resp. $\rho^\beta_\alpha\cdot\rho^\gamma_\beta=\rho^\gamma_\alpha$).*

**Definition 1.2.10.** *Let $X$, $Y$ be linear spaces, $\{X_\alpha,\rho^\alpha_\beta\}_{\alpha,\beta\in A}$ be an increasing family of linear spaces and $\rho^\alpha : X_\alpha \to X$ be linear maps such that:*

1. *for $\alpha\prec\beta$ it is $\rho^\beta\cdot\rho^\alpha_\beta=\rho^\alpha$.*

2. *if $\phi^\alpha : X_\alpha \to Y$ are linear maps such that $\phi^\alpha\cdot\rho^\beta_\alpha=\phi^\beta$ for $\beta\prec\alpha$, then there is a unique map $\phi: X\to Y$ such that $\phi^\alpha=\phi\cdot\rho^\alpha$.*

*The linear space $X$ is called the inductive limit of the family $\{X_\alpha\}$ and the linear maps $\rho^\alpha$ are called canonical maps. We will write $X=\varinjlim X_\alpha$.*

Analogously, we have:

**Definition 1.2.11.** *Let $X$, $Y$ be linear spaces, $\{X_\alpha,\rho^\beta_\alpha\}_{\alpha,\beta\in A}$ be an decreasing family of linear spaces, and let $\rho_\alpha : X\to X_\alpha$ be linear maps such that:*

1. *for $\alpha\prec\beta$ it is $\rho^\beta_\alpha\cdot\rho_\beta=\rho_\alpha$;*

2. *if $\phi_\alpha : Y\to X_\alpha$ are linear maps such that $\rho^\beta_\alpha\cdot\phi_\beta=\phi_\alpha$ for $\alpha\prec\beta$ there is a unique linear map $\phi : Y\to X$ such that $\phi_\alpha=\rho_\alpha\cdot\phi$.*

*Then $X$ is called the projective limit of $\{X_\alpha\}_{\alpha\in A}$ and we will write $X=\varprojlim X_\alpha$.*

When considering families of locally convex linear spaces, we are interested in describing the topology inherited by the linear space obtained as limit of a family:

**Definition 1.2.12.** *We say that* $\{X_\alpha, \rho_\beta^\alpha\}$ *with* $\alpha \prec \beta$ *is an increasing (resp. decreasing) family of locally convex linear spaces, on a directed set* $A$, *if it is an increasing (resp. decreasing) family and the linear maps* $\rho_\beta^\alpha : X_\alpha \to X_\beta$ *(resp.* $\rho_\alpha^\beta : X_\beta \to X_\alpha$*) are continuous.*

**Definition 1.2.13.** *Given an increasing family of locally convex spaces on a directed set* $A$, *the inductive limit* $X$ *of this family, denoted by* $X = \varinjlim X_\alpha$, *is defined as the linear space* $X$ *equal to the inductive limit of* $X_\alpha$ *equipped with the strongest locally convex topology for which all the canonical maps* $\rho^\alpha$ *are continuous.*

**Definition 1.2.14.** *Given a decreasing family of locally convex spaces on a directed set* $A$, *the projective limit* $X$ *of this family, denoted by* $X = \varprojlim X_\alpha$, *is defined as the linear space* $X$ *equal to the projective limit of* $X_\alpha$ *equipped with the weakest locally convex topology for which all the canonical maps* $\rho_\alpha$ *are continuous.*

Among the locally convex spaces arising as projective or injective limit there are those spaces coming from a family of Banach spaces described in the following definitions.

**Definition 1.2.15.** *A locally convex space* $X$ *is an FS space (Fréchet–Schwartz space) if there is a decreasing sequence of Banach spaces*

$$X_1 \overset{\rho_1^2}{\leftarrow} X_2 \overset{\rho_2^3}{\leftarrow} \dots X_j \overset{\rho_j^{j+1}}{\leftarrow} X_{j+1} \leftarrow \dots$$

*such that*

$$X = \varprojlim X_j$$

*and all the mappings* $\rho_j^{j+1} : X_{j+1} \to X_j$ *are compact, i.e., the sets* $\{\rho_j^{j+1}(x) : \|x\|_{X_{j+1}} \le 1\}$ *are relatively compact in* $X_j$.

Note that the topology on $X$ is given by the seminorms $\|\rho_j(x)\|_{X_j}$, $x \in X$ where $\rho_j : X \to X_j$ is the canonical map.

**Remark 1.2.1.** An important example of FS spaces is the space $\mathcal{E}(\Omega)$, where $\Omega$ is an open set in $\mathbb{R}^n$ with the topology given in Definition 1.2.18; another example is the space $\mathcal{O}(U)$, $U$ open set in $\mathbb{C}^n$, with the topology of the uniformly convergence on compact sets.

To complete the description, we conclude this section by recalling the following definition.

**Definition 1.2.16.** *A locally convex space $X$ is a DFS space (Dual Fréchet–Schwartz space) if there is an increasing sequence*

$$X_1 \overset{\rho_2^1}{\to} X_2 \overset{\rho_3^2}{\to} \dots X_j \overset{\rho_{j+1}^j}{\to} X_{j+1} \to \dots \tag{1.3}$$

*of Banach spaces such that all the continuous mappings $\rho_{j+1}^j : X_j \to X_{j+1}$ are injective and compact and $X = \varinjlim X_j$.*

Let (1.3) be an injective decreasing sequence of Banach spaces. If all the maps $\rho_{j+1}^j : X_j \to \rho_{j+1}^j(X_j)$ are homeomorphisms, then $X$ is called a strict inductive limit of $\{X_j\}$.

**Definition 1.2.17.** *The strict inductive limit of an injective increasing family of Fréchet spaces is called an LF space.*

### 1.2.2 Distributions

The theory of distributions was developed to extend the differential calculus to functions that are not differentiable in the classical sense of Newton and Leibniz. The first notion of such a "weak" derivative was introduced by Sobolev, even though Heaviside and then Dirac (among others) used these objects, like the famous Heaviside function and its "derivative", the Dirac function, long before Sobolev (see [58]). However, we have to wait until Schwartz's work in the 1950s to have a definitive setting for the theory of distributions from the topological point of view.

For the missing proofs of the results stated in the following we refer to [153], [209], [210]. We give the detailed proofs just for the results which are, in our opinion, significant in the context of this book.

We now introduce some topological spaces, essentially $\mathcal{E}(\Omega)$, $\mathcal{S}(\mathbb{R}^n)$, $\mathcal{D}(\Omega)$, which will be our spaces of "test functions" and whose duals, in a suitable sense, will be spaces of distributions.

**Definition 1.2.18.** *Let $\Omega \subseteq \mathbb{R}^n$ be an open set. We denote by $\mathcal{E}(\Omega)$ the topological linear space over $\mathbb{C}$ which, as a set, is $C^\infty(\Omega)$ (the set of infinitely differentiable functions $u : \Omega \to \mathbb{C}$) and whose topology is given by the family of seminorms defined as follows: for any compact set $K \subset \Omega$ and for any $m \in \mathbb{N} \cup \{0\}$ we set*

$$p_m(K, u) := \sup_{|\alpha| \le m} \sup_{x \in K} |D^\alpha u(x)|, \quad u \in C^\infty(\Omega).$$

**Remark 1.2.2.** The topology $\mathcal{T}_{p_m(K,u)}$ generated by the family of seminorms $p_m(K, u)$ makes $(\mathcal{E}(\Omega), \mathcal{T}_{p_m(K,u)})$ a metrizable space; in fact it is possible to prove that the family $p_m(K, u)$ is equivalent to a countable family of seminorms $p_m(K_j, u)$, for $j \in \mathbb{N}$. Moreover, with the topology generated by the seminorms $p_m(K_j, u)$, the space $\mathcal{E}(\Omega)$ turns out to be complete. The space $\mathcal{E}(\Omega)$ is metrizable and complete so it is a Fréchet space.

With the topology $\mathcal{T}_{p_m(K_j,u)}$, we get the following notion of convergence of a sequence in $\mathcal{E}(\Omega)$ to an element $u \in \mathcal{E}(\Omega)$: $u_j \to u$ if $u_j$ converges to $u$ uniformly with all its derivatives on compact sets in $\mathbb{R}^n$.

**Definition 1.2.19.** *The space $\mathcal{S}(\mathbb{R}^n)$ of rapidly decreasing functions contains those functions $u \in C^\infty(\mathbb{R}^n)$ such that*

$$\sup_{x\in\mathbb{R}^n} \left(1+|x|^2\right)^{m/2} |D^\alpha u(x)| < \infty, \text{ for any } m \in \mathbb{N}\cup\{0\} \text{ and for any } \alpha \in \mathbb{N}^n.$$

*The topology $\mathcal{T}_{p_m(u)}$ on $\mathcal{S}(\mathbb{R}^n)$ is given by the family of seminorms*

$$p_m(u) := \sup_{|\alpha|\le m} \sup_{x\in\mathbb{R}^n} \left(1+|x|^2\right)^{m/2} |D^\alpha u(x)| < \infty, \quad m \in \mathbb{N}\cup\{0\}.$$

The topology $\mathcal{T}_{p_m(u)}$ makes $\mathcal{S}(\mathbb{R}^n)$ a Fréchet space in which the notion of convergence (we will limit ourselves to the case of convergence to 0), is described as follows:

> $u_j \to 0$ in $\mathcal{S}(\mathbb{R}^n)$ if and only if $\left(1+|x|^2\right)^{m/2} D^\alpha u_j(x) \to 0$, as $j \to \infty$, uniformly in $\mathbb{R}^n$, for any $\alpha \in \mathbb{N}^n$ and any $m \in \mathbb{N}\cup\{0\}$.

We finally introduce the space of test functions $C_0^\infty(\Omega)$, recalling first the definition of support of a function (compare with the notion of support of a section in a sheaf):

**Definition 1.2.20.** *Let $\Omega$ is an open set in $\mathbb{R}^n$. The support of a function $u : \Omega \to \mathbb{C}$, denoted by* supp $u$, *is defined as the closure of the set $\{x \in \Omega : u(x) \neq 0\}$.*

**Definition 1.2.21.** *Let $\Omega \subseteq \mathbb{R}^n$ be an open set. We denote by $C_0^\infty(\Omega)$ the space of $C^\infty(\Omega)$ functions with compact support.*

**Example 1.2.4.** The function

$$\phi(x) = \begin{cases} \exp\dfrac{1}{|x|^2-1} & |x|<1 \\ 0 & |x|\ge 1. \end{cases} \tag{1.4}$$

belongs to $C_0^\infty(\mathbb{R}^n)$.

We observe that $C_0^\infty(\mathbb{R}^n)$ is strictly contained in $\mathcal{S}(\mathbb{R}^n)$. For example, the function $\exp(-|x|^2)$ belongs to $\mathcal{S}(\mathbb{R}^n)$ but it does not belong to $C_0^\infty(\mathbb{R}^n)$, because it does not have compact support. Moreover, $C_0^\infty(\mathbb{R}^n)$ is a dense set in $\mathcal{S}(\mathbb{R}^n)$, i.e., for every $\phi \in \mathcal{S}(\mathbb{R}^n)$ there exists a sequence $\phi_j \in C_0^\infty(\mathbb{R}^n)$ such that $\phi_j \to \phi$ in $\mathcal{S}(\mathbb{R}^n)$.

The linear space $C_0^\infty(\mathbb{R}^n)$ can be considered a (nonclosed) linear variety in $\mathcal{S}(\mathbb{R}^n)$; in fact it suffices to identify any $f \in C_0^\infty(\Omega)$ with its extension to $\mathbb{R}^n$ obtained by setting $f(x) = 0$ outside $\Omega$, and therefore $C_0^\infty(\mathbb{R}^n)$, can be endowed with the topology that is inherited from $\mathcal{S}(\mathbb{R}^n)$. With this topology, it becomes a linear metrizable space, but since it is not closed, it cannot be complete. The linear space $C_0^\infty(\Omega)$ is also a subspace of $\mathcal{E}(\Omega)$, therefore it can be endowed with the topology inherited by $\mathcal{E}(\Omega)$. Even in this case, $C_0^\infty(\Omega)$ becomes a metrizable space but not complete so, to make it a complete space, we need to introduce a stronger topology. For this, let us introduce the following family of seminorms:

**Definition 1.2.22.** *We define $\{p_{\chi_\alpha}(u)\}$ to be the family of seminorms*

$$p_{\chi_\alpha}(u) = \sup_{\alpha \in \mathbb{N}^n} \sup_{x \in \Omega} |\chi_\alpha(x) D^\alpha u(x)|, \quad u \in C_0^\infty(\Omega), \tag{1.5}$$

*where the continuous functions $\chi_\alpha$ are such that the family of their supports $\{\mathrm{supp}\chi_\alpha\}$ is locally finite (that is any compact set in $\Omega$ has nonempty intersection with at most a finite number of the supports $\mathrm{supp}\chi_\alpha$). We denote by $\mathcal{D}(\Omega)$ the vector space of $C_0^\infty(\Omega)$ endowed with the topology generated by the seminorms $p_{\chi_\alpha}$.*

**Remark 1.2.3.** Note that the condition on the family of supports $\{\mathrm{supp}\chi_\alpha\}$ makes $p_{\chi_\alpha}(u)$ well defined for any $u \in \mathcal{D}(\Omega)$. With the topology $\mathcal{T}_{p_{\chi_\alpha}}$ generated by the seminorms $p_{\chi_\alpha}$, the space $\mathcal{D}(\Omega)$ becomes a locally convex linear space which is Hausdorff and complete but not metrizable.

The topology $\mathcal{T}_{p_{\chi_\alpha}}$ can be described in an alternative way as the inductive limit topology of Fréchet spaces (see the important work of Dieudonné and Schwartz [57]). Let $\Omega$ be an open set in $\mathbb{R}^n$ and let $\{K_j\}_{j\in\mathbb{N}}$ be a sequence of compact sets in $\Omega$ such that $K_j$ lies in the interior of $K_{j+1}$ and $\Omega = \cup_{j\in\mathbb{N}} K_j$.

Denote by $\mathcal{D}(K_j)$ the space of all functions $u \in C^\infty(\Omega)$ whose support lies in $K_j$, endowed with the topology induced by the family of seminorms $p_m(K_j, u)$ defined by

$$p_m(K_j, u) = \sup_{|\alpha| \le m} \sup_{x \in K_j} |D^\alpha u(x)|, \quad u \in \mathcal{D}(K_j).$$

With the topology induced by the family of seminorms $\{p_m(K_j, u)\}$, the space $\mathcal{D}(K_j)$ turns out to be a Fréchet space. It is possible to verify that $\mathcal{D}(K_j) \subset \mathcal{D}(K_{j+1})$ and that $\mathcal{D}(K_j)$ is closed in $\mathcal{D}(K_{j+1})$. Using the strict inductive limit, we define the LF space

$$\mathcal{D}(\Omega) = \varinjlim \mathcal{D}(K_j). \tag{1.6}$$

By definition, the strict inductive topology is the strongest locally convex topology in which all the canonical inclusions $\mathcal{D}(K_j) \hookrightarrow \mathcal{D}(\Omega)$ are continuous. The reader may verify that the inductive limit topology in $\mathcal{D}(\Omega)$ coincides with the topology generated by the seminorms (1.5). This fact shows that even though the space $\mathcal{D}(\Omega)$ is not metrizable, it is a limit of Fréchet, i.e., metrizable spaces. The convergence in $\mathcal{D}(\Omega)$ can be described as follows:

> Let $\{u_j\}$ be a sequence of functions in $\mathcal{D}(\Omega)$, and let $u \in \mathcal{D}(\Omega)$. We say that $u_j \to u$ in $\mathcal{D}(\Omega)$ if and only if there exists a compact set $K$

in $\Omega$ such that supp $u \subset K$, supp $u_j \subset K$ for any $j \in \mathbb{N} \cup \{0\}$, and for every multiindex $\alpha \in \mathbb{N}^n$ we have, as $j \to \infty$

$$D^\alpha u_j(x) \to D^\alpha u(x), \quad \text{uniformly for } x \in K.$$

We now mention a proposition, usually known as the partition of the unity for functions in $\mathcal{D}(\Omega)$, which will be crucial in the proof of the Gluing Theorem.

**Proposition 1.2.3.** *Let $\{\Omega_k\}$ be locally finite covering of $\Omega$. Then there exists a set of functions $\{g_k\}$ such that*

- $g_k \in \mathcal{D}(\Omega_k)$, *for every* $k \in \mathbb{N}$,
- $\sum_{k \geq 1} g_k(x) = 1, \quad 0 \leq g_k(x) \leq 1, \quad x \in \Omega.$

We recall that the sum $\sum_{k \geq 1} g_k(x)$ consists of a finite number of addends for every $x \in \Omega$.

**Remark 1.2.4.** In the language of sheaves, Proposition 1.2.3 implies that once that one has proved that $\mathcal{D}$ is a sheaf, $\mathcal{D}$ is a fine sheaf.

Distributions, which are the main topic of this subsection, are nothing but linear continuous functionals on $\mathcal{D}$. To introduce them, we recall the notion of continuous linear functional. Note that in this discussion the field $\mathbb{K}$ is the complex field.

**Definition 1.2.23.** *Let $X$ be a topological linear space over $\mathbb{K}$ and let $T$ be a linear functional, i.e., a linear map $T : X \to \mathbb{K}$. Then $T$ is continuous on $X$ if*

$$\langle T, \phi_j \rangle \to \langle T, \phi \rangle$$

*for every $\phi_j \to \phi$ in $X$ as $j \to \infty$.*

**Definition 1.2.24.** *Let $X$ be a topological linear space over $\mathbb{K}$. The set of all linear and continuous functionals $T : X \to \mathbb{K}$ with the vector addition and scalar multiplication defined by*

$$\langle T_1 + T_2, \phi \rangle = \langle T_1, \phi \rangle + \langle T_2, \phi \rangle, \quad \text{for any } \phi \in X$$

$$\langle \mu T, \phi \rangle = \mu \langle T, \phi \rangle, \quad \text{for any } \phi \in X, \quad \text{for any } \mu \in \mathbb{K}$$

*is a linear space called dual of $X$ and it is denoted by $X'$.*

**Definition 1.2.25.** *The space of distributions, denoted by $\mathcal{D}'(\Omega)$, consists of all linear continuous functionals on $\mathcal{D}(\Omega)$.*

We point out that not all the linear functionals defined on $\mathcal{D}(\Omega)$ are continuous, but the existence of noncontinuous functionals can be proved only using the Zorn lemma, so none of them can be explicitly represented.

**Remark 1.2.5.** The inductive limit topology on $\mathcal{D}(\Omega)$ implies a characterization of continuous linear functionals on it: a linear functional on $\mathcal{D}(\Omega) = \varinjlim \mathcal{D}(K_j)$ (see (1.6)) is continuous if and only if its restriction to $\mathcal{D}(K_j)$ is continuous for any $j$.

Note that, up to now, the space $\mathcal{D}'(\Omega)$ is only a linear space, so we need to introduce the notion a convergence on it.

**Definition 1.2.26.** *Let $\{T_j\}$ be a sequence of functionals in $\mathcal{D}'(\Omega)$ and let $T \in \mathcal{D}'(\Omega)$. We say that $T_j \to T$ in $\mathcal{D}'(\Omega)$ if*

$$\langle T_j, \phi \rangle \to \langle T, \phi \rangle, \quad \text{for any } \phi \in \mathcal{D}(\Omega).$$

The topology associated to this notion of convergence is called weak $*$ topology (to be read weak star topology).

**Definition 1.2.27.** *Every linear continuous functional on the space of rapidly decreasing functions $\mathcal{S}(\mathbb{R}^n)$ is called a tempered distribution. We denote the space of such functionals by $\mathcal{S}'(\mathbb{R}^n)$.*

The convergence in $\mathcal{S}'(\mathbb{R}^n)$ is defined as follows:

**Definition 1.2.28.** *The sequence of distributions $\{T_j\} \in \mathcal{S}'(\mathbb{R}^n)$ converges to the distribution $T \in \mathcal{S}'(\mathbb{R}^n)$ if*

$$\langle T_j, \phi \rangle \to \langle T, \phi \rangle, \quad \text{for any } \phi \in \mathcal{S}(\mathbb{R}^n).$$

We have already observed that there is an inclusion $j: \mathcal{D}(\mathbb{R}^n) \hookrightarrow \mathcal{S}(\mathbb{R}^n)$ that is continuous and with dense image so, for the dual spaces, we have $\mathcal{S}'(\mathbb{R}^n) \subset \mathcal{D}'(\mathbb{R}^n)$.

Even though the spaces of distributions $\mathcal{D}'(\Omega)$ and $\mathcal{S}'(\mathbb{R}^n)$ contain objects which are quite unusual compared with classical functions, a surprising density theorem holds. It shows that we can approximate every distribution with suitable regular functions.

**Theorem 1.2.4.** *For any $T \in \mathcal{D}'(\Omega)$ there exists a sequence $\{\phi_j\} \in \mathcal{D}(\Omega)$ such that $\phi_j \to T$ in $\mathcal{D}'(\Omega)$.*

**Example 1.2.5.** (Regular distributions). The simplest examples of distributions are those generated by locally Lebesgue integrable functions $f \in L^1_{loc}(\Omega)$ and which are called regular distributions:

$$\langle T_f, \phi \rangle = \int_\Omega f\phi \, dx, \quad \text{for any } \phi \in \mathcal{D}(\Omega).$$

Thanks to the linearity property of the Lebesgue integral and the Lebesgue convergence theorem, we get that $T_f$ is a linear continuous functional on $\mathcal{D}(\Omega)$, that is $T_f \in \mathcal{D}'(\Omega)$. In what follows, we will denote by $T_f$ a regular distribution defined by a function $f \in L^1_{loc}(\Omega)$.

**Example 1.2.6.** The $\delta$ distribution, introduced by Dirac and also called Dirac delta, is defined by

$$\delta : \mathcal{S}(\mathbb{R}) \to \mathbb{R}, \qquad \langle \delta, \phi \rangle = \phi(0), \quad \text{for any } \phi \in \mathcal{S}(\mathbb{R}).$$

It is easy to show that $\delta$ is a singular (i.e., nonregular) distribution in the sense that there is no function $f \in L^1_{loc}(\mathbb{R})$ such that $\delta = T_f$.

**Example 1.2.7.** Another important example of distribution which is not a regular distribution is the Cauchy principal value, denoted by $P\frac{1}{x}$, and first introduced by Dirac. We make $P\frac{1}{x}$ act on $\mathcal{D}((-h,h))$ for $h > 0$, defining

$$P\frac{1}{x} : \mathcal{D}((-h,h)) \to \mathbb{R},$$

$$\langle P\frac{1}{x}, \phi \rangle := p.v. \int_{-h}^{h} \frac{\phi(x)}{x} dx := \lim_{\varepsilon \to 0} \left\{ \int_{-\infty}^{-\varepsilon} \frac{\phi(x)}{x} dx + \int_{\varepsilon}^{+\infty} \frac{\phi(x)}{x} dx \right\}.$$

The functional $P\frac{1}{x}$ is linear by definition and it is continuous thanks to the estimates

$$|\langle P\frac{1}{x}, \phi \rangle| \leq \left| p.v. \int_{-h}^{h} \frac{\phi(0) + x\phi'(x_0)}{x} dx \right|$$

$$\leq \left| \int_{-h}^{h} |\phi'(x_0) dx \right| \leq 2h \sup_{x_0 \in (-h,h)} |\phi'(x_0)|, \quad \text{for any } \phi \in \mathcal{D}((-h,h)),$$

showing that $P\frac{1}{x}$ is bounded. The continuity of $P\frac{1}{x}$ follows from the fact that every bounded linear functional is continuous.

The distributions $\delta$ and $P\frac{1}{x}$ are related by the following formulas:

$$\frac{1}{x + i0} = -i\pi\delta(x) + P\frac{1}{x}, \qquad \frac{1}{x - i0} = i\pi\delta(x) + P\frac{1}{x},$$

where

$$\frac{1}{x \pm i0} = \lim_{\epsilon \to 0} \frac{1}{x \pm i\epsilon}, \qquad \text{in the space } \mathcal{D}'(\Omega).$$

These relations (sometimes known by physicists as the Lippman–Schwinger relations) have to be understood in the sense of functionals.

A crucial point in the theory of distributions is the fact that $\mathcal{D}'(\Omega)$ is not a metric space, but it is sequentially complete in the weak $*$ topology. Before we can prove such a statement we need a preparation result (essentially the Banach–Steinhaus theorem, see [153]).

**Theorem 1.2.5.** *Consider a sequence $\{T_j\}$ in a set of weakly bounded functionals, i.e. of functionals $T_j \in \mathcal{D}'(\Omega)$ such that $|\langle T_j, \phi\rangle| \leq k(\phi)$ for any $\phi \in \mathcal{D}(\Omega)$, and $k(\phi)$ positive constants. Suppose that the sequence $\{\phi_j\} \subset \mathcal{D}(\Omega)$ be such that $\phi_j \to 0$ in $\mathcal{D}(\Omega)$. Then, as $j \to +\infty$, we have*

$$|\langle T_j, \phi_j\rangle| \to 0.$$

We now prove the following important theorem.

**Theorem 1.2.6.** *The space of distributions $\mathcal{D}'(\Omega)$ is a sequentially complete space in the weak $*$ topology, i.e., if a sequence $\{T_j\} \subset \mathcal{D}'(\Omega)$ is such that the sequence $\langle T_j, \phi\rangle$ is convergent for all $\phi \in \mathcal{D}(\Omega)$, then the functional $T$ defined on $\mathcal{D}(\Omega)$ by*

$$\langle T, \phi\rangle = \lim_{j\to+\infty} \langle T_j, \phi\rangle,$$

*belongs to $\mathcal{D}'(\Omega)$.*

*Proof.* Consider a sequence of distributions $\{T_j\} \subset \mathcal{D}'(\Omega)$ such that the sequence of numbers

$$\langle T_j, \phi\rangle$$

is convergent for any $\phi \in \mathcal{D}(\Omega)$. We have to prove that the limit functional $T$ defined by

$$\langle T, \phi\rangle = \lim_{j\to+\infty} \langle T_j, \phi\rangle$$

belongs to $\mathcal{D}'(\Omega)$, i.e., $T$ is linear and continuous on $\mathcal{D}(\Omega)$. The linearity is trivial, so we prove continuity: we show that if $\phi_j \to 0$, then $\langle T, \phi_j\rangle \to 0$. Suppose the contrary. Taking a subsequence, if necessary, still denoted by $\phi_j$, we have $|\langle T, \phi_j\rangle| \geq 2\varepsilon$ for some $\varepsilon > 0$. By definition, we also have $\langle T, \phi_j\rangle = \lim_j \langle T_j, \phi_j\rangle$, so $\langle T_{j_0}, \phi_{j_0}\rangle > \varepsilon$ for some $j_0$. Thanks to Theorem 1.2.5 we get a contradiction and the theorem is proved. □

The following theorem characterizes distributions and will allow the definition of order of a distribution. To state it, we recall the following notation: if $\Omega$ is an open set in $\mathbb{R}^n$, we say that $\Omega_0$ is compactly contained in $\Omega$, and we write $\Omega_0 \Subset \Omega$, if $\Omega_0$ is bounded and its closure $\overline{\Omega}_0$ is contained in $\Omega$.

**Theorem 1.2.7.** *Let $T$ be a linear functional on $\mathcal{D}(\Omega)$. Then $T \in \mathcal{D}'(\Omega)$ if and only if for every open set $\omega \Subset \Omega$ there exist two numbers $C = C(\omega) > 0$ and $m = m(\omega) \in \mathbb{N} \cup \{0\}$ such that*

$$|\langle T, \phi\rangle| \leq C\|\phi\|_{C^m(\bar{\omega})}, \quad \text{for any } \phi \in \mathcal{D}(\omega). \tag{1.7}$$

**Definition 1.2.29.** *Suppose that there exists $m$, independent of $\omega$, such that estimate (1.7) holds. We say that a distribution is of finite order $\nu$ if*

$$\nu = \min\{m \in \mathbb{N} \cup \{0\} \text{ such that (1.7) holds}\}.$$

**Example 1.2.8.** The $\delta$ distribution has order zero; in fact

$$|\langle \delta, \phi \rangle| = |\phi(0)| \leq \|\phi\|_{C^0(\Omega)}.$$

A distribution $T \in \mathcal{D}'(\Omega)$ takes value zero in the open set $\omega \subset \Omega$ if its restriction to $\omega$ is the null functional of $\mathcal{D}'(\omega)$, i.e., $\langle T, \phi \rangle = 0$, for any $\phi \in \mathcal{D}(\omega)$. In this case, we will write $T(x) = 0$ for $x \in \omega$.

Using an argument involving the partition of the unity in $\mathcal{D}(\Omega)$ it is possible to prove the following two results. We mention the first without proof.

**Theorem 1.2.8.** *Let $T \in \mathcal{D}'(\Omega)$ be such that $T(x) = 0$ in some neighborhood of every point of the open set $\Omega$. Then $T(x) = 0$, for any $x \in \Omega$.*

The following result, the so called Gluing Theorem, gives conditions for gluing together two distributions.

**Theorem 1.2.9.** (Gluing Theorem). *For every point $z \in \Omega$ let $V(z) \Subset \Omega$ be a neighborhood of $z$. Let $T_z$ be a distribution in $V(z)$ such that $T_{z_1}(x) = T_{z_2}(x)$ if $x \in V(z_1) \cap V(z_2) \neq \emptyset$. Then there exists a unique distribution $T \in \mathcal{D}'(\Omega)$ such that $T = T_z$ in $V(z)$ for every $z \in \Omega$.*

*Proof.* Since Euclidean spaces are paracompact, we can always extract from the open covering $V(z)$, $z \in \Omega$, a locally finite covering of $\Omega$ indicated by $\{\Omega_k\}$ with $\Omega_k \Subset \Omega$ corresponding to the partition of unity $\{g_k\}$. Thanks to Proposition 1.2.3 we have

$$\langle T, \phi \rangle = \sum_{k \geq 1} \langle T, \phi g_k \rangle, \quad \text{for any } \phi \in \mathcal{D}(\Omega). \tag{1.8}$$

Since the number of addends on the right-hand side of (1.8) is always finite and does not depend on $\phi \in \mathcal{D}(\Omega')$ for every $\Omega' \Subset \Omega$, the functional $T$ is linear and continuous on $\mathcal{D}(\Omega)$ i.e., $T \in \mathcal{D}'(\Omega)$. Moreover if $\phi \in \mathcal{D}(V(z))$, then $\phi g_k \in \mathcal{D}(V(z_k))$ and

$$\langle T_z, \phi g_k \rangle = \langle T_{z_k}, \phi g_k \rangle$$

in such a way that from (1.8) we get

$$\langle T, \phi \rangle = \sum_{k \geq 1} \langle T_{z_k}, \phi g_k \rangle = \langle T_z, \phi \sum_{k \geq 1} g_k \rangle = \langle T_z, \phi \rangle.$$

This means $T = T_z$ in $V(z)$. The uniqueness of the distribution $T$ follows from Theorem 1.2.8. This completes the proof. □

**Remark 1.2.6.** The results of Theorems 1.2.8 and 1.2.9 can be summarized by saying, in the language of the previous section, that $\mathcal{D}'$ is a soft sheaf on $\mathbb{R}^n$.

The definition of support of a distribution is natural, but we repeat it here for the convenience of the reader.

**Definition 1.2.30.** *We define the support of a distribution $T$ as the complement set in $\Omega$ of the union $U_T$ of all sets on which $T = 0$. We denote the support by supp $T$. It is a closed set and supp $T = \Omega \setminus U_T$.*

We now state the following result:

**Theorem 1.2.10.** *Let $T \in \mathcal{D}'(\Omega)$ and $\phi \in \mathcal{D}(\Omega)$.*

- *Assume that supp $T \cap$ supp $\phi$ is the empty set; then*

$$\langle T, \phi \rangle = 0.$$

- *$x \in$ supp $T$ if and only if $T \neq 0$ in any neighborhood of $x$.*

Among distributions, those ones with compact support constitute an important class and they are characterized in the following theorem:

**Theorem 1.2.11.** *Let $T_0 \in \mathcal{D}'(\mathbb{R}^n)$. Then $T_0$ is a distribution with compact support if and only if the functional $T_0$ can be extended to a linear and continuous functional $T$ on $\mathcal{E}(\mathbb{R}^n)$.*

From the characterization theorem of compactly supported distributions, we have that the distributions with compact support are elements in the dual space of $\mathcal{E}(\mathbb{R}^n)$ that will be denoted by $\mathcal{E}'(\mathbb{R}^n)$.
In the particular case in which the support of a distribution reduces to a single point we have a representation theorem.

**Theorem 1.2.12.** *Let $T \in \mathcal{E}'(\mathbb{R}^n)$ with* supp $T = \{0\}$. *Then $T$ admits the representation*

$$T(x) = \sum_{|\alpha| \leq \nu} k_\alpha D^\alpha \delta(x)$$

*where $k_\alpha \in \mathbb{R}$, $\nu$ is the order of the distribution $T$ and the equality must be interpreted in the sense of functionals.*

We now consider some operations on distributions.
Let $f \in L^1_{\text{loc}}(\Omega)$ and consider the regular distribution $T_f$ defined by

$$\langle T_f, \phi \rangle = \int_\Omega f \phi \, dx, \quad \text{for any } \phi \in \mathcal{D}(\Omega).$$

We define the product of a function $g \in C^\infty(\Omega)$ with the distribution $T_f$ by

$$\langle g T_f, \phi \rangle = \int_\Omega g f \phi \, dx, \quad \text{for any } \phi \in \mathcal{D}(\Omega).$$

Observe that the product is well defined because

$$\langle g T_f, \phi \rangle = \langle T_f, g\phi \rangle, \quad \text{for any } \phi \in \mathcal{D}(\Omega) \tag{1.9}$$

and the map $\phi \longrightarrow \phi g$ is linear and continuous on $\mathcal{D}(\Omega)$, therefore the functional $g T_f$ belongs to $\mathcal{D}'(\Omega)$.

**Definition 1.2.31.** *Let $T \in \mathcal{D}'(\Omega)$ and $g \in C^\infty(\Omega)$. The product $gT$ is defined by the equality*

$$\langle gT, \phi\rangle = \langle T, g\phi\rangle, \quad \text{for all } \phi \in \mathcal{D}(\Omega). \tag{1.10}$$

Things are not so easy when one attempts to multiply two distributions. In general it is not possible to define the product of two distributions (see the section on hyperfunctions for a discussion on this issue).

In the case of regular distributions, for example, if we take $f = g = |x|^{-1/2} \in L^1_{\rm loc}(\mathbb{R}^n)$ we see that $fg \notin L^1_{\rm loc}(\mathbb{R}^n)$.

To define the derivatives of a distribution we consider the integration by parts formula for $f \in C^k(\Omega)$, $|\alpha| \le k$ and $\phi \in \mathcal{D}(\Omega)$:

$$\langle D^\alpha f, \phi\rangle = \int_\Omega D^\alpha f \phi \, dx = (-1)^{|\alpha|} \int_\Omega f D^\alpha \phi \, dx = (-1)^{|\alpha|} \langle f, D^\alpha \phi\rangle,$$

which motivates the following definition.

**Definition 1.2.32.** *Let $f \in L^1_{\rm loc}(\Omega)$. We define $D^\alpha T_f$ by the equality*

$$\langle D^\alpha T_f, \phi\rangle = (-1)^{|\alpha|} \langle T_f, D^\alpha \phi\rangle, \quad \textit{for any } \phi \in \mathcal{D}(\Omega).$$

**Remark 1.2.7.** Observe that the map $\phi \to D^\alpha \phi$ is linear and continuous from $\mathcal{D}(\Omega) \to \mathcal{D}(\Omega)$ so the functional $D^\alpha T_f$ belongs to $\mathcal{D}'(\Omega)$. We assume this definition not only for regular distributions $T_f$ but for all distributions $T \in \mathcal{D}'(\Omega)$. The map $T \to D^\alpha T$ is linear and continuous from $\mathcal{D}'(\Omega)$ to $\mathcal{D}'(\Omega)$.

**Example 1.2.9.** Consider the function $f(x) = \log|x|$, for $|x| \neq 0$ and $f(0) = 0$. Then $f \in L^1_{\rm loc}(\mathbb{R})$. Such a function defines a regular distribution $T_f$ whose derivative is the Cauchy principal value $P\frac{1}{x}$. In fact we have, for all $\phi \in \mathcal{D}(\mathbb{R})$,

$$\begin{aligned}
\langle DT_f, \phi\rangle &= -\langle T_f, D\phi\rangle \\
&= -\int_{\mathbb{R}} f(x) D\phi(x)\, dx \\
&= -\lim_{\varepsilon\to 0} \int_{-\infty}^{-\varepsilon} f(x) D\phi(x)\, dx - \lim_{\varepsilon\to 0} \int_{\varepsilon}^{\infty} f(x) D\phi(x)\, dx \\
&= \lim_{\varepsilon\to 0} \int_{-\infty}^{-\varepsilon} \frac{\phi(x)}{x}\, dx - \lim_{\varepsilon\to 0} \int_{\varepsilon}^{\infty} \frac{\phi(x)}{x}\, dx \\
&= \langle P\frac{1}{x}, \phi\rangle.
\end{aligned}$$

We finally introduce the convolution between distributions. Let us suppose that $f$ and $g$ belong to $L^1_{\rm loc}(\mathbb{R}^n)$ and they are such that the integral $\int_{\mathbb{R}^n} f(y) g(x-y)\, dy$ exists for almost all $x \in \mathbb{R}^n$ and defines a function in $L^1_{\rm loc}(\mathbb{R}^n)$. Then we call this function the convolution of $f$ and $g$. Obviously the integral is not always convergent, so we give first a definition of convolution under a suitable hypothesis on the supports of the $f$ and $g$, and then we will extend the definition to the case of more general distributions.

**Definition 1.2.33.** *Let $f$ and $g \in L^1_{\text{loc}}(\mathbb{R}^n)$ with* $\operatorname{supp} f \subseteq A_1$ *and* $\operatorname{supp} g \subseteq A_2$*; suppose that the set*

$$B(\rho) := \{(x,y) : x \in A_1,\ y \in A_2,\ |x+y| \le \rho \ \text{ for } \rho > 0\}$$

*is bounded in $\mathbb{R}^{2n}$. We define the convolution product $f * g$ by*

$$f * g(x) := \int_{\mathbb{R}^n} f(y) g(x-y)\, dy.$$

It is not difficult to prove that, thanks to Fubini's theorem, $f * g \in L^1_{\text{loc}}(\mathbb{R}^n)$ and therefore the convolution $f*g$ defines a regular distribution $T_{f*g} \in \mathcal{D}'(\mathbb{R}^n)$. One can also show that

$$\langle T_{f*g}, \phi\rangle = \int_{\mathbb{R}^n} f * g(x)\, \phi(x)\, dx$$

$$= \int_{\mathbb{R}^{2n}} f(x) g(y)\, \phi(x+y)\, dx\, dy, \quad \text{for any } \phi \in \mathcal{D}(\mathbb{R}^n). \tag{1.11}$$

In fact, using again Fubini's theorem and a suitable change of variables, we have the chain of equalities

$$\int_{\mathbb{R}^n} f * g(x)\, \phi(x)\, dx = \int_{\mathbb{R}^n} \phi(x) \int_{\mathbb{R}^n} f(y) g(x-y)\, dx\, dy$$

$$= \int_{\mathbb{R}^n} f(y) \int_{\mathbb{R}^n} g(x-y)\phi(x)\, dx\, dy = \int_{\mathbb{R}^n} f(y) \int_{\mathbb{R}^n} g(\xi)\phi(y+\xi)\, d\xi\, dy.$$

It is also possible to prove that (1.11) can be rewritten as

$$\langle T_{f*g}, \phi\rangle = \lim_{j\to\infty} \int_{\mathbb{R}^{2n}} f(x) g(y)\, \chi_j(x,y) \phi(x+y)\, dx\, dy, \quad \text{for any } \phi \in \mathcal{D}(\mathbb{R}^n) \tag{1.12}$$

where the sequence of functions $\{\chi_j(x,y)\}_{j\in\mathbb{N}} \subset \mathcal{D}(\mathbb{R}^{2n})$ is chosen to converge to 1 in $\mathbb{R}^{2n}$ in the following sense

- for every compact set $K$ there exists $m(K) \in \mathbb{N}$ such that $\chi_j(x,y) = 1$ for $(x,y) \in K$, $j \ge m$,
- the derivatives satisfy the uniform estimates

$$|D^\alpha \chi_j(x,y)| \le c(\alpha), \quad c(\alpha) > 0, \quad \text{for any } j \in \mathbb{N}, \ \text{ for any } \alpha \in \mathbb{N}^n.$$

We use (1.12) to define the convolution in the general case. Let us first assume that $f \in L^1_{\text{loc}}(\Omega_1)$, $\Omega_1 \subset \mathbb{R}^n$ and $g \in L^1_{\text{loc}}(\Omega_2)$, $\Omega_2 \subset \mathbb{R}^n$. Since $fg = gf \in L^1_{\text{loc}}(\Omega_1 \times \Omega_2)$, the product $fg$ defines a regular distribution on $\mathcal{D}(\Omega_1 \times \Omega_2)$:

$$\langle T_{fg}, \phi\rangle = \langle T_f T_g, \phi\rangle$$

$$= \int_{\Omega_1 \times \Omega_2} f(x)g(y)\phi(x,y)dx\,dy, \quad \text{for any } \phi \in \mathcal{D}(\Omega_1 \times \Omega_2);$$

by Fubini's theorem we get

$$\langle T_f T_g, \phi \rangle = \langle T_f, T_g \phi \rangle, \quad \text{for any } \phi \in \mathcal{D}(\Omega_1 \times \Omega_2)$$

and

$$\langle T_g T_f, \phi \rangle = \langle T_g, T_f \phi \rangle, \quad \text{for any } \phi \in \mathcal{D}(\Omega_1 \times \Omega_2).$$

We use the above formulas to define the direct product between distributions.

**Definition 1.2.34.** *Let $T \in \mathcal{D}'(\Omega_1)$ and $G \in \mathcal{D}'(\Omega_2)$. We define the direct product of distributions (denoted by the symbol $\otimes$) by setting*

$$\langle T \otimes G, \phi \rangle := \langle T, G\phi \rangle, \quad \text{for any } \phi \in \mathcal{D}(\Omega_1 \times \Omega_2) \tag{1.13}$$

*and*

$$\langle G \otimes T, \phi \rangle := \langle G, T\phi \rangle, \quad \text{for any } \phi \in \mathcal{D}(\Omega_1 \times \Omega_2). \tag{1.14}$$

It is left to the reader to verify that the direct product is well defined.

**Definition 1.2.35.** *Consider $T$, $G \in \mathcal{D}'(\mathbb{R}^n)$ such that the direct product $T \otimes G$ admits an extension $\langle T \otimes G, \psi \rangle$ to the functions of the form $\psi(x,y) = \phi(x+y)$, for some $\phi \in \mathcal{D}(\mathbb{R}^n)$, in the sense that for any sequence $\{\chi_j\}_{j \in \mathbb{N}}$ of functions in $\mathcal{D}(\mathbb{R}^{2n})$ (with the properties stated above), converging to 1 in $\mathbb{R}^{2n}$, the limit*

$$\lim_{j \to \infty} \langle T \otimes G, \chi_j \psi \rangle = \langle T \otimes G, \psi \rangle$$

*exists and does not depend on the sequence $\chi_j$.*

*The convolution $T * G$ is defined by the functional*

$$\langle T * G, \phi \rangle = \langle T \otimes G, \psi \rangle = \lim_{j \to \infty} \langle T \otimes G, \chi_j \psi \rangle, \quad \text{for any } \phi \in \mathcal{D}(\mathbb{R}^n) \tag{1.15}$$

*where $\psi(x,y) = \phi(x+y)$.*

We observe that the function $\phi(x+y)$ does not belong to $\mathcal{D}(\mathbb{R}^{2n})$ since it does not have compact support on $\mathbb{R}^{2n}$, so the functional in Definition 1.2.35, i.e., $\langle T * G, \phi \rangle = \langle T \otimes G, \psi \rangle$ is not necessarily defined for arbitrary distributions $T$ and $G \in \mathcal{D}'(\mathbb{R}^n)$.

**Remark 1.2.8.** We end this section recalling some results about the convolution. This results are of crucial importance because the convolution is involved in the representation formula of the solutions of nonhomogeneous linear constant coefficients partial differential equations which will be introduced in the next section, see Remark 1.2.11.

**Theorem 1.2.13.** *Let $T$ and $G \in \mathcal{D}'(\mathbb{R}^n)$.*

- *Suppose that the convolution $T * G$ is defined. Then $G * T$ is defined and*

$$T * G = G * T.$$

- *The convolution $T * \delta$ is defined for every $T \in \mathcal{D}'(\mathbb{R}^n)$ and*

$$T * \delta = \delta * T = T.$$

- *Suppose that the convolutions $T * G$ and $G * T$ are defined. Then the convolutions $(D^\alpha T) * G$ and $T * (D^\alpha G)$ are defined and*

$$D^\alpha(T * G) = (D^\alpha T) * G = T * (D^\alpha G).$$

Let us finally recall a sufficient conditions to guarantee that the convolution of two distributions is defined.

**Theorem 1.2.14.** *Suppose that $T \in \mathcal{D}'(\mathbb{R}^n)$ and $G \in \mathcal{E}'(\mathbb{R}^n)$. Then the convolution $T * G$ is defined and it belongs to $\mathcal{D}'(\mathbb{R}^n)$. If $T \in \mathcal{S}'(\mathbb{R}^n)$ and $G \in \mathcal{E}'(\mathbb{R}^n)$, then the convolution $T * G$ is defined and it belongs to $\in \mathcal{S}'(\mathbb{R}^n)$.*

We introduce the space $\theta_M$ of functions $f \in C^\infty(\mathbb{R}^n)$ satisfying the condition:

$\forall\alpha$ there exist $K = K(\alpha) \geq 0$ and $m = m(\alpha) \in \mathbb{N}$ such that

$$|D^\alpha f(x)| \leq K(1 + |x|)^m, \ \forall x \in \mathbb{R}^n. \tag{1.16}$$

**Theorem 1.2.15.** *Let $T \in \mathcal{S}'(\mathbb{R}^n)$ and $g \in \mathcal{S}(\mathbb{R}^n)$, then $T * g$ is defined and belongs to $\in \theta_M$.*

We observe that Theorem 1.2.15 is important to deduce the properties of nonhomogeneous linear constant partial differential equations in the case the data belongs to $\mathcal{S}(\mathbb{R}^n)$ (see Theorem 1.2.24 and Remark 1.2.11).

### 1.2.3 Fourier transform and fundamental solutions

One of the most important reasons to define the space $\mathcal{S}(\mathbb{R}^n)$ is that the Fourier transform maps this space into itself and the inverse operator is also well defined. Moreover, since $\mathcal{S}(\mathbb{R}^n)$ is dense both in $L^2(\mathbb{R}^n)$ and in $\mathcal{S}'(\mathbb{R}^n)$, we can extend the Fourier transform to those spaces. We will only prove the results of particular interest in the context of the book, and for the missing proofs of the results stated in the following we refer the reader to [153], [209], [210].
Let $\lambda = (\lambda_1, \dots, \lambda_n)$ and $x = (x_1, \dots, x_n) \in \mathbb{R}^n$; the inner product in $\mathbb{R}^n$ will be denoted by

$$(\lambda, x) := \lambda_1 x_1 + \cdots + \lambda_n x_n.$$

Since $\mathcal{S}(\mathbb{R}^n) \subset L^1(\mathbb{R}^n)$ we can give the following definition.

**Definition 1.2.36.** *Let $\phi \in \mathcal{S}(\mathbb{R}^n)$. The linear integral operator defined by*

$$\mathcal{F}[\phi](\lambda) := \int_{\mathbb{R}^n} \phi(x) e^{-i(\lambda,x)}\, dx, \quad i = \sqrt{-1}$$

*is called the Fourier transform of $\phi(x)$.*

In the sections to follow, we sometimes use the notation $\widehat{\phi}$ for the Fourier transform $\mathcal{F}[\phi]$.

It is possible to prove that the inverse of the Fourier transform is defined, for every $\psi \in \mathcal{S}(\mathbb{R}^n)$, by

$$\mathcal{F}^{-1}[\psi](x) := \frac{1}{(2\pi)^n} \int_{\mathbb{R}^n} \psi(\lambda) e^{i(\lambda,x)}\, d\lambda. \tag{1.17}$$

We observe that the operator $\mathcal{F}^{-1}$ is well defined on $\mathcal{S}(\mathbb{R}^n)$ and for every $\psi \in \mathcal{S}(\mathbb{R}^n)$ we have

$$\psi = \mathcal{F}[\mathcal{F}^{-1}[\psi]] = \mathcal{F}^{-1}[\mathcal{F}[\psi]]. \tag{1.18}$$

A fundamental result is the following:

**Theorem 1.2.16.** *The Fourier transform $\mathcal{F}$ is an isomorphism between $\mathcal{S}(\mathbb{R}^n)$ and itself.*

*Proof.* We introduce in $\mathcal{S}(\mathbb{R}^n)$ the countable family of seminorms

$$p_m(\phi) = \sup_{|\alpha| \le m,\ x \in \mathbb{R}^n} (1 + |x|^2)^{m/2} |D^\alpha \phi(x)|, \quad m \in \mathbb{N} \cup \{0\}.$$

Take $\phi \in \mathcal{S}(\mathbb{R}^n)$ and observe that, since the test functions approach zero at infinity faster than every power of $|x|^{-1}$, we can differentiate under the integral infinitely many times so it follows that $\mathcal{F}[\phi] \in C^\infty(\mathbb{R}^n)$. We then get the formula

$$D^\alpha \mathcal{F}[\phi](\lambda) = \int_{\mathbb{R}^n} (-ix)^\alpha \phi(x) e^{-i(\lambda,x)}\, dx = \mathcal{F}[(-ix)^\alpha \phi](\lambda), \tag{1.19}$$

and the derivatives $D^\alpha \phi$ have the property

$$\mathcal{F}[D^\alpha \phi](\lambda) = \int_{\mathbb{R}^n} D^\alpha \phi(x) e^{-i(\lambda,x)}\, dx = (i\lambda)^\alpha \mathcal{F}[\phi](\lambda). \tag{1.20}$$

From equalities (1.19) and (1.20), for any multiindex $\alpha \in \mathbb{N}^n$ and $m \in \mathbb{N} \cup \{0\}$, we have the estimates

$$(1 + |\lambda|^2)^{m/2} |D^\alpha \mathcal{F}[\phi](\lambda)| \le (1 + |\lambda|^2)^{[(m+1)/2]} |D^\alpha \mathcal{F}[\phi](\lambda)|$$

$$\le \left| \int_{\mathbb{R}^n} (1 - \Delta)^{[(m+1)/2]} ((-ix)^\alpha \phi(x)) e^{-i(\lambda,x)}\, dx \right|$$

$$\leq C \sup_x (1+|x|^2)^{(n+1)/2}(1-\Delta)^{[(m+1)/2]}\big(x^\alpha \phi(x)\big)$$

where $[(m+1)/2]$ denotes the integer part, $C$ is a positive constant and $\Delta$ is the Laplace operator. From the chain of inequalities we get

$$p_m(\mathcal{F}[\phi]) \leq C_m p_{m+n+1}(\phi), \quad \text{for any } m \in \mathbb{N} \cup \{0\} \tag{1.21}$$

where the positive constants $C_m$ are independent of $\phi$. Estimates (1.21) show that $\phi \to \mathcal{F}[\phi]$ is continuous; moreover from (1.17) and (1.18); follows that every function $\phi \in \mathcal{S}(\mathbb{R}^n)$ is the Fourier transform of the function $\psi = \mathcal{F}^{-1}[\phi]$ in $\mathcal{S}(\mathbb{R}^n)$, where $\phi = \mathcal{F}[\psi]$, and if $\mathcal{F}[\phi] = 0$, then $\phi = 0$. This means that the map $\phi \to \mathcal{F}[\phi]$ is a bijection from $\mathcal{S}(\mathbb{R}^n)$ into itself. The same properties hold also for the inverse Fourier transform $\mathcal{F}^{-1}$ and this completes the proof. □

Thanks to Theorem 1.2.16 and to the density of $\mathcal{S}(\mathbb{R}^n)$ in the space of tempered distributions $\mathcal{S}'(\mathbb{R}^n)$, it is possible to define the operator $\mathcal{F}$ on $\mathcal{S}'(\mathbb{R}^n)$. First, we consider regular distributions $T_f$ defined by functions $f \in L^1_{\text{loc}}(\mathbb{R}^n)$ and we define

$$\begin{aligned} \langle \mathcal{F}[T_f], \phi \rangle &= \int_{\mathbb{R}^n} \mathcal{F}[T_f](\lambda)\phi(\lambda)\, d\lambda \\ &= \int_{\mathbb{R}^n} \mathcal{F}[f](\lambda)\phi(\lambda)\, d\lambda, \quad \text{for any } \phi \in \mathcal{S}(\mathbb{R}^n). \end{aligned}$$

By Fubini's theorem, we can change the order of integration to get

$$\langle \mathcal{F}[T_f], \phi \rangle = \langle T_f, \mathcal{F}[\phi] \rangle, \quad \text{for any } \phi \in \mathcal{S}(\mathbb{R}^n), \quad f \in L^1_{\text{loc}}(\mathbb{R}^n). \tag{1.22}$$

Since the relation (1.22) holds for regular distributions, we can assume the following definition of Fourier transform for tempered distributions.

**Definition 1.2.37.** *Let $T \in \mathcal{S}'(\mathbb{R}^n)$. We define the Fourier transform on the space of tempered distributions as*

$$\langle \mathcal{F}[T], \phi \rangle = \langle T, \mathcal{F}[\phi] \rangle, \qquad \textit{for any} \quad \phi \in \mathcal{S}(\mathbb{R}^n).$$

The Fourier transform on the space of tempered distributions is well defined. In fact, observe that since $\mathcal{F}$ is an isomorphism from $\mathcal{S}(\mathbb{R}^n)$ to itself the map $\phi \to \mathcal{F}[\phi]$ is linear and continuous from $\mathcal{S}(\mathbb{R}^n)$ to $\mathcal{S}(\mathbb{R}^n)$, the functional $\langle T, \mathcal{F}[\phi] \rangle$ represents a distribution in $\mathcal{S}'(\mathbb{R}^n)$ and moreover the map $T \to \mathcal{F}[T]$ is linear and continuous from $\mathcal{S}'(\mathbb{R}^n)$ to $\mathcal{S}'(\mathbb{R}^n)$.

The natural candidate to be the inverse of the Fourier transform in the space of tempered distributions is

$$\mathcal{F}^{-1}[T] = \frac{1}{(2\pi)^n}\mathcal{F}[\check{T}], \quad \text{for any } T \in \mathcal{S}'(\mathbb{R}^n) \tag{1.23}$$

where we have set $\check{T} := T(-x)$.

**Theorem 1.2.17.** *The Fourier transform $\mathcal{F}$ is an isomorphism between $\mathcal{S}'(\mathbb{R}^n)$ and itself.*

**Example 1.2.10.** We can easily verify that, if $a \in \mathbb{R}^n$,

$$\mathcal{F}[\delta(x-a)] = e^{-i(\lambda,a)} \tag{1.24}$$

in the sense of distributions. In particular

$$\mathcal{F}[\delta(x)] = 1$$

and

$$\mathcal{F}[1] = (2\pi)^n \delta(x).$$

We now state some properties of the Fourier transform in $\mathcal{S}'(\mathbb{R}^n)$.

**Theorem 1.2.18.** *Suppose that $T \in \mathcal{S}'(\mathbb{R}^n)$ and $\alpha \in \mathbb{N}^n$ is a multiindex and $a \in \mathbb{R}^n$. Then the following properties hold:*

- *Differentiation of the Fourier transform*

$$D^\alpha \mathcal{F}[T] = \mathcal{F}[(-ix)^\alpha T].$$

- *Fourier transform of the derivatives*

$$\mathcal{F}[D^\alpha T] = (i\lambda)^\alpha \mathcal{F}[T].$$

- *Translation of the Fourier transform*

$$\mathcal{F}[T](\lambda + a) = \mathcal{F}[e^{-i(a,x)}T](\lambda).$$

*Proof.* It is easy to prove the above relations in the case $T$ is a function in $\mathcal{S}(\mathbb{R}^n)$. Since $\mathcal{S}(\mathbb{R}^n)$ is dense in the space of tempered distributions, the above relations hold also in $\mathcal{S}'(\mathbb{R}^n)$. □

**Remark 1.2.9.** Some immediate consequences of Theorem 1.2.18 are the Fourier transforms of polynomials

$$\mathcal{F}[x^\alpha] = i^{|\alpha|} D^\alpha \mathcal{F}[1] = (2\pi)^n \, i^{|\alpha|} D^\alpha \delta(x)$$

and keeping in mind that $\mathcal{F}[\delta(x)] = 1$, we also obtain

$$\mathcal{F}[D^\alpha \delta] = (i\lambda)^\alpha \mathcal{F}[\delta] = (i\lambda)^\alpha.$$

Since $\mathcal{E}'(\mathbb{R}^n)$ is contained in $\mathcal{S}'(\mathbb{R}^n)$ it turns out that we can define the Fourier transform of distributions with compact support. We can state the following results, recalling (1.16):

**Theorem 1.2.19.** *Let $T \in \mathcal{E}'(\mathbb{R}^n)$. Then $\mathcal{F}[T] \in \theta_M$; moreover there exist two numbers $C_\alpha(T) \geq 0$ and $m \in \mathbb{N} \cup \{0\}$ such that*

$$|D^\alpha \mathcal{F}[T](\lambda)| \leq C_\alpha(T)(1+|\lambda|^2)^{m/2}, \quad \lambda \in (\mathbb{R}^n).$$

**Theorem 1.2.20.** *Let $T \in \mathcal{S}'(\mathbb{R}^n)$ and $G \in \mathcal{E}'(\mathbb{R}^n)$. Then $T * G \in \mathcal{S}'(\mathbb{R}^n)$, and the Fourier transform of the convolution is given by*

$$\mathcal{F}[T * G] = \mathcal{F}[T]\mathcal{F}[G].$$

**Remark 1.2.10.** The Fourier transform of any distribution $T \in \mathcal{E}'(\mathbb{R}^n)$ can be extended to an entire analytic function in $\mathbb{C}^n$ called the Fourier–Laplace transform of $T$, by virtue of Theorem 7.1.14 in Hörmander's book [91].

**Theorem 1.2.21.** (Paley–Wiener–Schwartz). *Let $K \subset \mathbb{R}^n$ be a convex compact set. Let $T$ be a distribution of order $m$ with support contained in $K$. Then*

$$|\mathcal{F}[T](\lambda)| \leq C(1+|\lambda|)^m e^{H_K(Im\lambda)}, \quad \lambda \in \mathbb{C}^n, \tag{1.25}$$

*where*

$$H_K(\zeta) = \sup_{x \in K}(x, \zeta)$$

*is the so-called supporting function of $K$. Conversely, every entire analytic function in $\mathbb{C}^n$ satisfying an estimate of type (1.25) is the Fourier–Laplace transform of a distribution with support in $K$.*

*If $\phi \in \mathcal{D}(K)$ there exists for every integer $m$ a constant $C_m$ such that*

$$|\mathcal{F}[\phi](\lambda)| \leq C_m(1+|\lambda|)^{-m} e^{H_K(\,Im\lambda)}, \quad \lambda \in \mathbb{C}^n. \tag{1.26}$$

*Conversely, every entire analytic function in $\mathbb{C}^n$ satisfying an estimate of type (1.26) for every $m$ is the Fourier–Laplace transform of a function in $\mathcal{D}(K)$.*

One of the most important results in the field of partial differential equations is the existence of a fundamental solution for partial differential equations with constant coefficients.

**Definition 1.2.38.** *A linear partial differential operator with constant coefficients of order $m \in \mathbb{N}$ is an operator of the form*

$$P(D) = \sum_{|\alpha| \leq m} c_\alpha D^\alpha \tag{1.27}$$

*where $D^\alpha$ are partial derivatives and the coefficients $c_\alpha$ are complex numbers such that*

$$\sum_{|\alpha|=m} |c_\alpha| \neq 0.$$

**Definition 1.2.39.** *Let $P(D)$ be a nonzero linear partial differential operator of order $m$ with constant coefficients. We call fundamental solution of the operator $P(D)$ the distribution $E$ such that*

$$P(D)E = \delta$$

*where $\delta$ is the Dirac distribution.*

Our next result, the Malgrange–Ehrenpreis theorem, is one of the most important in the early theory of differential equations in the space of distributions; as the reader will soon recognize, the result is actually a very special case of the more general study of syzygies of linear systems.

Before the Malgrange–Ehrenpreis theorem, we state an important result that is the core of its proof.

**Theorem 1.2.22.** *Let*

$$Q(\eta) = \sum_{\alpha \leq m} a_\alpha \eta^\alpha$$

*be a polynomial of degree $m \geq 1$ (i.e., $\sum_{\alpha=m} a_\alpha \eta^\alpha \neq 0$). Then there exists a real, nonsingular matrix $A \in M_{\mathbb{R}}(n \times n)$ such that if*

$$\eta = A\xi, \quad A = [a_{ij}]_{i,j=1,\dots,n}, \quad \xi, \eta \in \mathbb{R}^n,$$

*$Q(\eta)$ can be transformed into the form*

$$P(\xi) = a\xi_1^m + \sum_{0 \leq k \leq m-1} Q_k(\xi_2, \dots, \xi_n)\xi_1^k, \quad a \neq 0 \tag{1.28}$$

*where $P(\xi) = Q(A\xi)$ and $Q_k$ are polynomials independent of $\xi_1$. Moreover, there exists a constant $h > 0$, depending only on $m$, such that for every $\xi \in \mathbb{R}^n$ there exists an integer $j \in [0, m]$ such that the following inequality holds*

$$|P(\xi_1 + i\tau j/m, \xi_2, \dots, \xi_n)| \geq ha, \quad |\tau| = 1. \tag{1.29}$$

**Theorem 1.2.23.** (Malgrange–Ehrenpreis). *Every nontrivial linear partial differential operator with constant coefficients $P(D)$ has a fundamental solution in $\mathcal{D}'(\Omega)$.*

*Proof.* Since a real nonsingular linear transformation maps $\mathcal{D}'(\Omega)$ into itself we assume that the polynomial $P(i\xi)$ is in the form (1.28). Consider $m+1$ non negative, measurable functions $\{f_j\}_{j=0,1,\dots,m}$ such that

$$\sum_{j=0}^{m} f_j(\xi) = 1, \quad \xi \in \mathbb{R}^n$$

and $f_j(\xi) = 0$ when

$$\min_{|\tau|=1} |P(i\xi_1 - \tau j/m, i\xi_2, \dots, i\xi_n)| < ha.$$

Define the distribution $E$ by setting, for any $\phi \in \mathcal{D}(\mathbb{R}^n)$,

$$\langle E, \phi\rangle = \frac{1}{(2\pi)^n}\sum_{j=0}^{m}\int_{\mathbb{R}^n} f_j(\xi)\frac{1}{2\pi i}\int_{|\tau|=1}\frac{L[\phi](\xi_1 + i\tau\frac{j}{m}, \xi_2, ..., \xi_n)}{P(i\xi_1 - \tau\frac{j}{m}, i\xi_2, ..., i\xi_n)}\frac{d\tau}{\tau}\, d\xi$$

where $L[\phi]$ is the Laplace transform (see [91]) of $\phi$. We have to prove that $\langle E, \cdot\rangle$ is a linear and continuous functional on $\mathcal{D}(\mathbb{R}^n)$, i.e., that $E \in \mathcal{D}'(\mathbb{R}^n)$. Consider the estimates

$$|\langle E, \phi\rangle| \leq \frac{1}{(2\pi)^n}\sum_{j=0}^{m}\int_{\mathbb{R}^n} f_j(\xi)\frac{\max_{|\tau|=1}\left|L[\phi](\xi_1 + i\tau\frac{j}{m}, \xi_2, ..., \xi_n)\right|}{\min_{|\tau|=1}\left|P(i\xi_1 - \tau\frac{j}{m}, i\xi_2, ..., i\xi_n)\right|}\, d\xi$$

$$\leq \frac{1}{(2\pi)^n\, ha}\sum_{j=0}^{m}\max e^{|Re(\tau)|\frac{j}{m}}\int_{\mathbb{R}^n}\frac{1}{(1 + |\xi_1 + i\tau\frac{j}{m}|^2 + \xi_2^2 + ... + \xi_n^2)^N}\, d\xi$$

$$\times \int_{|x|<R}|(1-\Delta)^N\phi(x)|\, dx,$$

where $\Delta$ denotes the Laplacian. This implies that for every integer $N > n/2$ and for any $\phi \in \mathcal{D}(U_R)$, $U_R$ being the disc $|x| < R$,

$$|\langle E, \phi\rangle| \leq K_N\int_{|x|<R}|(1-\Delta)^N\phi(x)|\, dx,$$

where in the last estimate we used Paley–Wiener–Schwartz's theorem.

We finally have to prove that $E$ is a solutions to the equation $P(D)E = \delta$.

$$\langle P(D)E, \phi\rangle = \langle E, P(-D)\phi\rangle$$

$$= \frac{1}{(2\pi)^n}\sum_{j=0}^{m}\int_{\mathbb{R}^n} f_j(\xi)\frac{1}{2\pi i}\int_{|\tau|=1}\frac{L[P(-D)\phi](\xi_1 + i\tau\frac{j}{m}, \xi_2, \ldots, \xi_n)}{P(i\xi_1 - \tau\frac{j}{m}, i\xi_2, \ldots, i\xi_n)}\frac{d\tau}{\tau}\, d\xi$$

$$= \frac{1}{(2\pi)^n}\sum_{j=0}^{m}\int_{\mathbb{R}^n} f_j(\xi)\frac{1}{2\pi i}\int_{|\tau|=1} L[\phi](\xi_1 + i\tau\frac{j}{m}, \xi_2, \ldots, \xi_n)\frac{d\tau}{\tau}\, d\xi$$

$$= \frac{1}{(2\pi)^n}\sum_{j=0}^{m}\int_{\mathbb{R}^n} f_j(\xi)\mathcal{F}[\phi](\xi)\, d\xi$$

$$= \frac{1}{(2\pi)^n}\int_{\mathbb{R}^n}\mathcal{F}[\phi](\xi)\, d\xi$$

$$= \phi(0) = \langle\delta, \phi\rangle, \qquad \text{for any } \phi \in \mathcal{D}(\Omega).$$

This completes the proof. □

**Remark 1.2.11.** The fact that the fundamental solution always exists in $\mathcal{D}'$, thanks to the Malgrange–Ehrenpreis theorem, has important consequences. In fact, consider the nonhomogeneous partial differential equation with constant coefficients

$$P(D)u = g, \tag{1.30}$$

where $g$ is a given distribution with compact support. Let $E$ be the fundamental solution; then the function

$$u = E * g$$

is a solution of equation (1.30) because

$$P(D)(E * g) = (P(D)E) * g = \delta * g = g$$

by Theorem 1.2.13. The existence of a fundamental solution thus leads to a general existence theorem for the solution to the equation (1.30). Note also that every solution of (1.30) differs from $E * g$ by a solution of the homogeneous equation $P(D)u = 0$. For the properties of the solution $u = E * g$, see Remark 1.2.8.

In particular, one can ask if there is a result analogous to the Malgrange–Ehrenpreis theorem when particular conditions on the solutions are required, for example conditions on the smoothness, support, and growth. Here we mention only the case of the existence of a solution of the differential equation (1.30) in the space of tempered distributions:

**Theorem 1.2.24.** (Hörmander). *For every $g \in \mathcal{S}'(\mathbb{R}^n)$ the differential equation*

$$P(D)u = g,$$

*with $P(D) \neq 0$, has a solution in $\mathcal{S}'(\mathbb{R}^n)$. In particular it has a fundamental solution $E \in \mathcal{S}'(\mathbb{R}^n)$ since $\delta \in \mathcal{S}'(\mathbb{R}^n)$.*

**Definition 1.2.40.** *We say that the operator $P(D)$ is hypoelliptic if its fundamental solution $E$ is of class $C^\infty$ for $x \neq 0$. We say that the operator $P(D)$ is elliptic if its fundamental solution $E$ is real analytic for $x \neq 0$.*

**Remark 1.2.12.** Every elliptic operator is hypoelliptic. We recall that important operators like the Laplace operator, Cauchy–Riemann operator, Cauchy–Fueter operator are elliptic, while the heat operator is hypoelliptic. Many of the operators studied in this book are elliptic, although we will also deal with some significant examples of hypoelliptic operators.

We recall analytic conditions to characterize elliptic and hypoelliptic operators.

**Theorem 1.2.25.** *The operator $P(D)$ is elliptic if and only if for every open set $\Omega$ every solution $u \in \mathcal{D}'(\Omega)$ to the equation*

$$P(D)u = 0$$

*is a real analytic function in* $\Omega$.

*The operator* $P(D)$ *is hypoelliptic if and only if for every open set* $\Omega$ *every solution* $u \in \mathcal{D}'(\Omega)$ *of the equation*

$$P(D)u = f, \quad \textit{where} \quad f \in C^{\infty}(\Omega)$$

*belongs to* $C^{\infty}(\Omega)$.

It is possible to give algebraic conditions to characterize elliptic and hypoelliptic operators with constant coefficients.

**Theorem 1.2.26.** *An operator* $P(D)$ *is elliptic if and only if its principal part*

$$P_m(D) := \sum_{|\alpha|=m} c_\alpha D^\alpha$$

*satisfies the condition*

$$P_m(\xi) \neq 0, \quad \text{for } \xi \neq 0.$$

*An operator* $P(D)$ *is hypoelliptic if and only if*

$$\lim_{|\xi|\to\infty} \frac{P^{(\alpha)}(-i\xi)}{P(-i\xi)} = 0, \quad \textit{for any } \alpha \textit{ with } |\alpha| \geq 1.$$

## 1.3 Elements of hyperfunction theory

Back in the late 1950s, the Japanese mathematician M. Sato set out [168], [170] to construct a space of generalized functions which would be the "analytic" equivalent of Schwartz's distributions. His inspiration came from some work in theoretical physics which was done by his mentor S. I. Tomonaga, who eventually received the Nobel prize for physics, jointly with R. Feynman, for his work on the scattering phenomenon. This work showed the necessity of dealing with boundary values of holomorphic functions, and even though the first few papers of Sato are relatively simple, it quickly turned out that the only way to build such a theory is through the use of sheaf theory and derived functors. We provide here some basic material on this theory, and we refer the reader to [98], [102], [137] for all the missing proofs.

### 1.3.1 Hyperfunctions in one variable

In this section we will give a very elementary treatment of hyperfunction theory in one variable.

Let us consider an open set $\Omega \subset \mathbb{R}$; an open set $U \subset \mathbb{C}$ such that $\Omega$ is a closed subset of $U$, is said to be a complex neighborhood of $\Omega$.
Let us consider the complex vector space $\mathcal{O}(U \setminus \Omega)$, its subspace $\mathcal{O}(U)$ and their quotient $\mathcal{O}(U \setminus \Omega)/\mathcal{O}(U)$. We will define a hyperfunction $f(x)$ as an equivalence

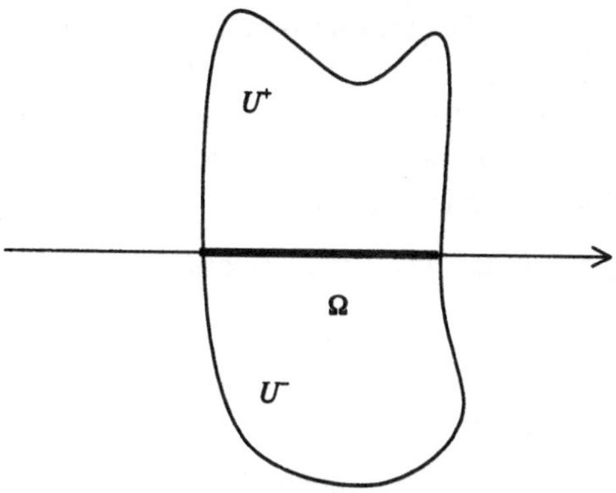


**Figure 1.1.**

class $f(x) = [F(z)]$ in this quotient. Any function $F(z)$ in the equivalence class is said to be a defining function for $f(x)$.

The quotient $\mathcal{O}(U \setminus \Omega)/\mathcal{O}(U)$ depends a priori on the choice of the open set $U \subseteq \mathbb{C}$, but the next proposition, which relies on the Mittag–Leffler theorem, shows that this is not the case.

**Proposition 1.3.1.** *Let $U$ and $U'$ be two complex neighborhoods of the same open set $\Omega \subseteq \mathbb{R}$. Then, there is a vector space isomorphism*

$$\frac{\mathcal{O}(U \setminus \Omega)}{\mathcal{O}(U)} \cong \frac{\mathcal{O}(U' \setminus \Omega)}{\mathcal{O}(U')}.$$

*Proof.* First, note that we can suppose $U' \subset U$. Consider the homomorphism given by the restriction map

$$\rho : \frac{\mathcal{O}(U \setminus \Omega)}{\mathcal{O}(U)} \to \frac{\mathcal{O}(U' \setminus \Omega)}{\mathcal{O}(U')}.$$

The map $\rho$ is injective since two holomorphic functions $F$ and $G$ that coincide on $U' \setminus \Omega$ must coincide on $U \setminus \Omega$. We now have to show that $\rho$ is also surjective. The Mittag–Leffler theorem asserts that if $V$ and $V'$ are two open sets in $\mathbb{C}$, then for any function $F \in \mathcal{O}(V \cap V')$ there exist $F_1 \in \mathcal{O}(V)$, $F_2 \in \mathcal{O}(V')$ such that $F(z) = F_1(z) - F_2(z)$ on $V \cap V'$. Applying this result to the pair of open sets $U \setminus \Omega$ and $U'$, we get that any $F \in \mathcal{O}(U' \setminus \Omega)$ can be expressed as $F(z) = F_1(z) - F_2(z)$ with $F_1 \in \mathcal{O}(U \setminus \Omega)$ and $F_2 \in \mathcal{O}(U')$. Then $\rho(F_1) = F$. □

**Definition 1.3.1.** *Let $\Omega$ be an open set in $\mathbb{R}$. The vector space of hyperfunctions on $\Omega$ is defined as*

$$\mathcal{B}(\Omega) = \frac{\mathcal{O}(U \setminus \Omega)}{\mathcal{O}(U)}, \tag{1.31}$$

*where $U$ is any complex neighborhoods of $\Omega$.*

**Remark 1.3.1.** Let $F \in \mathcal{O}(U \setminus \Omega)$ and denote by $f = [F]$ the hyperfunction $f$ defined by the quotient (1.31). If the function $F$ is holomorphic at every point of $\Omega$, then $f$ is the zero hyperfunction. Note, however, that it is not possible to speak about the value of a hyperfunction at a given point, so it is not correct to think of $f(x) = 0$ as a numerical value.

**Theorem 1.3.1.** *Let $\Omega \subseteq \mathbb{R}$. Then the correspondence*

$$\Omega \to \mathcal{B}(\Omega),$$

*defines a flabby sheaf $\mathcal{B}$ on $\mathbb{R}$.*

*Proof.* Let $\Omega$, $\Omega' \subseteq \mathbb{R}$ be open sets such that $\Omega' \subset \Omega$ and let $U$, $U'$ be open complex neighborhoods such that $U' \setminus \Omega' \subset U \setminus \Omega$. The restriction map $\mathcal{O}(U \setminus \Omega) \to \mathcal{O}(U' \setminus \Omega')$ induces a map $\mathcal{B}(\Omega) \to \mathcal{B}(\Omega')$ which does not depend on the choice of $U$ and $U'$. It is immediate to verify that $\mathcal{B}$ with this restriction map is a presheaf. Let us now consider an open covering $\{\Omega_i\}_{i \in I}$ of $\Omega$. We have to prove that

(i) if $f \in \mathcal{B}(\Omega)$ satisfies $f_{|\Omega_i} = 0$, for every $i \in I$, then $f = 0$ in $\mathcal{B}(\Omega)$;

(ii) if $\{f_j\}_{j \in I}$ is a family of hyperfunctions, $f_j \in \mathcal{B}(\Omega_j)$ satisfying

$$f_{j|\Omega_j \cap \Omega_k} = f_{k|\Omega_j \cap \Omega_k},$$

for every $j, k \in I$ such that $\Omega_j \cap \Omega_k \neq \emptyset$, then there is $f \in \mathcal{B}(\Omega)$ such that for every $j \in I$ it is $f_{|\Omega_j} = f_j$.

It is easy to prove that (i) is satisfied. Indeed, if $f = [F] \in \mathcal{B}(\Omega)$ then the defining function $F$ is holomorphic through the real axis at every point of $\Omega_i$, $\forall i$. Since $\{\Omega_i\}$ is a covering of $\Omega$, $F$ is holomorphic at every point of $\Omega$ and therefore $f$ vanishes on $\Omega$.

To prove (ii), let us consider $f_j \in \mathcal{B}(\Omega_j)$ and an open set $V_j$ in $\mathbb{C}$ such that $V_j \cap \mathbb{R} = \Omega_j$ and let $F_j \in \mathcal{O}(V_j \setminus \Omega_j)$ be a representative of $f_j$. Set $V = \cup_{j \in I} V_j$, so that $V \cap \mathbb{R} = \Omega$. Since $\mathbb{R}$ and $\mathbb{C}$ are paracompact spaces, we can assume that $\{\Omega_j\}$ and $\{V_j\}$ are locally finite. Consider the functions $G_{ij} = F_j - F_i$. They are holomorphic on $V_i \cap V_j$ and obviously satisfy

$$G_{ij} + G_{ji} = 0, \qquad \text{on } V_i \cap V_j,$$

$$G_{ij} + G_{jk} + G_{ki} = 0, \qquad \text{on } V_i \cap V_j \cap V_k.$$

These equalities imply that $\{G_{ij}\}$ is a 1-cocycle on $V$. By the Mittag–Leffler theorem there exist functions $G_i \in \mathcal{O}(V_i)$, such that

$$G_{ij} = G_j - G_i, \qquad \text{on } V_i \cap V_j.$$

We may then write

$$F_j - F_i = G_j - G_i, \qquad \text{on } V_i \cap V_j$$

i.e.,

$$F_i - G_i = F_j - G_j, \qquad \text{on } (V_i \backslash \Omega_i) \cap (V_j \backslash \Omega_j).$$

We define a holomorphic function $F \in \mathcal{O}(V \backslash \Omega)$, by setting

$$F = F_j - G_j, \qquad \text{on } V_j \setminus \Omega_j.$$

Since it is $[F_j - G_j] = [F_j] = f_j$ on $\mathcal{B}(\Omega_j)$, we have that $[F] = f$ is the hyperfunction such that $f_{|\Omega_j} = f_j$ and (ii) is proved. We now prove that $\mathcal{B}$ is a flabby sheaf. Let $\Omega$ and $\Omega'$ be two open sets in $\mathbb{R}$ such that $\Omega' \subseteq \Omega$. We have to show that the restriction map $\mathcal{B}(\Omega) \to \mathcal{B}(\Omega')$ is surjective. Let $f \in \mathcal{B}(\Omega')$, $\partial\Omega'$ be the boundary of $\Omega'$ and set $V = \mathbb{C} \backslash \partial\Omega'$. The set $\Omega'$ is a closed subset of $V$, so $f$ admits a representative $F \in \mathcal{O}(V \backslash \Omega') = \mathcal{O}(\mathbb{C} \backslash \overline{\Omega'})$. Since $\Omega' \subseteq \Omega$, we have that $F \in \mathcal{O}(\mathbb{C} \backslash \overline{\Omega})$. Since $W = \mathbb{C} \backslash \partial\Omega$ is a complex neighborhood of $\Omega$ we have that $F \in \mathcal{O}(W \backslash \Omega)$ and $[F]$ defines a hyperfunction $g \in \mathcal{B}(\Omega)$. Obviously $g_{|\Omega'} = f$, and the sheaf $\mathcal{B}$ is flabby. □

It is possible to give an important alternative definition of $\mathcal{B}(\Omega)$. In fact the relative cohomology sequence of the pair $(U, U \setminus \Omega)$ with coefficients in the sheaf $\mathcal{O}$ gives

$$0 \to H^0_\Omega(U, \mathcal{O}) \to H^0(U, \mathcal{O}) \to H^0(U \setminus \Omega, \mathcal{O})$$
$$\to H^1_\Omega(U, \mathcal{O}) \to H^1(U, \mathcal{O}) \to \dots .$$

Because of the Mittag–Leffler theorem we know that $H^1(U, \mathcal{O}) = 0$ and the unique continuation property for holomorphic functions implies $H^0_\Omega(U, \mathcal{O}) = 0$. By the previous exact sequence we get

$$\mathcal{B}(\Omega) = \frac{\mathcal{O}(U \setminus \Omega)}{\mathcal{O}(U)} \cong H^1_\Omega(U, \mathcal{O}).$$

If we define the support of a hyperfunction $f \in \mathcal{B}(\Omega)$ as the complement in $\Omega$ of the largest open subset on which $f$ vanishes, and we denote by $\mathcal{B}[K]$ the space of hyperfunctions supported by the compact set $K$, we have the following result.

**Theorem 1.3.2.** *Let $K$ be a compact set in $\mathbb{R}$ and let $U$ be any complex neighborhood of $K$. We have the following isomorphism*

$$\mathcal{B}[K] \cong \frac{\mathcal{O}(U \setminus K)}{\mathcal{O}(U)}. \tag{1.32}$$

*Proof.* Let $f \in \mathcal{B}[K]$ and let $\Omega \subset \mathbb{R}$ be any open set containing $K$; then $f$ can be represented by an element in $\mathcal{B}(\Omega) = \mathcal{O}(U \setminus \Omega)/\mathcal{O}(U)$, where $U$ is any neighborhood of $\Omega$. Using this representation, one sees that (1.32) corresponds to hyperfunctions with support in $K$. Since $f$ can be thought of as an element in $\mathcal{B}(\Omega)$ by extending it to zero outside $K$, the inclusion $\mathcal{B}[K] \hookrightarrow \mathcal{B}(\Omega)$ is well defined for any $\Omega$ containing $K$ and the discussion does not depend on the choice of $\Omega$. □

Before giving some examples of hyperfunctions, it is convenient to define some elementary operations on them, besides the operation of sum and multiplication by a complex number that are naturally implied by the vector space structure.

Note that, since a hyperfunction is determined by the equivalence class of a function $F \in \mathcal{O}(U \setminus \Omega)$, we can set $U \setminus \Omega = U^+ \cup U^-$ with $U^\pm = U \cap \{\pm \mathrm{Im} z > 0\}$ and $F = (F^+, F^-)$, $F^\pm \in \mathcal{O}(U^\pm)$ so that the hyperfunction $f$ can be represented by the pair $(F^+, F^-)$.

**Definition 1.3.2.** (Multiplication of a hyperfunction by a real analytic function). *Let $\phi(x)$ be a real analytic function on $\Omega$ and let $f = [F] \in \mathcal{B}(\Omega)$. We define $\phi(x)f(x) = [\phi(z)F(z)]$ where $\phi(z)$ is an analytic continuation of $\phi(x)$.*

**Definition 1.3.3.** (Differentiation). *We define*

$$\frac{d}{dx}f(x) = \frac{d}{dx}([F(z)]) = \left[\frac{d}{dz}F(z)\right]$$

*and, in general,*

$$\frac{d^n}{dx^n}f(x) = \left[\frac{d^n}{dz^n}F(z)\right].$$

It is possible to define the notion of definite integral for hyperfunctions: let $f$ be a hyperfunction defined on a neighborhood of the interval $[a, b]$ and let $f$ be real analytic in a neighborhood of each of the two endpoints of the interval. Let $F = (F^+, F^-)$ be a defining function for $f$: by definition both $F^+$ and $F^-$ can be analytically continued to some neighborhoods of the points $a$ and $b$. Let $\gamma^\pm \subset U^\pm$ be piecewise smooth arcs connecting the points $a$, $b$, then (see Figure 1.2)

$$\int_a^b f(x)dx = \int_{\gamma^+} F^+(z)dz - \int_{\gamma^-} F^-(z)dz.$$

The definition is not affected by the choices made.

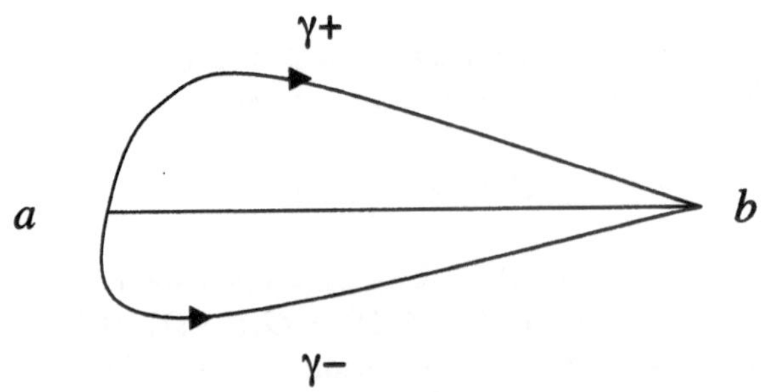


**Figure 1.2.**

Let us now consider compactly supported hyperfunctions.

**Definition 1.3.4.** (Integration). *Let $F(z) \in \mathcal{O}(U \setminus K)$ be a defining function for a hyperfunction $f(x)$ supported in $K$, where $U$ is any complex neighborhood of $K$. Let $\gamma \subset U$ be a closed, piecewise smooth curve encircling once $K$. We will assume $\gamma$ oriented counterclockwise. We define*

$$\int_{\mathbb{R}} f(x)dx = -\oint_{\gamma} F(z)dz.$$

Using Cauchy's integral theorem, it is immediate to verify that the notion is independent of the choices of $F$, $U$ and $\gamma$.

**Example 1.3.1.** The Dirac delta function can be defined as the hyperfunction

$$\delta(x) = \left[-\frac{1}{2\pi i}\frac{1}{z}\right].$$

Since $\frac{1}{z}$ is holomorphic on $\mathbb{C} \setminus \{0\}$, $\delta(x) = 0$ on $\mathbb{R} \setminus \{0\}$ so its support is the origin. The chain of equalities

$$\int_{\mathbb{R}} \phi(x)\delta(x)dx = \frac{1}{2\pi i}\int_{\gamma} \frac{\phi(z)}{z}dz = \phi(0)$$

proves, at least when $\phi(x)$ is a real analytic function, that the $\delta$ hyperfunction behaves like the $\delta$-distribution. In particular, we have

$$\int_{\mathbb{R}} \delta(x)dx = 1.$$

Moreover, we also have

$$\frac{d}{dx}\delta(x) = \left[\frac{1}{2\pi i}\frac{1}{z^2}\right]$$

and, in general,

$$\frac{d^n}{dx^n}\delta(x) = \left[-\frac{(-1)^n n!}{2\pi i}\frac{1}{z^{n+1}}\right].$$

**Example 1.3.2.** The Heaviside function $H(x)$, whose derivative (in $\mathcal{D}'(\mathbb{R})$) is the Dirac delta, is defined as the function

$$H(x) = \begin{cases} 0 \text{ for } x < 0 \\ 1 \text{ for } x \geq 0. \end{cases}$$

In this setting the Heaviside hyperfunction is defined as

$$H(x) = \left[-\frac{1}{2\pi i}\log(-z)\right],$$

where we consider the principal value of the logarithmic function. It is immediate to verify that, even within the theory of hyperfunctions, $H'(x) = \delta(x)$ as one expects from the theory of distributions.

**Example 1.3.3.** Consider the following hyperfunctions:

$$\varepsilon(z) = \begin{cases} 1 \text{ for } \mathrm{Im} z > 0 \\ 0 \text{ for } \mathrm{Im} z < 0 \end{cases} \qquad \bar{\varepsilon}(z) = \begin{cases} 0 \text{ for } \mathrm{Im} z > 0 \\ -1 \text{ for } \mathrm{Im} z < 0. \end{cases}$$

The hyperfunction associated to $\varepsilon$ is defined on $\mathbb{R}$ and it can be seen as the unit hyperfunction 1 if we think a hyperfunction as the difference of boundary values of holomorphic functions. We obviously have $[\varepsilon] = [\bar{\varepsilon}]$.

**Remark 1.3.2.** Any real analytic function $\phi(x)$ can be thought of as a hyperfunction. In other words, denoting by $\mathcal{A}(\Omega)$ the vector space of real analytic functions on $\Omega$, we have that $\mathcal{A}(\Omega) \hookrightarrow \mathcal{B}(\Omega)$, and $\mathcal{A}$ is a subsheaf of $\mathcal{B}$. In fact, since $\phi \in \mathcal{A}(\Omega) = \varinjlim_{U \supset \Omega} \mathcal{O}(U)$, we can set

$$\phi(z) \in \mathcal{O}(U) \to [\phi(z)\varepsilon(z)] \in \mathcal{B}(\Omega).$$

**Example 1.3.4.** A function $F$ holomorphic on $U \setminus \Omega$ defines a hyperfunction $f = [F]$ that can be realized as the boundary value of $F$ as follows. Let us set

$$F(x + i0) = [F \cdot \varepsilon], \qquad F(x - i0) = [F \cdot \bar{\varepsilon}].$$

We have

$$f(x) = F(x + i0) - F(x - i0)$$

which gives formal meaning to the notion of boundary value.

More generally, let $F(z)$ be a meromorphic function in a neighborhood of $\Omega$. We define the hyperfunction called the finite part of $F$ as follows:

$$f.p.\ F(x) = \frac{1}{2}(F(x + i0) + F(x - i0)).$$

When $F(z) = \dfrac{1}{z}$, this hyperfunction is referred to as Cauchy's Principal Value.

**Remark 1.3.3.** From the previous example, it follows that the Dirac delta hyperfunction can be rewritten as

$$\delta(x) = \frac{-1}{2\pi i}\left(\frac{1}{x + i0} - \frac{1}{x - i0}\right) = \frac{i}{2\pi} f.p. \frac{1}{x},$$

while

$$\frac{d^n}{dx^n}\delta(x) = (-1)^{n+1}\frac{n!}{2\pi i}\left(\frac{1}{(x + i0)^{n+1}} - \frac{1}{(x - i0)^{n+1}}\right).$$

Another useful result involving the Dirac delta is the following:

**Proposition 1.3.2.** *Let $\phi \in \mathcal{A}(\{0\})$. We have*

$$\phi(x)\frac{d^n}{dx^n}\delta(x) = \sum_{j=0}^{n}(-1)^j\binom{n}{j}\frac{d^j}{dx^j}\phi(0)\ \frac{d^{n-j}}{dx^{n-j}}\delta(x);$$

*in particular*

$$\phi(x)\delta(x) = \phi(0)\delta(x).$$

*Proof.* The proof is based on the following computations

$$\begin{aligned}
\phi(x)\frac{d^n}{dx^n}\delta(x) &= (-1)^{n+1}\frac{n!}{2\pi i}\left[\frac{\phi(z)}{z^{n+1}}\right] \\
&= (-1)^{n+1}\frac{n!}{2\pi i}\left[\frac{1}{z^{n+1}}\sum_{j=0}^{+\infty}\frac{z^j}{j!}\frac{d^j}{dz^j}\phi(0)\right] \\
&= \sum_{j=0}^{n}(-1)^{n+1}\frac{n!}{j!(n-j)!}\frac{(n-j)!}{2\pi i}\left[\frac{1}{z^{n-j+1}}\frac{d^j}{dz^j}\phi(0)\right] \\
&= \sum_{j=0}^{n}(-1)^j\binom{n}{j}\frac{d^j}{dx^j}\phi(0)\,\frac{d^{n-j}}{dx^{n-j}}\delta(x). \qquad \square
\end{aligned}$$

**Proposition 1.3.3.** *Let $\Omega$ be an open interval in $\mathbb{R}$ and let $f \in \mathcal{B}(\Omega)$. Then $\frac{d^n}{dx^n}f = 0$ if and only if $f$ is a polynomial of degree less than $n$.*

*Proof.* Let us consider a representative $F \in \mathcal{O}(U \setminus \Omega)$ of $f$ and let us suppose $\frac{d}{dx}f = 0$. By definition $\frac{d}{dz}F \in \mathcal{O}(U \setminus \Omega)$ and it is not reductive to suppose that $U$, $U^+$ and $U^-$ are simply connected. By the monodromy principle there exists $g \in \mathcal{O}(U)$ such that $\frac{d}{dz}g = \frac{d}{dz}F$ on $U$. Since

$$\frac{d^n}{dz^n}(F(z) - g(z)) = 0$$

on $U \setminus \Omega$ then there are two polynomials $p$ and $q$ of degree less than $n$ such that

$$F(z) - g(z) = \varepsilon(z)p(z) + \bar{\varepsilon}(z)q(z)$$

hence $f = p - q$. The other implication is trivial. $\square$

**Proposition 1.3.4.** *Let $\Omega$ be an open interval in $\mathbb{R}$ and let $g \in \mathcal{B}(\Omega)$. Then the ordinary differential equation*

$$\frac{d^n}{dx^n}f = g$$

*has a hyperfunction solution $f$ on $\Omega$ unique up to polynomials of degree less than $n$.*

*Proof.* Let us consider a representative $G \in \mathcal{O}(U \setminus \Omega)$ of $g$; it is not reductive to suppose that $U$, $U^+$ and $U^-$ are simply connected. By the monodromy principle there exists $F \in \mathcal{O}(U \setminus \Omega)$ such that $\frac{d^n}{dx^n}F = G$ on $U \setminus \Omega$. It is clear that $f = [F]$ is a solution to the differential equation. The uniqueness up to polynomials of degree less than $n$ follows by the previous proposition. $\square$

**Proposition 1.3.5.** *Let $\Omega$ be a locally closed set in $\mathbb{R}$ containing 0. A hyperfunction $f \in \mathcal{B}(\Omega)$ satisfies the equation*

$$x^n f(x) = 0$$

*if and only if*

$$f(x) = \sum_{j=0}^{n-1} c_j \frac{d^j}{dx^j} \delta(x), \tag{1.33}$$

*for some $c_j \in \mathbb{C}$, $j = 0, \ldots, n-1$.*

*Proof.* The condition (1.33) is trivially sufficient by Proposition 1.3.2. Let us now suppose that $f = [F]$, $F \in \mathcal{O}(U \setminus \Omega)$. By hypothesis there exists $G(z) \in \mathcal{O}(U)$ such that $G(z) = z^n F(z)$ so that we have

$$[F] = \left[\frac{G(z)}{z^n}\right]$$

and computations similar to the ones in the proof of Proposition 1.3.2 show that $f$ has the form (1.33). □

**Proposition 1.3.6.** *Let $\Omega$ be a locally closed set in $\mathbb{R}$ and $0 \in \Omega$. The equation*

$$x^n f(x) = g(x)$$

*has a solution $f \in \mathcal{B}(\Omega)$ for any $g \in \mathcal{B}(\Omega)$. The solution is unique up to linear combinations of the delta function and its derivatives up to order $n-1$.*

*Proof.* Let $g = [G]$, $G \in \mathcal{O}(U \setminus \Omega)$. The hyperfunction $f$ defined by $\dfrac{G(z)}{z^n} \in \mathcal{O}(U \setminus \Omega)$ is clearly a solution and the uniqueness follows by Proposition 1.3.5. □

More generally, let us consider the operator

$$P\left(x, \frac{d}{dx}\right) = \sum_{i=0}^{m} a_i(x) \frac{d^i}{dx^i} \tag{1.34}$$

where $a_i \in \mathcal{A}(\Omega)$ for $i = 0, \ldots, m$, $a_m(x) \not\equiv 0$ and let $f = [F] \in \mathcal{B}(\Omega)$. Let $U$ be an open set to which all the $a_i$ can be holomorphically extended, and let $F \in \mathcal{O}(U \setminus \Omega)$. We define

$$P\left(x, \frac{d}{dx}\right) f = \left[P\left(z, \frac{d}{dz}\right) F\right]$$

where

$$P\left(z, \frac{d}{dz}\right) = \sum_{i=0}^{m} a_i(z) \frac{d^i}{dz^i}.$$

It is easy to check that this definition does not depend on the choice of the open set $U$, the defining function $F$ and the extensions $a_i(z)$ of $a_i(x)$.

We have the following result about the existence of hyperfunction solutions to the differential equation

$$P\left(x, \frac{d}{dx}\right) f = g. \tag{1.35}$$

**Theorem 1.3.3.** *For any $g \in \mathcal{B}(\Omega)$ there exists a solution $f \in \mathcal{B}(\Omega)$ to the differential equation (1.35).*

*Proof.* Let $U$ be any complex neighborhood of $\Omega$ such that $U$, $U^+$ and $U^-$ are simply connected. Let $a_i \in \mathcal{O}(U)$ be the holomorphic extension of $a_i(x)$ to $U$ and $a_m(z) \not\equiv 0$ on $U \backslash \Omega$. We can consider $F, G \in \mathcal{O}(U \backslash \Omega)$ two defining functions for $f$, $g$, respectively. By the Cauchy existence theorem the differential equation

$$P\left(z, \frac{d}{dz}\right) F(z) = G(z)$$

admits a local solution which, by the monodromy principle, can be extended to the simply connected domains $U^+$ and $U^-$. Then we obtain a solution $\widetilde{F} \in \mathcal{O}(U \backslash \Omega)$. The hyperfunction $f = [\widetilde{F}]$ is the hyperfunction solution of (1.35). □

If we consider the homogeneous equation

$$P\left(x, \frac{d}{dx}\right) f = 0 \tag{1.36}$$

it is possible to determine the dimension of the space $\mathcal{B}^P$ of its hyperfunction solutions on $\Omega$.

**Theorem 1.3.4.** *Let $\Omega$ be an open interval in $\mathbb{R}$; then*

$$\dim(\mathcal{B}^P(\Omega)) = m + \sum_{x \in \Omega} \mathrm{ord}_x a_m(x)$$

*where $\mathrm{ord}_x a_m(x)$ denotes the order of the zero of $a_m$ at the point $x$.*

For the proof of this result see [107].

The following result characterizes ellipticity.

**Theorem 1.3.5.** *Let $\Omega$ be an open set in $\mathbb{R}$. The following conditions are equivalent:*

*(i) any solution $f \in \mathcal{B}(\Omega)$ to the equation (1.36) belongs to $\mathcal{A}(\Omega)$;*

*(ii) if $P\left(x, \frac{d}{dx}\right) f \in \mathcal{A}(\Omega)$, then $f \in \mathcal{A}(\Omega)$;*

*(iii) $a_m(x) \neq 0$ for any $x \in \Omega$.*

For the proof, see [108] or [137].

Now we wish to consider differential operators more general than (1.34), i.e., the so-called local operators. As is well known, every local operator on distributions is a finite sum of convolutions with the Dirac delta and its derivatives. On the other hand, one of the great advantages of the hyperfunctions is that they allow the use of a larger class of local operators in which infinitely many derivatives can be considered. More generally, we can consider the operator

$$P\left(x, \frac{d}{dx}\right) = \sum_{i=0}^{+\infty} a_i(x) \frac{d^i}{dx^i} \tag{1.37}$$

where $\{a_i(x)\}_{i\in\mathbb{N}}$ is a sequence of functions. In order for this operator to act on a hyperfunction $f = [F]$ supported at the origin, we need to ensure that when $F(z)$ is holomorphic in a neighborhood of $\{0\}$, then

$$P\left(z, \frac{d}{dz}\right) F = \sum_{i=0}^{+\infty} a_i(z) \frac{d^i}{dz^i} F(z) \tag{1.38}$$

is holomorphic in $U$, so that we can define

$$P\left(x, \frac{d}{dx}\right) f = \left[P\left(z, \frac{d}{dz}\right) F(z)\right].$$

The series on the right-hand side of (1.38) converges under suitable growth conditions on the coefficients $a_i(z)$.

**Theorem 1.3.6.** *Consider the operator defined by*

$$P\left(z, \frac{d}{dz}\right) = \sum_{i=0}^{+\infty} a_i(z) \frac{d^i}{dz^i}$$

*and let $a_i(z)$ be holomorphic functions on an open set $U$. Then $P(z, \frac{d}{dz})$ is a sheaf homomorphism from $\mathcal{O}$ to itself if and only if, for any compact set $K \subset U$,*

$$\lim_{i\to+\infty} \sqrt[i]{\sup_K |a_i(z)| i!} = 0. \tag{1.39}$$

*Proof.* In order to prove the theorem, we will show that $P\left(z, \frac{d}{dz}\right)$ is a homomorphism at the stalk level $\mathcal{O}_{z_0}$. Let us consider an element in $\mathcal{O}_{z_0}$, i.e., a germ of a holomorphic function $F$ at a point $z_0$. In particular, $F$ is holomorphic on a closed disc $C$ around $z_0$ having radius $r$ and reaches a maximum value $M$ on it. The condition on the coefficients $a_i(z)$ implies that for every $\varepsilon > 0$ there exists a constant $\Gamma_\varepsilon$ such that for every $z \in C$, $|a_i(z)| i! \leq \Gamma_\varepsilon \varepsilon^i$ and, by the Cauchy inequality, we have that for every $z \in C$ it is

$$\left|\frac{d^i F(z)}{dz^i}\right| \leq \frac{Mi!}{r^{i+1}}.$$

Then, if we set $\varepsilon = r/2$, we obtain

$$\left|\sum_{i=0}^{+\infty} a_i(z)\frac{d^iF(z)}{dz^i}\right| \leq \sum_{i=0}^{+\infty} |a_i(z)| \left|\frac{d^iF(z)}{dz^i}\right| \leq \sum_{i=0}^{+\infty} \Gamma_\varepsilon \varepsilon^i \frac{M}{r^i} \leq \Gamma_\varepsilon M \sum_{i=0}^{+\infty} \left(\frac{1}{2}\right)^i.$$

This proves that the series on the left-hand side is uniformly convergent and defines a holomorphic function in a neighborhood of any point $z_0$ in which $F(z)$ is holomorphic. Now we have to show the converse, i.e., we have to prove that if the series converges then (1.39) holds. Suppose (1.39) does not hold; then for some $\varepsilon > 0$ there exists a sequence $\{z_j\} \subset K$, convergent to $z_0$, and such that

$$\sqrt[i_j]{\sup_K |a_{i_j}(z_j)| i_j!} \geq 2\varepsilon, \qquad j = 1, 2, \ldots, \quad i_j \to +\infty.$$

The function

$$F(z) = \frac{1}{z - z_0 - \varepsilon}$$

is holomorphic on a disc with center $z_0$ and radius $\varepsilon$ so we can apply to it the operator $P(z, d/dz)$ and we get

$$\sum_{i=0}^{+\infty} a_i(z)\frac{d^i}{dz^i}\left(\frac{1}{z - z_0 - \varepsilon}\right) = \sum_{i=0}^{+\infty} (-1)^i \frac{a_i(z) i!}{(z - z_0 - \varepsilon)^{i+1}};$$

if we set $z = z_j$, we obtain

$$\left|\frac{a_i(z_j) i_j!}{(z_j - z_0 - \varepsilon)^{i_j+1}}\right| \geq \frac{(2\varepsilon)^{i_j}}{(2\varepsilon)^{i_j+1}} = \frac{1}{2\varepsilon}.$$

This inequality implies that the series does not converge locally uniformly which contradicts our assumption. □

**Definition 1.3.5.** *An operator $P(z, d/dz)$ is said to be a local operator if it satisfies (1.39).*

In particular, when the coefficients are constant, i.e., $a_i(z) = a_i$ for every index $i$, local operators allow a characterization of hyperfunctions supported at one point (we consider here the case of the origin).

**Theorem 1.3.7.** *Let $f \in \mathcal{B}[\{0\}]$. Then there is a sequence $\{a_i\}_{i \in \mathbb{N}} \subset \mathbb{C}$ satisfying*

$$\lim_{i \to +\infty} \sqrt[i]{|a_i| i!} = 0 \qquad (1.40)$$

*such that*

$$f(x) = \sum_{i=0}^{+\infty} a_i \frac{d^i \delta}{dx^i}(x).$$

*Proof.* Let $f = [F(z)]$ with $F(z) \in \mathcal{O}(\mathbb{C} \setminus \{0\})$. Then $F$ admits a Laurent expansion at the origin of the form

$$F(z) = \sum_{i=-\infty}^{+\infty} b_i z^i = \sum_{i=-\infty}^{-1} b_i z^i + G(z).$$

The function $G(z)$ is entire, so that we can consider as a representative of $f$ the function $\tilde{F}(z) = \sum_{i=-\infty}^{-1} b_i z^i$ and any other representative of $f$ differs from $\tilde{F}$ by an entire function. Moreover $\tilde{F}$ converges on $\mathbb{C} \setminus \{0\}$, so that its radius of convergence is infinite, i.e., $\lim_{i\to+\infty} \sqrt[i]{|b_{-i}|} = 0$. Rewriting $1/z^{j+1}$ as

$$\frac{1}{z^{j+1}} = (-1)^j \frac{1}{j!} \frac{d^j}{dz^j} \left(\frac{1}{z}\right)$$

and setting $a_j = (-1)^{j+1} 2\pi i \dfrac{b_{-j-1}}{j!}$ one has that

$$F(z) = \sum_{j=0}^{+\infty} a_j \frac{d^j}{dz^j} \left(\frac{-1}{2\pi i z}\right).$$

By the definition of delta-hyperfunction and its derivatives, we obtain

$$f(x) = \sum_{j=0}^{+\infty} a_j \frac{d^j}{dx^j} \delta(x)$$

and the coefficients satisfy the condition $\lim_{j\to+\infty} \sqrt[j]{|a_j| j!} = 0$. □

In the next section we will see that a hyperfunction on a compact set $K$ is nothing but a locally analytic functional on $K$; it is then natural to compare distributions and hyperfunctions. Since there is a canonical injection $i : \mathcal{A}(\mathbb{R}) \hookrightarrow \mathcal{E}(\mathbb{R})$ with dense image, there is also an injective dual map $i^* : \mathcal{E}'(\mathbb{R}) \hookrightarrow \mathcal{A}'(\mathbb{R})$ preserving the supports by virtue of the following proposition (see [137]).

**Proposition 1.3.7.** *Let $K$ be a compact set in $\mathbb{R}$. The support of a functional $T \in \mathcal{E}'(\mathbb{R})$ is contained in $K$ if and only if $i^*(T)$ can be extended to a continuous linear functional on $\mathcal{A}(K)$.*

We now prove that there is a proper inclusion of the space of distributions into the space of hyperfunctions defined on the same open set.

**Proposition 1.3.8.** *Let $\Omega$ be an open set in $\mathbb{R}$. There is a canonical injection*

$$j : \mathcal{D}'(\Omega) \hookrightarrow \mathcal{B}(\Omega)$$

*preserving the support.*

*Proof.* Any distribution $T \in \mathcal{D}'(\Omega)$ can be written as a locally finite sum of distributions $T_\ell$ with compact support as a consequence of the existence of a partition of the unity (see Proposition 1.2.3). By the previous proposition, $i^*T_\ell$ is a hyperfunction with compact support; moreover supp $T_\ell =$ supp $i^*T_\ell$. The hyperfunction $\sum i^*(T_\ell)$ is a locally finite sum of hyperfunctions, and we can define $j(T) = \sum i^*(T_\ell) \in \mathcal{B}(\Omega)$. The map $j$ is the desired injection and preserves the support by construction. □

As a consequence we can see the sheaf of distributions as a subsheaf of hyperfunctions.

**Theorem 1.3.8.** *The sheaf $\mathcal{D}'$ is a subsheaf of $\mathcal{B}$.*

*Proof.* Let $\mu$ be a distribution in $\mathcal{D}'$. Since every distribution is a locally finite sum of compactly supported distributions, we can assume $\mu$ itself to be compactly supported. Let $K \subset \mathbb{R}$ be the support of $\mu$. The so-called indicatrix of $\mu$, namely $\tilde{\mu}(\zeta) = \mu[\frac{1}{z-\zeta}]$, is a holomorphic function on $\mathbb{C} \setminus K$. As such $\tilde{\mu}(\zeta)$ defines a hyperfunction supported in $K$. One can then show that the map $\mu \to [\tilde{\mu}]$ is an injection of sheaves. □

**Remark 1.3.4.** In general, it is possible to show that the following inclusions of sheaves

$$\mathcal{A} \hookrightarrow \mathcal{L}^1_{\text{loc}} \hookrightarrow \mathcal{D}' \hookrightarrow \mathcal{B}$$

hold (see [98]).

We mention, without the proof that can be found in [109], the following result characterizing hyperfunctions that are distributions.

**Theorem 1.3.9.** *Let $\Omega$ be an open set in $\mathbb{R}$ and let $U$ be a complex neighborhood of $\Omega$. Let $f = [F] \in \mathcal{B}(\Omega)$. The hyperfunction $f$ is a distribution on $\Omega$ if and only if for any compact set $K$ in $U$ there are positive constants $C$, $\eta$ and $N \in \mathbb{N}$ such that, if $0 < |y| < \eta$, then*

$$\sup_{x \in K} |F(x+iy)| \le C|y|^{-N}.$$

**Example 1.3.5.** The hyperfunction supported at the origin

$$f = [\exp(-1/z)]$$

can be obtained using infinite order local operators as

$$\left[\exp(-\frac{1}{z})\right] = \sum_{j=0}^{+\infty} \frac{2\pi i}{j!(j+1)!} \frac{d^j}{dx^j} \delta(x).$$

This hyperfunction is not a distribution.

Finally, we wish to mention a theorem relating the distribution solution and the hyperfunction solution to a linear ordinary differential equation.

**Theorem 1.3.10.** *Let $\Omega$ be an open set in $\mathbb{R}$ and let*

$$P\left(x, \frac{d}{dx}\right) f(x)$$

$$= a_n(x)\frac{d^n}{dx^n} f(x) + a_{n-1}(x)\frac{d^{n-1}}{dx^{n-1}} f(x) + \cdots + a_0(x) f(x) = 0 \qquad (1.41)$$

*where we assume that the functions $a_j \in \mathcal{A}(\Omega)$ for any $j$ and $a_n(x) \not\equiv 0$. The following conditions are equivalent.*

*(i) Any solution $f \in \mathcal{B}(\Omega)$ to the equation (1.41) belongs to $\mathcal{D}'(\Omega)$.*

*(ii) If $P\left(x, \frac{d}{dx}\right) f \in \mathcal{D}'(\Omega)$, then $f \in \mathcal{D}'(\Omega)$.*

For the proof see [107].

This result is typical of differential operators with "regular" singularities in the sense of Section 3 in [108].

### 1.3.2 Cohomological properties of the sheaf of holomorphic functions

In this section we will collect some useful cohomology vanishing theorems that will allow us to interpret hyperfunctions in terms of analytic functionals. Let us begin with the following definition.

**Definition 1.3.6.** *An open set $U \subset \mathbb{C}^n$ is said to be a domain of holomorphy if for every $z$ on the boundary of $U$ there exists a function $F \in \mathcal{O}(U)$ that cannot be analytically continued to any neighborhood of this point. An open set $U \subseteq \mathbb{C}^n$ is called a Stein open set if every connected component of $U$ is a domain of holomorphy.*

It is useful to point out that the intersection of two Stein open sets is a Stein open set. More generally, if $\{U_\alpha\}_{\alpha\in A}$ is an infinite family of Stein open sets, the interior of the intersection $\cap_{\alpha\in A} U_\alpha$ is a Stein open set.

**Remark 1.3.5.** Any open set $U \subseteq \mathbb{C}$ is a domain of holomorphy as shown by the Mittag–Leffler theorem. An open set in $\mathbb{C}^n$, $n > 1$, is not necessarily a domain of holomorphy since, by Hartogs' theorem 2.1.12, it may occur that all functions can be analytically continued beyond the boundary. For example if $B$ is a closed ball in $\mathbb{C}^n$, the open set $\mathbb{C}^n \setminus B$ is not a domain of holomorphy in $\mathbb{C}^n$.

The key results in this direction are the following (see, for example [98]).

**Theorem 1.3.11.** (Oka–Cartan–Serre). *Let $U$ be a Stein open set in $\mathbb{C}^n$. Then $H^j(U, \mathcal{O}) = 0$ for $j \geq 1$.*

**Theorem 1.3.12.** (Malgrange). *Let $U$ be an open set in $\mathbb{C}^n$; then*

$$H^j(U, \mathcal{O}) = 0, \qquad j \geq n.$$

The following results are useful to compute relative cohomology groups with coefficients in $\mathcal{O}$. We will consider an open set $U \subseteq \mathbb{C}^n$ and a closed set $A$ in $U$.

**Theorem 1.3.13.** *Let $U \subseteq \mathbb{C}^n$ be an open set and let $A$ be a closed set in $U$. Let $(\mathcal{V}, \mathcal{V}')$ be an open covering of the pair $(U, U \setminus A)$ such that $\mathcal{V}$ is a Stein open covering. Then*

$$H^j(U, \mathcal{O}) \cong H^j(\mathcal{V}, \mathcal{O}), \qquad j \geq 0, \tag{1.42}$$

$$H^j_A(U, \mathcal{O}) \cong H^j(\mathcal{V}, \mathcal{V}', \mathcal{O}), \qquad j \geq 0. \tag{1.43}$$

*Proof.* Let $U_i \in \mathcal{V}$ be Stein open sets. The intersection $U_{i_0} \cap U_{i_1} \cap \ldots \cap U_{i_j}$ is a Stein open set. Therefore Theorem 1.3.11 implies

$$H^j(U_{i_0} \cap U_{i_1} \cap \ldots \cap U_{i_j}, \mathcal{O}) = 0, \qquad j \geq 1,$$

and by Leray's theorem 1.1.10, we get (1.43). Setting $A = U$, from (1.43) we get (1.42). □

**Theorem 1.3.14.** *Let $U$ be a Stein open set in $\mathbb{C}^n$ and let $A$ be a closed set in $U$. Then we have the exact sequence*

$$0 \to H^0_A(U, \mathcal{O}) \to H^0(U, \mathcal{O}) \to H^0(U \setminus A, \mathcal{O}) \to H^1_A(U, \mathcal{O}) \to 0.$$

*Moreover, we have*

$$H^{j-1}(U \setminus A, \mathcal{O}) \cong H^j_A(U, \mathcal{O}), \qquad j \geq 2.$$

*Proof.* Consider the following exact sequence of relative cohomology groups

$$\begin{aligned} 0 &\to H^0_A(U, \mathcal{O}) \to H^0(U, \mathcal{O}) \to H^0(U \setminus A, \mathcal{O}) \\ &\to H^1_A(U, \mathcal{O}) \to H^1(U, \mathcal{O}) \to H^1(U \setminus A, \mathcal{O}) \\ &\to H^2_A(U, \mathcal{O}) \to H^2(U, \mathcal{O}) \to \ldots . \end{aligned}$$

By assumption, $U$ is a Stein open set, so by Theorem 1.3.11 it is

$$H^j(U, \mathcal{O}) = 0, \quad \text{for} \quad j \geq 1,$$

and we immediately get the result. □

**Corollary 1.3.1.** *Let $U$ be a Stein open set in $\mathbb{C}^n$ and let $A$ be a closed set in $U$. Then we have*

$$H^j(U \setminus A, \mathcal{O}) = 0, \quad j \geq n+1.$$

*Proof.* By the Malgrange Theorem 1.3.12, we have $H^{j-1}(U \setminus A, \mathcal{O}) = 0$ for $j > n$ and we get the statement. □

Now we give some important definitions and results that will be a key tool in the theory of hyperfunctions of several variables.

**Definition 1.3.7.** *A closed set $A$ in a topological space $X$ is purely $n$-codimensional with respect to the sheaf $\mathcal{F}$ if*

$$\mathcal{H}^j_A(\mathcal{F}) = 0 \quad \textit{for} \quad j \neq n.$$

**Definition 1.3.8.** *Let $X$ be a topological space and let $\mathcal{F}$ be a sheaf on $X$. A flabby resolution of $\mathcal{F}$ is an exact sequence*

$$0 \to \mathcal{F} \to \mathcal{L}^0 \to \mathcal{L}^1 \to \dots$$

*where $\mathcal{L}^i$ are flabby sheaves. The minimum of the length of the flabby resolution of $\mathcal{F}$ is called the flabby dimension of $\mathcal{F}$ and it will be denoted by fl.dim $\mathcal{F}$.*

**Remark 1.3.6.** A sheaf $\mathcal{F}$ is flabby if and only if fl.dim $\mathcal{F} = 0$.

**Theorem 1.3.15.** *Let $X$ be a topological space and let $\mathcal{F}$ be a sheaf on $X$. If fl.dim $\mathcal{F} \leq m$ and a closed set $A$ in $X$ is purely $m$-codimensional with respect to $\mathcal{F}$, then $\mathcal{H}^m_A(\mathcal{F})$ is flabby sheaf.*

**Theorem 1.3.16.** (Sato). *$\mathbb{R}^n$ is purely $n$-codimensional in $\mathbb{C}^n$ with respect to the sheaf $\mathcal{O}$.*

We now introduce the topological structure of some functional spaces. Let $U$ be an open set in $\mathbb{C}^n$. The space $\mathcal{C}^\infty(U)$ with the topology of uniform convergence on the compact sets is a Fréchet space. Since the space $\mathcal{O}(U)$ is a linear subspace of $\mathcal{C}^\infty(U)$ that turns out to be closed (as any subspace of all $\mathcal{C}^\infty$ solutions of a partial differential equation with constant coefficients) then also $\mathcal{O}(U)$ is a Fréchet space with the topology induced by the topology of $\mathcal{C}^\infty(U)$. Let us consider an increasing sequence of compact sets $\{K_j\}_{j\in\mathbb{N}}$, such that

$$K_1 \Subset K_2 \Subset \dots, \qquad U = \cup_{j=1}^\infty K_j.$$

Let us consider the norm on $K_j$ given by

$$p_j(f) := \sup_{z\in K_j} |f(z)|, \quad f(z) \in \mathcal{O}(U).$$

This norm makes the space $\mathcal{O}_B(K_j)$ of bounded functions on $K_j$, which are holomorphic in the interior of $K_j$, a Banach space; since the restriction maps $\mathcal{O}_B(K_{j+1}) \to \mathcal{O}_B(K_j)$ are compact for all $j \geq 1$, the space

$$\mathcal{O}(U) = \varprojlim \mathcal{O}_B(K_j)$$

with the topology of the uniform converge on the compact sets is an FS space. Since the strong dual of an FS space is a DFS space, as a consequence we get the following classical result from functional analysis.

**Theorem 1.3.17.** *The space $\mathcal{O}'(U)$ is a DFS space.*

Recalling the definition of inductive limit of linear topological spaces, we give the following definition:

**Definition 1.3.9.** *Let $K$ be any compact set in $\mathbb{C}^n$ and let $\mathcal{U}$ be the totality of open sets containing $K$. We define*

$$\mathcal{O}(K) = \varinjlim_{U \in \mathcal{U}} \mathcal{O}(U).$$

**Definition 1.3.10.** *Let $K$ be any compact set in $\mathbb{C}^n$. A locally analytic functional on $K$ is a continuous linear functional on the locally convex space $\mathcal{O}(K)$. The dual space $\mathcal{O}'(K)$ of $\mathcal{O}(K)$ is the space of locally analytic functionals on $K$ equipped with the strong topology.*

**Theorem 1.3.18.** *Let $K$ be any compact set in $\mathbb{C}^n$; then $\mathcal{O}(K)$ is a DFS space while $\mathcal{O}'(K)$ is a FS space.*

We say that an analytic functional $\mu \in \mathcal{O}'(\mathbb{C}^n)$ is carried by the compact set $K$ (and $K$ is said to be a carrier) if for every open neighborhood $U$ of $K$ there is a positive constant $c_U$ such that for every $f \in \mathcal{O}(\mathbb{C}^n)$ it is

$$|\mu(f)| \leq c_U \sup_{z \in U} |f(z)|.$$

**Remark 1.3.7.** Every analytic functional has a minimal carrier by Zorn's Lemma, but it does not have a minimum carrier, i.e., if $\mu$ is carried by $K_1$ and $K_2$, it is not necessarily carried by $K_1 \cap K_2$. On the other hand, if $\mu$ is carried by some compact $K$ in $\mathbb{R}^n$, then there exists a smallest compact set in $\mathbb{R}^n$ which carries $\mu$: such a set is said to be a support for $\mu$. This remark will acquire a deeper significance after we define hyperfunctions and we realize that they are a sheaf.

**Definition 1.3.11.** *Let $\mu \in \mathcal{O}'(K)$ with $K$ a compact convex subset of $\mathbb{C}^n$. The Fourier–Borel transform of $\mu$ is defined by*

$$\hat{\mu}(\zeta) = \mu(\exp(-(z, \zeta))).$$

**Theorem 1.3.19.** (Martineau, [128]). *The space $\mathcal{O}'(K)$ is topologically isomorphic, via the Fourier–Borel transform, to the space $Exp(K)$ of entire functions $f$ such that for any $\varepsilon > 0$ there exists $C_\varepsilon > 0$ such that*

$$|f(z)| \leq C_\varepsilon \exp(H_K(z) + \varepsilon |z|)$$

*where $H_K(z) = \sup_{\zeta \in K} \operatorname{Re}(z, \zeta)$ is the supporting function of $K$.*

It is immediate to observe that the Fourier–Borel transform also provides a topological isomorphism between $\mathcal{O}'(\mathbb{C}^n)$ and the space $\operatorname{Exp}(\mathbb{C}^n)$ of those entire functions $f$ for which there are positive constant $A$, $B$ such that

$$|f(z)| \leq A \exp(B|z|).$$

The space $\mathrm{Exp}(\mathbb{C}^n)$ is known as the space of an entire function of exponential type.

**Definition 1.3.12.** *When $K = \{0\}$ the space $Exp(\{0\})$ is also denoted by $\mathrm{Exp}_0(\mathbb{C}^n)$ and its elements are the entire functions of minimal type i.e., for every $\varepsilon > 0$ there exists $A_\varepsilon > 0$ such that*

$$|f(z)| \leq A_\varepsilon \exp(\varepsilon |z|).$$

*In the terminology of Sato, the functions in $\mathrm{Exp}_0(\mathbb{C}^n)$ are called functions of infraexponential type.*

The description of the space $\mathcal{O}'(K)$ can be done in cohomological terms thanks to the Martineau–Harvey's theorem (see [98] or [137]):

**Theorem 1.3.20.** (Martineau–Harvey). *Let $K$ be a compact set in $\mathbb{C}^n$ satisfying*

$$H^j(K, \mathcal{O}) = 0, \qquad \text{for all } j \geq 1,$$

*and let $U$ be any open neighborhood of $K$. We have*

$$H^j_K(U, \mathcal{O}) = 0, \qquad j \neq n,$$

*and*

$$H^n_K(U, \mathcal{O}) \cong \mathcal{O}'(K).$$

If $K$ is a real compact set we have the isomorphism $\mathcal{O}(K) \cong \mathcal{A}(K)$, and therefore we obtain the following result that we mention in the case of hyperfunctions in one variable, but can be naturally generalized to the case of several variables:

**Theorem 1.3.21.** *Let $K$ be a compact set in $\mathbb{R}$. Then we have*

$$\mathcal{A}'(K) \cong \mathcal{B}[K].$$

*Proof.* We know that $\mathcal{B}[K] \cong H^1_K(U, \mathcal{O})$ and, by Martineau–Harvey's Theorem 1.3.20 whose assumption is satisfied by any real compact set, $H^1_K(U, \mathcal{O})$ is isomorphic to $\mathcal{A}'(K)$. □

**Remark 1.3.8.** By the previous duality theorem, it follows that the space $\mathcal{B}[K]$ can be endowed with the structure of an FS space.

### 1.3.3 Hyperfunctions of several variables

In this section we sketch the basic notions in the theory of hyperfunctions on $\mathbb{R}^n$. Here, a different level of complexity is needed (for a history of its developement we refer the reader to [102]): the case of a single variable can be handled with elementary tools in complex analysis, while the case of several variables requires knowledge of sheaf theory and of relative cohomology with coefficients in a sheaf. For example, the definition of hyperfunctions in one variable supported by the compact set $K$ as the quotient $\mathcal{O}(U \setminus K)/\mathcal{O}(U)$, where $U$ is any complex neighborhood of $K$, cannot be extended to several variables, since, by Hartogs' theorem on the removability of compact singularities of functions in several variables, the quotient vanishes. In fact, we have observed (see Remark 1.3.5) that there exist open sets in $\mathbb{C}^n$, $n > 1$, which are not domain of holomorphy or, in cohomological terms, it is not true that $H^1(U, \mathcal{O}) = 0$. The result that will play the role of the Mittag–Leffler theorem in this setting is the pure codimensionality of $\mathbb{R}^n$ in $\mathbb{C}^n$ with respect to $\mathcal{O}$. By this result and the definition of derived sheaf we can provide the definition for the sheaf of hyperfunctions that generalizes the one given in the case of a single variable. We will also show that the idea of boundary value, which was the guiding notion in the case of one variable, can be introduced also in this case, even though it is less intuitive because of the many directions to take into account.

**Definition 1.3.13.** *The sheaf*

$$\mathcal{B} = \mathcal{H}^n_{\mathbb{R}^n}(\mathcal{O})$$

*is called the sheaf of hyperfunctions on* $\mathbb{R}^n$.

This definition can be rewritten in an equivalent way as follows:

**Theorem 1.3.22.** *Let $\Omega$ be an open set in $\mathbb{R}^n$ and let $U \subseteq \mathbb{C}^n$ be a complex neighborhood of $\Omega$. Then*

$$\mathcal{B}(\Omega) = H^n_\Omega(U, \mathcal{O}) = H^n_\Omega(\mathbb{C}^n, \mathcal{O}). \tag{1.44}$$

*Proof.* The left-hand side equality in (1.44) follows from the definition of derived sheaf, Theorem 1.1.13 and from Sato's Theorem 1.3.16. The right-hand side equality follows by excision (see Theorem 1.1.12). □

**Remark 1.3.9.** It is immediate to verify that $\mathcal{B}(\Omega)$ is an $\mathcal{A}(\Omega)$-module.

The most important feature of the sheaf of hyperfunctions is its flabbiness stated in the next theorem.

**Theorem 1.3.23.** *The sheaf $\mathcal{B}$ is flabby.*

*Proof.* A simple way to see the flabbiness makes use of Malgrange's theorem, which implies that fl.dim $\mathcal{O} \leq n$, and Theorems 1.3.15 and 1.3.16. Another way to show that the sheaf $\mathcal{B}$ is flabby is the following. Let $\Omega$ be an open bounded set in $\mathbb{R}^n$ and let $\partial\Omega = \overline{\Omega} \setminus \Omega$. Now we consider the exact sequence

$$\ldots \longrightarrow H^n_{\partial\Omega}(\mathbb{C}^n, \mathcal{O}) \longrightarrow H^n_{\overline{\Omega}}(\mathbb{C}^n, \mathcal{O}) \longrightarrow$$

$$\longrightarrow H^n_\Omega(\mathbb{C}^n \setminus \partial\Omega, \mathcal{O}) \longrightarrow H^{n+1}_{\partial\Omega}(\mathbb{C}^n, \mathcal{O}) \longrightarrow \ldots.$$

Theorem 1.3.20 implies $H^{n+1}_{\partial\Omega}(\mathbb{C}^n, \mathcal{O}) = 0$, so the restriction map

$$H^n_{\overline{\Omega}}(\mathbb{C}^n, \mathcal{O}) \longrightarrow H^n_\Omega(\mathbb{C}^n \setminus \partial\Omega, \mathcal{O})$$

i.e., the map $\mathcal{B}_{\overline{\Omega}}(\mathbb{C}^n, \mathcal{O}) \to \mathcal{B}(\Omega)$, is surjective. Since the flabbiness is a local property, that is if $\{\Omega_j\}$ is any covering of $\mathbb{R}^n$ if each sheaf $\mathcal{B}_{\Omega_i}$ is flabby so is $\mathcal{B}$, and the statement follows. □

Now we wish to establish a comparison between hyperfunctions and suitable equivalence classes of holomorphic functions. To this purpose, let us recall from Grauert [80] that any open set $\Omega \subseteq \mathbb{R}$ has a fundamental system of neighborhoods whose elements are Stein. This allows us to think hyperfunctions as families of holomorphic functions.

**Proposition 1.3.9.** *Let $\Omega \subseteq \mathbb{R}$ be an open set and let $U$ be a Stein open set such that $U \cap \mathbb{R} = \Omega$. Let $(\mathcal{V}, \mathcal{V}')$ be a relative Stein open covering of the pair $(U, U \setminus \Omega)$. We have*

$$\mathcal{B}(\Omega) \cong H^n(\mathcal{V}, \mathcal{V}', \mathcal{O}).$$

*Proof.* It immediately follows from Theorems 1.3.22 and 1.3.13. □

**Proposition 1.3.10.** *Let $z = (z_1, z_2, \dots, z_n)$ be a point in $\mathbb{C}^n$, let $\Omega$ be an open set in $\mathbb{R}^n$, and let $U$ be a Stein open set such that $U \cap \mathbb{R}^n = \Omega$. Define*

$$U_j = \{z \in U \; : \; Imz_j \neq 0 \}, \quad j = 1, 2, \dots, n, \quad U_0 = U$$

*and*

$$\mathcal{V} = \{U_0, U_1, \dots, U_n\}, \qquad \mathcal{V}' = \{U_1, \dots, U_n\}.$$

*Then $(\mathcal{V}, \mathcal{V}')$ is a relative Stein open covering of the pair $(U, U \setminus \Omega)$. Moreover we have*

$$\mathcal{B}(\Omega) \cong \frac{\mathcal{O}(U_1 \cap \ldots \cap U_n)}{\sum_{j=1}^n \mathcal{O}(U_1 \cap \ldots \cap \widehat{U_j} \cap \ldots \cap U_n)}, \tag{1.45}$$

*where the symbol $\widehat{U_j}$ denotes the omission of the element $U_j$.*

In fact, the space of hyperfunctions can be expressed as a Cech cohomology group as

$$H^n(\mathcal{V}, \mathcal{V}', \mathcal{O}) = \frac{\ker\{\delta^n \; : \mathcal{C}^n(\mathcal{V}, \mathcal{V}', \mathcal{O}) \to \mathcal{C}^{n+1}(\mathcal{V}, \mathcal{V}', \mathcal{O})\}}{\operatorname{im}\{\delta^{n-1} \; : \mathcal{C}^{n-1}(\mathcal{V}, \mathcal{V}', \mathcal{O}) \to \mathcal{C}^n(\mathcal{V}, \mathcal{V}', \mathcal{O})\}}.$$

Since the cardinality of $\mathcal{V}$ is $n+1$, then $\mathcal{C}^{n+1}(\mathcal{V}, \mathcal{V}', \mathcal{O}) = 0$ so the kernel of $\delta^n$ coincides with $\mathcal{C}^n(\mathcal{V}, \mathcal{V}', \mathcal{O})$ which is isomorphic to

$$\mathcal{O}(U_0 \cap U_1 \cap \ldots \cap U_n).$$

Moreover, the image of the map $\delta^{n-1}$ by definition is

$$\operatorname{im}(\delta^{n-1}) = \sum_{j=1}^n \mathcal{O}(U_0 \cap \ldots \cap \widehat{U_j} \cap \ldots \cap U_n).$$

**Remark 1.3.10.** In analogy with what happens for distributions, where $\mathcal{D}'(\mathbb{R}^2)$ is the completion of $\mathcal{D}'(\mathbb{R}) \otimes \mathcal{D}'(\mathbb{R})$, one would expect $\mathcal{B}(\mathbb{R}^2)$ to be related to $\mathcal{B}(\mathbb{R}) \otimes \mathcal{B}(\mathbb{R})$. However, the space of hyperfunctions needed to construct a tensor product which cannot be endowed with any natural locally convex topology and so the idea to take the completion of $\mathcal{B}(\mathbb{R}) \otimes \mathcal{B}(\mathbb{R})$ seems to fail. Still, if one looks at

$$\mathcal{B}(\mathbb{R}) \otimes \mathcal{B}(\mathbb{R}) = \left(\frac{\mathcal{O}(\mathbb{C} \setminus \mathbb{R})}{\mathcal{O}(\mathbb{C})}\right) \otimes \left(\frac{\mathcal{O}(\mathbb{C} \setminus \mathbb{R})}{\mathcal{O}(\mathbb{C})}\right)$$

which in turn is equal to

$$\frac{\mathcal{O}(\mathbb{C} \setminus \mathbb{R}) \otimes \mathcal{O}(\mathbb{C} \setminus \mathbb{R})}{\mathcal{O}(\mathbb{C} \setminus \mathbb{R}) \otimes \mathcal{O}(\mathbb{C}) + \mathcal{O}(\mathbb{C}) \otimes \mathcal{O}(\mathbb{C} \setminus \mathbb{R})} \tag{1.46}$$

and

$$\mathcal{B}(\mathbb{R}^2) \cong \frac{\mathcal{O}(U_1 \cap U_2)}{\mathcal{O}(U_2) + \mathcal{O}(U_1)} = \frac{\mathcal{O}((\mathbb{C} \setminus \mathbb{R}) \times (\mathbb{C} \setminus \mathbb{R}))}{\mathcal{O}((\mathbb{C} \setminus \mathbb{R}) \times \mathbb{C}) + \mathcal{O}(\mathbb{C} \times (\mathbb{C} \setminus \mathbb{R}))} \tag{1.47}$$

(by the previous proposition with $\Omega = \mathbb{R}^2$ and $U = \mathbb{C}^2$), one realizes that the numerator and the denominator in (1.47) are the completion of those of (1.46). Although this is a correct definition for hyperfunctions in two variables and it can be generalized to several variables, it strongly depends on the coordinate system.

**Example 1.3.6.** The description of hyperfunctions in (1.45) allows the introduction of a defining function of a hyperfunction $f$. For example, the Dirac delta $\delta \in \mathcal{B}[\{0\}]$ is given by

$$\delta(x_1, \ldots, x_n) = \left(\frac{-1}{2\pi i}\right)^n \left[\frac{1}{z_1 \cdot \ldots \cdot z_n}\right].$$

In the case of hyperfunctions in one variable, we have another useful description of them in terms of boundary values of holomorphic functions. We wish to generalize this fact to the case of several variables. Note that in the one variable case, we have two directions to approach the real axis, while working with $n$ variables one has $2^n$ ways to approach an open set $\Omega \subseteq \mathbb{R}^n$. For example, in the case of two variables, one has four possibilities as shown in the following figure. A different way to represent hyperfunctions can be obtained using $n+1$ suitable angular domains. In this description a hyperfunction can be represented by $n+1$ holomorphic functions (see for example [137]).

Therefore, we need to introduce some suitable definitions that will allow us to take in consideration all such boundary values.

**Definition 1.3.14.** *Let $\Omega \subseteq \mathbb{R}^n$ be an open set and let us denote by $\mathbb{R}^n_y$ the space of purely imaginary coordinates. Let $\Gamma \subseteq \mathbb{R}^n_y$ be an open cone with vertex in the origin. A set of the form $\mathbb{R}^n + i\Gamma$ is called a wedge.*

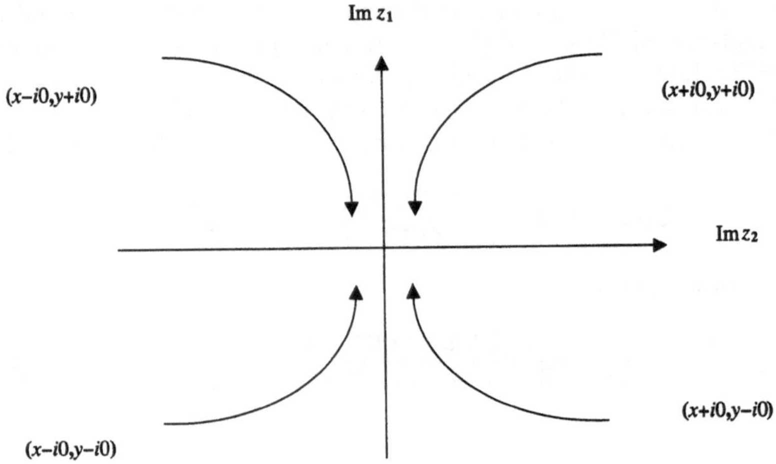


**Figure 1.3.**

The notion of infinitesimal wedge, introduced below, generalizes the expression $x \pm i0$ introduced in the case of one variable.

**Definition 1.3.15.** *Let $\Omega$ an open set in $\mathbb{R}^n$ and let $\Gamma$ be an open cone in $\mathbb{R}^n_y$ (see Figure 1.4). A complex open set $U \subset \mathbb{C}^n$ is called an infinitesimal wedge of type $\Omega + i\Gamma 0$ if $U \subset \Omega + i\Gamma$ and for every proper subcone $\Gamma'$ of $\Gamma$ and for every $\varepsilon > 0$ there is $\delta > 0$ such that*

$$U \supset \Omega_\varepsilon + i(\Gamma' \cap \{|y| < \delta\})$$

*where we have set $\Omega_\varepsilon = \{x \in \Omega : \operatorname{dist}(x, \partial\Omega) > \varepsilon\}$. $\Omega$ is said to be the edge of the infinitesimal wedge.*

We now can introduce the notion of hyperfunctions using a more intuitive definition than the one based on cohomology. All these definitions turn out to be equivalent as proved in [98], Chapter 7.

**Definition 1.3.16.** (Boundary value representation of hyperfunctions). *Let $\{F_\ell(z)\}_{\ell=1,\dots,m}$ be a set of holomorphic functions defined on infinitesimal wedges $\Omega + i\Gamma_\ell 0$, $\ell = 1, \dots, m$. The formal sum*

$$f(x) = \sum_{\ell=1}^{m} F_\ell(x + i\Gamma_\ell 0)$$

*is a hyperfunction on $\Omega$.*

**Remark 1.3.11.** Whenever $\Gamma_r \cap \Gamma_s \neq \emptyset$ the sum reads as

$$F_r(x + i\Gamma_r 0) + F_s(x + i\Gamma_s 0) = (F_r + F_s)(x + i(\Gamma_0 \cap \Gamma_s)0).$$

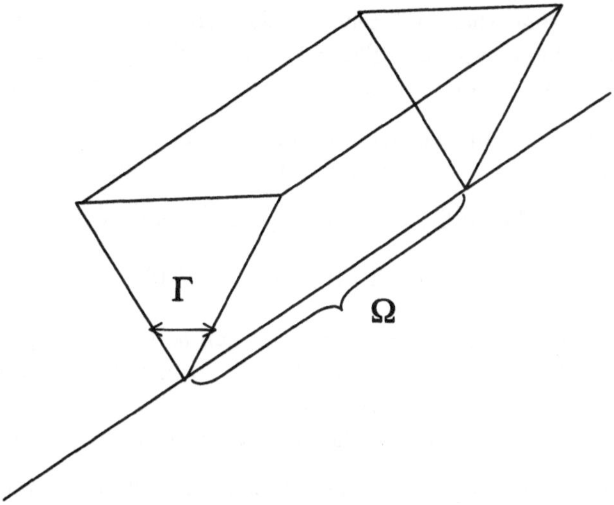


**Figure 1.4.**

Note also that we may suppose that each $\Gamma_\ell$ is convex since any function holomorphic in $\Omega + i\Gamma_\ell 0$ is holomorphic also in $\Omega + i\Gamma'_\ell 0$ where $\Gamma'_\ell$ is the convex hull of $\Gamma_\ell$.

**Remark 1.3.12.** The operations defined for hyperfunctions of one variable can be extended in a natural way to the case of several variables. It is worth repeating the notion of definite integration. Let $f \in \mathcal{B}(\Omega)$ and let $K \in \Omega$ be a compact set with piecewise smooth boundary. Let

$$f(x) = \sum_{\ell=1}^{m} F_\ell(x + i\Gamma_\ell 0)$$

be a boundary value representation of $f$ near $K$. For each wedge $K + i\Gamma_\ell 0$ we define the integration path $K + i\gamma_\ell = \{x + i\gamma_\ell(x) \ : \ x \in K\}$ such that $\gamma_\ell(x) = 0$ for $x \in \partial K$ and $x + i\gamma_\ell(x) \in K + i\Gamma_\ell 0$ for $x \in K \setminus \partial K$, where $\gamma_\ell$ is a continuous piecewise smooth function. Under suitable regularity conditions on $f$, $F$, we can define

$$\int_K f(x)\,dx = \sum_{\ell=1}^{m} \int_{K+i\gamma_\ell} F_\ell(z)\,dz.$$

For example, the integral makes sense if $f$ is real analytic near $\partial K$ and each $F_\ell$ is analytically continuable to the real axis where $f$ is real analytic. It can be shown that the value of the integral does not depend on the choice we have made. In particular, the fact that it is independent of the choice of the integrating path is a consequence of Cauchy's theorem, while the independence of the boundary value representation follows from a classical result called "Edge of the Wedge Theorem." If we consider two different boundary value representations of the hyperfunction $f$, then their difference is the zero hyperfunction and it must satisfy some suitable conditions. The Edge of the Wedge Theorem allows us to determine when a hyperfunction coincides with the zero hyperfunction.

We will state the following weaker version of the Edge of the Wedge Theorem. With the notation $\Gamma_1 + \Gamma_2$ we will mean the convex hull of $\Gamma_1 \cup \Gamma_2$.

**Theorem 1.3.24.** (Local Edge of the Edge Theorem). *Let us suppose that*

$$F_1(x + i\Gamma_1 0) = F_2(x + i\Gamma_2 0)$$

*as elements in $\mathcal{B}(\Omega)$. Then $F_1(z)$ and $F_2(z)$ can be analytically continued to a function $F(z)$ holomorphic on $\Omega + i(\Gamma_1 + \Gamma_2)0$. In particular, if the intersection of the interior of $\Gamma_1$ and the interior of $\Gamma_2$ is empty then $F_1(z)$ and $F_2(z)$ can be glued together as a holomorphic function through the real axis.*

The Edge of the Wedge Theorem can be written in different ways and has an interesting history going back to the work of Bogoliubov, whose interests were in theoretical physics, in particular in the Wightman function and dispersion relations. There are also important early contributions by Malgrange, Martineau and Zerner. For historical remarks on the history of this theorem we refer the reader to the book [102].

**Theorem 1.3.25.** *The following inclusions of sheaves hold:*

$$\mathcal{A} \hookrightarrow \mathcal{L}^1_{\mathrm{loc}} \hookrightarrow \mathcal{D}' \hookrightarrow \mathcal{B}.$$

As in the case of hyperfunctions in one variable we can characterize hyperfunctions with support contained in a compact set $K$, in a theorem that is a generalization of Theorem 1.3.21 and essentially a restatement of Martineau–Harvey's Theorem 1.3.20.

**Proposition 1.3.11.** *Let $K$ be a compact set in $\mathbb{R}^n$. The space $\mathcal{B}[K]$ of hyperfunctions with support contained in $K$ is isomorphic to $H^n_K(\mathbb{C}^n, \mathcal{O})$. Moreover we have*

$$\mathcal{B}[K] \cong \mathcal{A}'(K).$$

**Remark 1.3.13.** As a consequence of the previous theorem, the space $\mathcal{B}[K]$ can be endowed with a topology that makes it a FS space. An interesting fact is that also the converse of the previous theorem holds true: the dual of $\mathcal{B}[K]$ is topologically isomorphic to the space $\mathcal{A}(K)$. The inner product between them is given by the definite integral

$$\langle f, \phi \rangle = \int_{\mathbb{R}^n} f(x)\phi(x)dx$$

for $f \in \mathcal{B}[K]$ and $\phi \in \mathcal{A}(K)$. We will not prove this result here since we will deduce it from a general result we will give in Section 2.1 (see Theorem 2.1.11).

**Remark 1.3.14.** Let us point out another interesting feature of hyperfunctions. Let $\Omega$ be an open and bounded set in $\mathbb{R}^n$. The duality theorem yields $\mathcal{B}[\overline{\Omega}] \cong \mathcal{A}'(\overline{\Omega})$ and $\mathcal{B}[\partial\Omega] \cong \mathcal{A}'(\partial\Omega)$. In general, if $\Omega$ is an open set in $\mathbb{R}^n$ (not necessarily bounded), we have the exact sequence

$$0 \to \mathcal{B}[\partial\Omega] \to \mathcal{B}[\overline{\Omega}] \to \mathcal{B}(\Omega) \to 0,$$

from which it follows that

$$\mathcal{B}(\Omega) \cong \frac{\mathcal{B}[\overline{\Omega}]}{\mathcal{B}[\partial\Omega]} \cong \frac{\mathcal{A}'(\overline{\Omega})}{\mathcal{A}'(\partial\Omega)}.$$

Note that this last isomorphism was taken by Martineau as the definition of hyperfunctions.

### 1.3.4 Fourier transform

In a previous section, we introduced the Fourier transform for functions in the space of rapidly decreasing functions $\mathcal{S}(\mathbb{R}^n)$, where the operator $\mathcal{F}$ is defined by the convergent integral

$$\mathcal{F}[f](\lambda) := \int_{\mathbb{R}^n} f(x)e^{-i(x,\lambda)}\,dx, \quad (x,\lambda) = \sum_{j=1}^{n} x_j\lambda_j. \tag{1.48}$$

Since the space $\mathcal{S}(\mathbb{R}^n)$ is dense in the space of tempered distribution $\mathcal{S}'(\mathbb{R}^n)$, we were able to define the Fourier transform also in that space of generalized functions. Here, we have to give a meaning to the integral (1.48) in the sense of hyperfunctions. Analogous to what we did for distributions we will assume that the operator (1.48) is the Fourier transform of a hyperfunction $f$ with compact support $K$ and that

$$\mathcal{F}^{-1}[g](x) := \frac{1}{(2\pi)^n}\int_{\mathbb{R}^n} g(\lambda)e^{i(x,\lambda)}\,d\lambda \tag{1.49}$$

is the inverse Fourier transform. To avoid confusion with the notation, we will denote the Fourier transform of a hyperfunction $f$ by $\hat{f}$.

**Definition 1.3.17.** (One variable case). *Recalling the definition of integral of a hyperfunction we define the Fourier transform of a hyperfunction $f = [F]$ with compact support $K \subset \mathbb{R}$ as*

$$\hat{f}(\lambda) = -\oint_{\gamma} F(z)e^{-i(z,\lambda)}\,dz,$$

*where $\gamma$ is an arbitrary integral path which encircles $K$ once in the positive direction.*

For hyperfunctions in more than one variable, with compact support $K$, we use the representation (see Definition 1.3.16)

$$f(x) = \sum_{j=1}^{m} F_j(x + i\Gamma_j 0)$$

where every defining function $F_j(z)$ can be analytically continued to a neighborhood of the real axis $\mathbb{R}^n \setminus K$ such that their sum is equal to zero there, in the ordinary sense.

**Definition 1.3.18.** (Several variables). *We define the Fourier transform of $f$ as*

$$\hat{f}(\lambda) = \sum_{j=1}^{m} \int_{\gamma_j} F_j(z) e^{-i\langle z,\lambda\rangle}\, dz$$

*where each $\gamma_j$ is an integral path obtained by fixing the boundary of a real neighborhood $\gamma$ of $K$ and deforming its interior inside the infinitesimal wedge $\mathbb{R}^n + i\Gamma_j 0$, which is the domain of analyticity of $F_j(z)$.*

Analogous to the case of distributions, we have a version of the Paley–Wiener theorem:

**Theorem 1.3.26.** *The Fourier transform $\hat{f}$ of a hyperfunction $f$ with compact support contained in a compact set $K$ is an entire function of $\lambda$. Moreover, for any $\varepsilon > 0$ there exists a positive constant $C_\varepsilon$ such that*

$$|\hat{f}(\lambda)| \le C_\varepsilon \exp\left(\varepsilon|\lambda| + H_K(Im(\lambda))\right) \tag{1.50}$$

*where $H_K(\zeta) = \sup_{x\in K}\langle x, \zeta\rangle$ is the supporting function of $K$. Conversely, an entire function which satisfies estimate (1.50) for any choice of $\varepsilon > 0$ is necessarily the Fourier transform of a unique hyperfunction whose support is contained in the convex hull of $K$.*

In the following result, by $P(D)$ we denote a linear partial differential operator with constant coefficients of the form (1.27) where $D_j = \frac{1}{i}\frac{\partial}{\partial x_j}$:

**Theorem 1.3.27.** *Let $f$ be a compactly supported hyperfunction and let $P(D)$ be a nontrivial linear partial differential operator with constant coefficients. Then we have*

$$\widehat{P(D)f} = P(\lambda)\hat{f}.$$

Observe that in particular when $P(D) = D_j$ we have that the Fourier transform maps the derivative $D_j$ into a multiplication by $\lambda_j$.

If we consider a hyperfunction with support at the origin, we know that it can be written as $P(D)\delta(x)$ with $P(D)$ a local operator with constant coefficients. In this case we get that $P(\lambda)$ is an entire function, called the total symbol of the operator, which satisfies the estimate of Theorem 1.3.26: for all $\varepsilon > 0$ there exists a positive constant $C_\varepsilon$ such that $|P(\lambda)| \le c_\varepsilon \exp(\varepsilon|\lambda|)$. Functions of this type will be called slowly decreasing according to the definition below in which we introduce some function spaces that turn out to be natural spaces for the operator $\mathcal{F}$. In fact, the Paley–Wiener theorem, in its several versions according to the type of (generalized) functions considered, establishes the possible "growth order" of the Fourier transform.

**Definition 1.3.19.** (Space of holomorphic functions decreasing exponentially.)

- *Let $\eta > 0$ and let $F(z)$ be an holomorphic function in a tubular domain $\mathbb{R}^n + iI$ ($I$ is a connected open set in $\mathbb{R}^n$). Let us suppose that for every compact set $K \Subset I$ and for every $\varepsilon > 0$ there exists a positive constant $C_{K,\varepsilon}$ such that*
$$|F(z)| \leq C_{K,\varepsilon} e^{-(\eta-\varepsilon)|Re(z)|}$$
*uniformly for $z \in \mathbb{R}^n + iK$. We say that $F$ is a holomorphic function decreasing exponentially with type $-\eta$ along the real axis. The set of these functions is denoted by*
$$\mathcal{O}_{\text{exp}}^{-\eta}(\mathbb{R}^n + iI).$$

Space of slowly increasing functions.

- *It contains the functions that satisfy the previous estimate for $\eta = 0$, that is*
$$\mathcal{O}_{\text{exp}}^{0}(\mathbb{R}^n + iI) = \{F \in \mathcal{O}(\mathbb{R}^n + iI) \ : \ |F(z)| \leq C_{K,\varepsilon} e^{\varepsilon|Re(z)|} \}.$$

Space of rapidly-decreasing real analytic functions.

- *It is defined as*
$$\mathcal{O}_{\text{exp}} = \varinjlim_{I \ni 0} \varinjlim_{\eta \to 0+} \mathcal{O}_{\text{exp}}^{-\eta}(\mathbb{R}^n + iI).$$

The spaces just introduced are natural with respect to the Fourier transform, as it appears clearly in the following theorem

**Theorem 1.3.28.** *Let $\eta$ and $\delta$ be two positive real constants. Then, denoting by $y$ the imaginary part of $z$, we get*
$$\mathcal{F}\mathcal{O}_{\text{exp}}^{-\eta}(\mathbb{R}^n + i\{|y| < \delta\}) = \mathcal{O}_{\text{exp}}^{-\delta}(\mathbb{R}^n + i\{|y| < \eta\}).$$
*Moreover, we have*
$$\mathcal{F}\mathcal{O}_{\text{exp}} = \mathcal{O}_{\text{exp}}.$$

This theorem can be further generalized.

**Theorem 1.3.29.** *Let $F \in \mathcal{O}^{-\eta}_{\exp}(\mathbb{R}^n + iI)$, $y \in I$ and consider the Fourier transform*

$$\hat{F}(\lambda) = \int_{\text{Im} z = y} e^{-i(z,\lambda)} F(z) dz. \tag{1.51}$$

*Then $\hat{F}(\lambda)$ is holomorphic in the neighborhood $|\text{Im}\lambda| < \eta$ of the real axis. Moreover, for any $K \Subset I$ and $\varepsilon > 0$, there is a positive constant $C_{K,\varepsilon}$ such that*

$$|\hat{F}(\lambda)| \leq C_{K,\varepsilon} e^{-H_{-K}(\text{Re}(\lambda))}$$

*uniformly for $|\text{Im}(\lambda)| \leq \eta - \varepsilon$. Conversely, any function satisfying the previous conditions is the Fourier transform of an element in $\mathcal{O}^{-\eta}_{\exp}(\mathbb{R}^n + iI)$.*

For further references on the topic see also [23], [38].

## 1.4 Appendix: category theory

**Definition 1.4.1.** *A category $\mathcal{C}$ consists of a collection of objects $\text{Ob}(\mathcal{C})$ and, for each pair of objects $A, B \in \text{Ob}(\mathcal{C})$, a set $\text{Mor}(A, B)$ (whose elements $f : A \to B$ are called morphisms) with a composition law*

$$\text{Mor}(A, B) \times \text{Mor}(B, C) \to \text{Mor}(A, C), \quad (f, g) \mapsto gf$$

*such that:*

1. *the composition is associative,*
2. *for any object $A$ there is a distinguished morphism $1_A \in \text{Mor}(A, A)$ such that, whenever defined, $f1_A = f$, $1_A g = g$.*

*A morphism $f \in \text{Mor}(A, B)$ is an isomorphism if there is $g \in \text{Mor}(B, A)$ such that $fg = 1_B$ and $gf = 1_A$.*

Other examples of categories, besides the category of $R$-modules, are the category of abelian groups with group homomorphisms, the category of vector spaces with linear maps and the category of topological spaces with continuous maps. Note that it is possible to have different categories having the same set of objects, as long as different morphisms are considered. For example, one could consider the category of topological spaces with equivalence classes of homotopic maps.

It is possible also to define homomorphisms between categories: these are called functors.

**Definition 1.4.2.** *If $\mathcal{C}$ and $\mathcal{C}'$ are two categories, a functor $F : \mathcal{C} \to \mathcal{C}'$ associates to each object $A \in \text{Ob}(\mathcal{C})$ an object $F(A) \in \text{Ob}(\mathcal{C}')$ and for every morphism $f : A \to B$, $F$ associates a morphism $F(f) : F(A) \to F(B)$ (in which* *case we say that the functor is covariant) or a morphism* $F(f) : F(B) \to F(A)$ *(in this case we say that the functor is contravariant). Moreover, the correspondence* $f \to F(f)$ *preserves composition and identity elements, i.e.,* $F(1_A) = 1_{F(A)}$, $F(fg) = F(f)F(g)$ *if* $F$ *is covariant or* $F(fg) = F(g)F(f)$ *if* $F$ *is controvariant.*

Given two functors, we can define a notion of transformation between them (we will limit ourselves to the covariant case, the contravariant case can be obtained with minor modifications):

**Definition 1.4.3.** *Given two covariant functors* $F, G : \mathcal{C} \to \mathcal{C}'$, *a natural transformation* $\tau : F \to G$ *is a rule assigning to each* $A \in \mathrm{Ob}(\mathcal{C})$ *an arrow* $\tau_A : F(A) \to G(A)$ *in* $\mathcal{C}'$ *such that for every* $f \in \mathrm{Mor}(A, B)$ *there is a commutative diagram*

$$\begin{array}{ccc} F(A) & \xrightarrow{\tau_A} & G(A) \\ {\scriptstyle F(f)}\downarrow & & \downarrow{\scriptstyle G(f)} \\ F(B) & \xrightarrow{\tau_B} & G(B). \end{array}$$

**Remark 1.4.1.** Covariant functors from a category $\mathcal{C}$ to a category $\mathcal{C}'$ are the objects of another category in which the morphisms are the natural transformations.

Now we wish to introduce a suitable class of categories in which it is defined the notion of "exact sequence": such categories are particular additive categories.

**Definition 1.4.4.** *A category* $\mathcal{C}$ *is said to be additive if the following conditions are satisfied:*

1. *for any two objects* $A$, $B \in \mathrm{Ob}(\mathcal{C})$, *the set* $\mathrm{Mor}(A, B)$ *can be endowed with a structure of abelian group such that for any* $A$, $B$, $C \in \mathrm{Ob}(\mathcal{C})$, *the composition law*

$$\mathrm{Mor}(A, B) \times \mathrm{Mor}(B, C) \to \mathrm{Mor}(A, C)$$

   *is bilinear;*

2. *there is an object* $0$ *in* $\mathcal{C}$ *such that* $\mathrm{Mor}(0, 0) = \{0\}$;

3. *for each ordered pair* $A, B \in \mathrm{Ob}(\mathcal{C})$ *there is an object denoted by* $A \oplus B$ *and an isomorphism of functors* $\mathrm{Mor}(A, -) \oplus \mathrm{Mor}(B, -) \cong \mathrm{Mor}(A \oplus B, -)$.

Let $\mathcal{C}$, $\mathcal{C}'$ be two additive categories. A covariant functor $F : \mathcal{C} \to \mathcal{C}'$ is called additive if for any two objects $X$, $Y \in \mathrm{Ob}(\mathcal{C})$, the map

$$\mathrm{Mor}(X, Y) \to \mathrm{Mor}(F(X), F(Y))$$

is a homomorphism of abelian groups. In an additive category $\mathcal{C}$, one can define the notions of kernel and cokernel of a morphism $f : A \to B$. Let $K$ be an object and let $j \in \text{Mor}(K, A)$ be such that every morphism $\alpha : C \to A$ with $f\alpha = 0$ factors as $j\alpha'$ for a suitable $\alpha' : C \to K$. If the correspondence sending $\alpha$ to $\alpha'$ is an isomorphism of functors in $\mathcal{C}$, then $K$ is said a kernel of $f$ and it is denoted by $\ker(f)$. Note that $\ker(f)$ is a functor of the morphism $f$. In a similar way one can define the notion of cokernel of $f$: if $N$ is an object and $p \in \text{Mor}(B, N)$ is such that every morphism $\alpha : B \to C$ such that $\alpha f = 0$ factors as $\alpha' p$. If the correspondence sending $\alpha \in \{\alpha \in \text{Mor}(B, C) \ : \alpha f = 0\}$ to $\alpha' \in \text{Mor}(N, C)$ is an isomorphism of functors in $\mathcal{C}$, then $N$ is said a kernel of $f$ and it is denoted by $\text{coker}(f)$. It is possible to show that both the kernel and the cokernel, if they exist, are unique up to isomorphisms. Moreover, if $f : A \to B$ has a kernel and $\ker(f) \to A$ has a cokernel, this cokernel is called coimage of $f$ and is denoted by $\text{coim}(f)$. Analogously, if $B \to \text{coker}(f)$ has a kernel, it is called the image of $f$. We say that $f \in \text{Mor}(A, B)$ is a monomorphism if $\ker(f) = 0$, while $f$ is called an epimorphism if $\text{coker}(f) = 0$.

**Definition 1.4.5.** *An additive category $\mathcal{C}$ is said to be an abelian category if* $\ker(f)$ *and* $\text{coker}(f)$ *exist for every morphism $f$ and the natural morphism sending coim($f$) to* $\text{im}(f)$ *is an isomorphism for every $f$.*

In an abelian category $\mathcal{C}$ a sequence of two morphisms

$$A \xrightarrow{f} B \xrightarrow{g} C$$

is called an exact sequence if $gf = 0$ and the induced morphism $\text{im}(f) \to \ker(g)$ is an isomorphism. In general, a sequence is called exact if any pair of adjacent morphisms is exact.

**Definition 1.4.6.** *Let $\mathcal{C}$, $\mathcal{C}'$ be two abelian categories. An additive (covariant) functor $F$ is said to be left exact if for every exact sequence in $\mathcal{C}$*

$$0 \to X \to Y \to Z$$

*the sequence in $\mathcal{C}'$*

$$0 \to F(X) \to F(Y) \to F(Z)$$

*is exact as well. Similarly, an additive functor $F$ is said to be right exact if for every exact sequence in $\mathcal{C}$*

$$Z \to Y \to X \to 0$$

*the sequence in $\mathcal{C}'$*

$$F(Z) \to F(Y) \to F(X) \to 0$$

*is exact as well.*

In the same hypotheses as above, we say (according to [79]) that an additive controvariant functor is left exact if for every exact sequence in $\mathcal{C}$

$$0 \to X \to Y \to Z$$

the sequence in $\mathcal{C}'$

$$F(Z) \to F(Y) \to F(X) \to 0$$

is exact as well, and similarly an additive controvariant functor is right exact if for every exact sequence in $\mathcal{C}$

$$Z \to Y \to X \to 0$$

the sequence in $\mathcal{C}'$

$$0 \to F(X) \to F(Y) \to F(Z)$$

is exact as well.

**Warning**. We point out that in the literature it is possible to find different definitions based on the exactness of the corresponding covariant functors in the dual category.

**Definition 1.4.7.** *An object $U$ in a category $\mathcal{C}$ is said to be a universal repelling object if, for each $A \in \mathrm{Ob}(\mathcal{C})$ there is a unique morphism $U \to A$ while $U$ is said to be a universal attracting object if there is a unique morphism $A \to U$. We will say that $U$ is an universal object to denote either a universal repelling or an universal attracting object when no confusion arises.*

**Remark 1.4.2.** It can be shown that universal objects are unique up to isomorphisms.

We can also give the definition of universal arrow.

**Definition 1.4.8.** *Let $F : \mathcal{C}' \to \mathcal{C}$ be a functor and let $A \in \mathrm{Ob}(\mathcal{C})$. A universal arrow from $A$ to $F$ is a pair $(U, u)$, where $U \in \mathrm{Ob}(\mathcal{C}')$ and $u : A \to F(U)$ is a morphism such that for any pair $(D, d)$ with $D \in \mathrm{Ob}(\mathcal{C}')$ and $d : A \to F(D)$ there is a unique morphism $f : U \to D$ of $\mathcal{C}'$ with $F(f)u = d$.*

Using this definition we can introduce the notion of inductive limit in a category (also called a colimit or direct limit) and of its dual notion, i.e., the projective limit (or limit or inverse limit). Let us consider two categories $\mathcal{C}$ and $J$. Let us consider the category $\mathcal{C}^J$ whose objects are the functors $F : J \to \mathcal{C}$. The so-called diagonal functor $\Delta : \mathcal{C} \to \mathcal{C}^J$ sends each object $A \in \mathrm{Ob}(\mathcal{C})$ to the constant functor $\Delta(A)$, i.e. the functor sending each object $j \in \mathrm{Ob}(J)$ to $A$ and each morphism of $J$ to $1_A$. If $f \in \mathrm{Mor}(A, B)$, then $\Delta(f)$ is the natural transformation $\Delta(f) : \Delta(A) \to \Delta(B)$ having the same value $f$ on each object of $J$.

**Definition 1.4.9.** *A universal arrow $(U, u)$, $U \in \mathrm{Ob}(\mathcal{C})$ from $F$ to $\Delta$ is called an inductive limit for the functor $F$. We will write $U = \underrightarrow{F}$. A universal arrow $(U', u')$ from $\Delta$ to $F$ is called a projective limit for $F$ and it is denoted by $U' = \underleftarrow{F}$.*

# 2

# Computational Algebraic Analysis for Systems of Linear Constant Coefficients Differential Equations

## 2.1 A primer of algebraic analysis

In this section we provide the background on those aspects of algebraic analysis which will be necessary in the rest of the book. Historically, we believe that Euler was the first major mathematician to use the term "algebraic analysis" in connection with his important work on general solutions to linear ordinary differential equations with constant coefficients, [71]. Currently, the term "algebraic analysis" refers to the work of the Japanese school of Kyoto (Sato, Kashiwara, Kawai, and their coworkers) which founded and developed methods to analyze algebraically systems of linear partial differential equations with real analytic coefficients [102]. Their results, however, rest on some preliminary work, in which algebra was used to study general properties of systems of linear differential equations with constant coefficients.

In addition to some fundamental early work of Malgrange and Martineau [122], [127], [129] and Ehrenpreis [63], these techniques were developed first in the works of Ehrenpreis [68] and Palamodov [142], in the 1960s. Because of the complexity of the machinery developed by Ehrenpreis and Palamodov, their results were (almost) never applied to specific systems of differential equations, but were used mainly to provide very general results on solutions of systems of linear partial differential equations with constant coefficients (the first being Ehrenpreis' remarkable new interpretation of Hartogs' theorem [66], and the most impressive being the so-called Ehrenpreis–Palamodov Fundamental Principle [64], [142]). Since this book is concentrated on systems that are variations of the Dirac system, which has constant coefficients, we will limit the main part of our overview of algebraic analysis to this specific case.

Let us set the notations for this section (which we will also maintain throughout the entire book). We will denote by $R$ the ring of polynomials in $n$ variables $z_1, \dots, z_n$ with complex coefficients, i.e.,

$$R = \mathbb{C}[z_1, \dots, z_n].$$

This ring can be thought of as the ring of linear constant coefficients differential operators in the following sense. If $P$ is a polynomial in $R$, then a differential operator can be obtained by replacing the variables $z_1, \dots, z_n$ with the derivatives $-i\partial_{x_1}, \dots, -i\partial_{x_n}$ so that $P(-i\partial_{x_1}, \dots, -i\partial_{x_n})$ is now a partial differential operator. We will often write $D$ to represent the $n$-tuple of derivatives $(-i\partial_{x_1}, \dots, -i\partial_{x_n})$, so that the differential operator associated to the polynomial $P(z)$ can be referred to as $P(D)$. Therefore we can say that $R$ is the ring of symbols of such operators. More generally, if $P = [P_{ij}]$ is an $r_1 \times r_0$ matrix of polynomials in $R$, then the corresponding matrix $P(D) = [P_{ij}(D)]$ represents a system of linear partial differential operators with constant coefficients. We point out that when we start from a matrix $P(D)$, we associate to it a matrix $P$ of polynomials by replacing, via the Fourier transform, $-i\partial_{x_j}$ by $z_j$.

Even though in the chapters to follow we will operate with the ring of polynomials, we wish to treat more general cases by replacing $R$ with the Weyl algebra $A_n(\mathbb{C})$ (in which case we are working with linear differential operators with polynomial coefficients), or with the sheaf $\mathscr{D}$ of differential operators with real analytic or holomorphic coefficients, or even with the sheaf of differential operators with real analytic or holomorphic coefficients and infinite order (this is the full context for algebraic analysis in the sense of the Japanese school).

Let us show how to associate a finitely generated left $A_n(\mathbb{C})$-module to a system of linear partial differential equations when the operator has polynomial coefficients.

Any operator $P_j(D)$ in the Weyl algebra $A_n(\mathbb{C})$ can be written in the form $P_j(D) = \sum_\alpha p_{j\alpha} D^\alpha$ where $\alpha = (\alpha_1, \dots, \alpha_n) \in \mathbb{N}^n$ is a multiindex, $p_{j\alpha} \in \mathbb{C}[z_1, \dots, z_n]$, $j = 1, \dots, r_1$, and $D^\alpha$ denotes $\partial_{z_1}^{\alpha_1} \dots \partial_{z_n}^{\alpha_n}$. Then we consider the following system of differential equations:

$$P_1(D)f = \dots = P_{r_1}(D)f = 0, \tag{2.1}$$

where

$$P_j(D)f = \sum_\alpha p_{j\alpha} D^\alpha f = 0, \quad j = 1, \dots, r_1,$$

and $f$ belongs to a vector space or to a sheaf that will be specified later.

**Definition 2.1.1.** *The left $A_n(\mathbb{C})$-module associated to the system (2.1) is defined as*

$$\mathcal{M} = A_n(\mathbb{C}) / \sum_{j=1}^{r_1} A_n(\mathbb{C}) P_j(D). \tag{2.2}$$

This definition is motivated by the fact that it allows the description of the solutions of the system. For example, the set of all the polynomial solutions of the system (2.1) forms a $\mathbb{C}$-vector space, denoted by $R^P$, described as follows:

**Proposition 2.1.1.** *Let $\mathcal{M}$ be the left $A_n(\mathbb{C})$-module defined in (2.2). Then the vector space $R^P$ is isomorphic to* $\mathrm{Hom}_{A_n(\mathbb{C})}(\mathcal{M}, R)$.

*Proof.* Let $f$ be a polynomial solution to the system (2.1). Consider the homomorphism

$$\sigma_f : A_n(\mathbb{C}) \to R$$

which maps 1 to $f$. It is clear that if an operator $Q(D)$ belongs to the vector space generated by $P_1(D), \dots, P_{r_1}(D)$ then $Q(D)f = 0$ so that $\sigma_f$ really induces a homomorphism (which we still denote by $\sigma_f$) from $\mathcal{M}$ to $R$. This construction defines a map $f \mapsto \sigma_f$ from the space of solutions to the system to $\mathrm{Hom}_{A_n(\mathbb{C})}(\mathcal{M}, R)$. It is easy to verify, by the linearity of the $P_j(D)$, that this is in fact a linear map. One can also explicitly construct its inverse as follows: if $\tau$ is an element of $\mathrm{Hom}_{A_n(\mathbb{C})}(\mathcal{M}, R)$, then $\tau(1)$ is an element $g \in R$ which is easily seen to be a solution of (2.1). Finally, one verifies that $\tau$ is the inverse of $\sigma$ and this concludes the proof. □

More generally, we give the following definition:

**Definition 2.1.2.** *Let $S$ be a left $A_n(\mathbb{C})$–module and let $\mathcal{M}$ be a finitely generated left $A_n(\mathbb{C})$–module. We call* $\mathrm{Hom}_{A_n(\mathbb{C})}(\mathcal{M}, S)$ *the vector space of solutions of $\mathcal{M}$ in $S$.*

Note that this definition allows us to look for solutions in various finitely generated $A_n(\mathbb{C})$-modules, even in modules of generalized functions. For example, one can consider $\mathcal{C}^\infty$ solutions to the system (2.1) and, repeating the argument in the proof of the previous proposition, one gets that the vector space of solutions over an open set $\Omega \subset \mathbb{R}^n$ is isomorphic to $\mathrm{Hom}_{A_n(\mathbb{C})}(\mathcal{M}, \mathcal{C}^\infty(\Omega))$.

One can further generalize the discussion and describe the solutions of a system associated to a matrix $P(D)$ not only as a vector space but even as a sheaf.

Let $\mathscr{D}$ be the sheaf of rings of linear partial differential operators with holomorphic coefficients on $\mathbb{C}^n$. The stalk $\mathscr{D}_z$ at the point $z = (z_1, \dots, z_n)$ contains elements of the type

$$P_i(z, D) = \sum_\alpha a_{i\alpha}(z) D^\alpha, \quad i = 1, \dots, r_1,$$

where $\alpha$ runs on a finite set of $n$-tuples of nonnegative integers, and $a_{i\alpha}(z)$ are germs of holomorphic functions at $z$. Locally, for a small enough neighborhood $U$ of $z$, we have

$$\mathscr{D}(U) = \mathcal{O}(U) \otimes_{\mathbb{C}} \mathbb{C}[\partial_{z_1}, \dots, \partial_{z_n}].$$

More generally, we can consider a system of differential equations of the form

$$P_i(D)f = \sum_{j=1}^{r_0} P_{ij}(D)f_j = 0, \qquad i = 1, \ldots, r_1,$$

we can associate to it a sheaf of germs of $\mathscr{D}$-modules, by setting

$$\mathcal{M} = \mathscr{D}^{r_0}/\mathscr{D}^{r_1}P(D).$$

Also the converse it true: if $\mathcal{M}$ is a sheaf of germ of $\mathscr{D}$-modules (in the sequel we will refer to it as a $\mathscr{D}$-module, for short) and if $\mathcal{M}$ is finitely generated over $\mathscr{D}$, it can be associated to system of linear partial differential equation. Let us suppose that $u_1, \ldots, u_{r_0}$ are generators for $\mathcal{M}$. We have an exact sequence of $\mathscr{D}$-modules

$$\mathscr{D}^{r_0} \overset{\bullet u}{\to} \mathcal{M} \to 0$$

where the so called augmentation map from $\mathscr{D} \oplus \ldots \oplus \mathscr{D}$ to $\mathcal{M}$ is defined as

$$(A_1U_1 \oplus \ldots \oplus A_{r_0}U_{r_0}) \bullet u = A_1u_1 + \ldots + A_{r_0}u_{r_0},$$

and where $U_j = [0, \ldots, 0, 1, 0, \ldots, 0]$ with the element equal to 1 sitting at the $j$-th place for $j = 1, \ldots, r_0$. Since $\mathscr{D}$ is left noetherian, the kernel of the augmentation map $\ker(\bullet u)$ is a finitely generated submodule of $\mathscr{D}^{r_0}$ with generators $v_1, v_2, \ldots, v_{r_1}$. We can define an epimorphism

$$\mathscr{D}^{r_1} \overset{\bullet v}{\to} \ker(\bullet u) \to 0$$

where the map $\bullet v$ is defined by

$$(B_1V_1 \oplus \ldots \oplus B_{r_1}V_{r_1}) \bullet v = B_1v_1 + \ldots + B_{r_1}v_{r_1},$$

where $V_i = [0, \ldots, 0, 1, 0, \ldots, 0]$ and the element equal to 1 sits at the $i$-th place for $i = 1, \ldots, r_1$. Note that each $v_\ell$ can be written as

$$v_\ell = P_{\ell 1}U_1 + \cdots + P_{\ell r_0}U_{r_0}, \quad \ell = 1, \ldots, r_1,$$

so we obtain the diagram

$$\mathscr{D}^{r_1} \xrightarrow{\bullet P} \mathscr{D}^{r_0} \xrightarrow{\bullet u} \mathcal{M} \to 0 \tag{2.3}$$

where the map $P = P_{ij}$ is the composition of the inclusion map $\ker(\bullet u) \hookrightarrow \mathscr{D}^{r_0}$ and of $\bullet v : \mathscr{D}^{r_1} \to \ker(\bullet u)$. By construction, it follows that

$$([B_1, \ldots, B_{r_1}] \bullet P) \bullet u = 0.$$

Observe now that if $P$ is an $r_1 \times r_0$ matrix with entries in $\mathscr{D}$ and if we write the augmentation map $\bullet u$ in matrix form using the vector $u = [u_1, \ldots, u_{r_0}]^t$, then the composition of the maps $P$ and $u$ becomes a system of $\mathscr{D}$-linear relations among the generators $u_1, \ldots, u_{r_0}$ of the $\mathscr{D}$-module $\mathcal{M}$:

$$\begin{cases} P_{11}u_1 + P_{12}u_2 + \cdots + P_{1r_0}u_{r_0} = 0 \\ P_{21}u_1 + P_{22}u_2 + \cdots + P_{2r_0}u_{r_0} = 0 \\ \vdots \\ P_{r_1 1}u_1 + P_{r_1 2}u_2 + \cdots + P_{r_1 r_0}u_{r_0} = 0. \end{cases} \tag{2.4}$$

This is nothing but a system of partial differential equations. Note, with a similar construction, one can extend the exact sequence (2.3) to get a free projective resolution of $\mathcal{M}$.

Let us now describe the solutions of the system associated to the $\mathscr{D}$-module $\mathcal{M}$ belonging to a sheaf $\mathcal{S}$. Let $f$ belong to $\mathcal{H}om_{\mathscr{D}}(\mathcal{M}, \mathcal{S})$, so that $f$ maps each generator $u_i$ of $\mathcal{M}$ to an element $f_i \in \mathcal{S}$. Let us apply $f$ to the system (2.4): we get

$$0 = f\left(\sum_{j=1}^{r_0} P_{ij}u_j\right) = \sum_{j=1}^{r_0} f(P_{ij}u_j) = \sum_{j=1}^{r_0} P_{ij}f(u_j) = \sum_{j=1}^{r_0} P_{ij}f_j,$$

implying that $(f_1, \ldots, f_{r_0})$ is a solution to the system.

Now we come back to the case of linear systems of partial differential equations with constant coefficients $P(D)f = 0$. We have associated to this system the matrix $P(z)$ with polynomial entries. It is at this point where the traditional tools from commutative algebra come into the picture: let us consider the matrix

$$P'(z) := P^t(-z), \tag{2.5}$$

that we will denote by $P'$ for short, and the $R$-homomorphism $P' : R^{r_1} \to R^{r_0}$. We introduce the $R$-module

$$M = R^{r_0}/P'R^{r_1} = R^{r_0}/\langle P' \rangle,$$

where $\langle P' \rangle$ is the submodule of $R^{r_0}$ generated by the columns of $P'$.

The first important step is the use of the Hilbert Syzygy Theorem 1.1.1 that shows that the module $M$ has a minimal finite free resolution of the form

$$0 \longrightarrow R^{r_m} \xrightarrow{P^t_{m-1}} R^{r_{m-1}} \longrightarrow \ldots \xrightarrow{P^t_1} R^{r_1} \xrightarrow{P^t} R^{r_0} \longrightarrow M \longrightarrow 0. \tag{2.6}$$

Let us now apply the functor $\mathrm{Hom}(-, R)$ to (2.6): we obtain the dual of the previous sequence. It is a complex

$$0 \longrightarrow R^{r_0} \xrightarrow{P} R^{r_1} \xrightarrow{P_1} \ldots \longrightarrow R^{r_{m-1}} \xrightarrow{P_{m-1}} R^{r_m} \longrightarrow 0 \tag{2.7}$$

which is not necessarily exact and its cohomology groups (see Example 1.1.12)

$$H^j(M, R) = \frac{\ker(P_j)}{\mathrm{im}(P_{j-1})}, \tag{2.8}$$

denoted by $\mathrm{Ext}^j(M, R)$, are uniquely determined by $M$ (note that the resolution (2.6) or complex (2.7) are not uniquely determined). The knowledge of these cohomology groups, which will be studied in what follows, is the fundamental tool for the theory developed by Ehrenpreis and Palamodov, and therefore it is crucial to be able to compute the resolution (2.6) for all the systems we are interested in. Let us provide a simple, but important, example in order to acquaint the reader who is not familiar with these concepts. We state first the following definition:

**Definition 2.1.3.** *An ordered sequence of elements $f_1, \ldots, f_m \in R$ is called a regular sequence for the R-module $M$ if $(f_1, \ldots, f_m)M \neq M$ and $f_\ell$ is a non zero divisor for $M/(f_1, \ldots, f_{\ell-1})M$ for $\ell = 1, \ldots, m$.*

**Example 2.1.1.** Let

$$P = \begin{bmatrix} Q_1 \\ Q_2 \\ Q_3 \end{bmatrix}$$

be a $3 \times 1$ matrix of polynomials; assume these polynomials have no common factors. Then

$$M = R/(Q_1, Q_2, Q_3)$$

where $(Q_1, Q_2, Q_3)$ denotes the ideal generated by $Q_1, Q_2, Q_3$ in $R$. It is obvious that elements such as

$$(-Q_2, Q_1, 0)^t, \quad (Q_3, 0, -Q_1)^t, \quad (0, -Q_3, Q_2)^t$$

belong to the kernel of $P^t$ in $R^3$, but in fact, under the previous assumptions they generate all of it, since $Q_1$, $Q_2$, $Q_3$ form a regular sequence according to the previous definition. So, in this case, the beginning of resolution (2.6) looks like

$$R^3 \xrightarrow{P_1^t} R^3 \xrightarrow{P^t} R \longrightarrow M \longrightarrow 0$$

where

$$P_1^t = \begin{bmatrix} -Q_2 & Q_3 & 0 \\ Q_1 & 0 & -Q_3 \\ 0 & -Q_1 & Q_2 \end{bmatrix}.$$

Using again the property of regularity, one finds out that the next map is given by the matrix $P_2^t = [Q_3, Q_2, Q_1]^t$. To build the kernel of this last map, we argue as above, and we show that the kernel is zero, so that the resolution has its final form

$$0 \longrightarrow R \xrightarrow{P_2^t} R^3 \xrightarrow{P_1^t} R^3 \xrightarrow{P^t} R \longrightarrow M \longrightarrow 0.$$

This construction is known as the Koszul complex, and can be replicated for any number of polynomials as well as for matrices with more than one column (in this case the complex is known as the generalized Koszul complex, see [32]).

**Remark 2.1.1.** First, we point out that the same exact process would work for any number of polynomials, so that the case $n = 3$ is really quite general. Second, we note an interesting feature in the Koszul complex: the different maps are all of degree one in the polynomials $Q_j$; the order of the matrices provides for a symmetrical complex, whose Betti numbers (the rank of the free modules $R^m$ appearing in the minimal resolution) are given by the binomial coefficients; finally, the last matrix of the complex is (up to signs and ordering) exactly the transpose of the matrix we started with. In particular, if we consider the associated differential operators, we have that the first and the last operator are the same in terms of their solution spaces.

We now particularize this discussion to the case in which the matrix $P$ represents the Cauchy–Riemann system in three complex variables $z_1, z_2, z_3$. Thus, if $\mathcal{S}$ is the sheaf of infinitely differentiable functions, the solution sheaf $\mathcal{S}^P$ of $P(D)\vec{f} = 0$, is the sheaf of germs of holomorphic functions. We should point out that the traditional treatment of this system is done without using these tools, because of the simplicity of the case. Nevertheless, looking at the multivariable Cauchy–Riemann system in this fashion is an excellent way to understand the way in which algebraic analysis can be used. In this particular case, the polynomials $Q_j$, $j = 1, 2, 3$, are given by

$$Q_1 = i\xi_1 - \eta_1, \quad Q_2 = i\xi_2 - \eta_2, \quad Q_3 = i\xi_3 - \eta_3$$

and

$$Q_1(D) = \partial/\partial\bar{z}_1, \quad Q_2(D) = \partial/\partial\bar{z}_2, \quad Q_3(D) = \partial/\partial\bar{z}_3.$$

The matrix $P_1$ gives the first syzygies that, for any pair of operators $\partial/\partial\bar{z}_j$, $\partial/\partial\bar{z}_i$, are given by

$$\frac{\partial}{\partial\bar{z}_j}\frac{\partial}{\partial\bar{z}_i} - \frac{\partial}{\partial\bar{z}_i}\frac{\partial}{\partial\bar{z}_j} = 0. \tag{2.9}$$

The construction of the syzygies is very simple because of several circumstances: the number of polynomials is small, the polynomials have no common factors, and the matrix $P$ has only one column. In general, however, a celebrated paper by Bayer and Stillman [15] has shown that the syzygy problem is of very high complexity, so that it is not possible to find for it a quick algorithm, which will be applicable to all the systems one may want to study. Part of the purpose of this book is to show how to successfully compute the syzygies for a large class of first order systems.

The general case, in this algebraic discussion, is the study of the inhomogeneous system

$$P(D)f = g, \tag{2.10}$$

where $P(D)$ is a matrix of differential operators with constant coefficients, for $f$ and $g$ with components in some given sheaf of (generalized) functions. As we have seen above, the archetype of all first order differential systems is the Cauchy–Riemann system, for which the compatibility conditions are easy to find and are well understood, see (2.9). They imply for any pair of indices $i \neq j$,

$$\frac{\partial}{\partial \bar{z}_j} g_i - \frac{\partial}{\partial \bar{z}_i} g_j = 0.$$

But the problem for more general systems is more complicated. The result obtained by algebraic analysis is the following theorem, whose proof is due to several authors: Ehrenpreis in the case of the sheaves $\mathcal{E}$ and $\mathcal{D}'$ see [64], Malgrange, Harvey and Komatsu for the sheaf $\mathcal{O}$ (see [106]) and Komatsu in the case of $\mathcal{A}$ and $\mathcal{B}$ [106].

**Theorem 2.1.1.** *Let $\mathcal{S}$ be one of the sheaves $\mathcal{E}$, $\mathcal{A}$, $\mathcal{O}$, $\mathcal{D}'$, $\mathcal{B}$. The system (2.10) has a solution $f \in \mathcal{S}(U)^{r_0}$ on a convex open (or convex compact) set $U$, if and only if $g \in \mathcal{S}(U)^{r_1}$ and it satisfies the compatibility system $P_1(D)g = 0$.*

*Proof.* To prove the statement, in the various cases, it is necessary to provide several technical results that are beyond our scope, so we limit ourselves to the proof in the case of $\mathcal{S} = \mathcal{B}$, assuming that the theorem holds for $\mathcal{S} = \mathcal{O}$ (see [106]).

Let us consider $\mathcal{S} = \mathcal{B}$, $U \subseteq \mathbb{R}^n$ and an open convex set $V \subseteq \mathbb{C}^n$ such that $V \cap \mathbb{R}^n = U$. Define

$$V_j = \{ z \in V \; : \; \operatorname{Im} z_j \neq 0 \}, \quad j = 1, 2, \dots, n, \quad V_0 = V$$

and let

$$\mathcal{V} = \{V_0, V_1, \dots, V_n\}, \qquad \mathcal{V}' = \{V_1, \dots, V_n\}$$

be an open covering of the pair $(V, V \setminus U)$. Then $\mathcal{V}$ is a Stein open covering. In what follows, we will respectively denote by $Z^n(\mathcal{V}, \mathcal{V}', \mathcal{O}^{r_j})$ and by $B^n(\mathcal{V}, \mathcal{V}', \mathcal{O}^{r_j})$ the group of relative $n$-cocycles and $n$-coboundaries with coefficients in $\mathcal{O}^{r_j}$. An element $g \in \mathcal{B}(U)$ can be represented by a relative $n$-cocycle $G \in Z^n(\mathcal{V}, \mathcal{V}', \mathcal{O}^{r_1})$ (see Proposition 1.3.10). The compatibility condition $P_1(D)G = 0$ implies that

$$P_1(D)G \in B^n(\mathcal{V}, \mathcal{V}', \mathcal{O}^{r_2}) = \delta^{n-1} \mathcal{C}^{n-1}(\mathcal{V}, \mathcal{V}', \mathcal{O}^{r_2}),$$

so that

$$P_1(D)G = \delta^{n-1} G_1,$$

with $G_1 \in \mathcal{C}^{n-1}(\mathcal{V}, \mathcal{V}', \mathcal{O}^{r_2})$. Coboundary and differentiation operators commute, so we have

$$\delta^{n-1}(P_2(D)G_1) = P_2(D)P_1(D)G = 0,$$

and since $H^{n-1}_U(V, \mathcal{O}^{r_2}) = 0$, there exists a cochain

$$G_2 \in \mathcal{C}^{n-2}(\mathcal{V}, \mathcal{V}', \mathcal{O}^{r_3})$$

such that

$$P_2(D)G_1 = \delta^{n-2}G_2.$$

By iterating this argument we get a sequence of cochains

$$G_j \in \mathcal{C}^{n-j}(\mathcal{V}, \mathcal{V}', \mathcal{O}^{r_{j+1}}), \quad \text{for } j = 1, 2, \dots, m-1$$

such that

$$P_j(D)G_{j-1} = \delta^{n-j}G_j. \tag{2.11}$$

The system represented by the last matrix, i.e., $P_{m-1}(D)$ has trivial compatibility condition so each component of $G_{m-1}$ is a holomorphic function on a convex open set in $\mathbb{C}^n$. By the validity of the result in the case of the sheaf $\mathcal{S} = \mathcal{O}$ (see [106]), there exists a cochain

$$H_{m-1} \in \mathcal{C}^{n-m+1}(\mathcal{V}, \mathcal{V}', \mathcal{O}^{r_{m-1}})$$

such that

$$G_{m-1} = P_{m-1}(D)H_{m-1}.$$

Substituting this equality in (2.11), it follows that

$$P_{m-1}(D)(G_{m-2} - \delta^{n-m+1}H_{m-1}) = 0,$$

so that each component of $G_{m-2} - \delta^{n-m+1}H_{m-1}$ satisfies the compatibility condition for $P_{m-2}(D)$. Then there exists a cochain

$$H_{m-2} \in \mathcal{C}^{n-m+2}(\mathcal{V}, \mathcal{V}', \mathcal{O}^{r_{m-2}})$$

such that

$$G_{m-2} - \delta^{n-m+1}H_{m-1} = P_{m-2}(D)H_{m-2}.$$

Iterating the procedure, we find cochains

$$H_j \in \mathcal{C}^{n-j}(\mathcal{V}, \mathcal{V}', \mathcal{O}^{r_j})$$

such that

$$G_j - \delta^{j+1}H_{j+1} = P_j(D)H_j, \quad \text{for} \quad j = m-1, m-2, \dots, 1.$$

Finally, there is a cochain $H \in \mathcal{C}^n(\mathcal{V}, \mathcal{V}', \mathcal{O}^{r_0})$ such that

$$G - \delta^{n-1}H_1 = P(D)H.$$

Denoting by $f$ the cohomology class of $H$, we have that $f$ is a solution to $P(D)f = g$. □

**Remark 2.1.2.** Notice that Theorem 2.1.1 is a far reaching generalization of the surjectivity results proved in Chapter 1, Section 1.2.3 for the case $r_0 = r_1 = 1$. In that case the first syzygy $P_1$ is trivial but the explicit construction of the solution is far from being obvious.

One can continue and find the compatibility conditions $P_2(D)$ for the system associated to $P_1(D)$ and, in general, the compatibility conditions $P_{j+1}(D)$ for $P_j(D)$. This amounts to translating the algebraic resolution (2.6) into an analytical resolution as in the following result.

**Theorem 2.1.2.** *Let $\mathcal{S}$ be one of the sheaves $\mathcal{E}$, $\mathcal{A}$, $\mathcal{O}$, $\mathcal{D}'$, $\mathcal{B}$ on $\mathbb{R}^n$ (or $\mathbb{C}^n$) and let $U$ be a convex open (or convex compact) set in $\mathbb{R}^n$ (or $\mathbb{C}^n$). Then the sequence*

$$0 \longrightarrow \mathcal{S}^P(U) \longrightarrow \mathcal{S}(U)^{r_0} \overset{P(D)}{\longrightarrow} \mathcal{S}(U)^{r_1} \overset{P_1(D)}{\longrightarrow} \ldots$$
$$\ldots \overset{P_{m-1}(D)}{\longrightarrow} \mathcal{S}(U)^{r_m} \longrightarrow 0$$

*is exact.*

Since convex sets form a fundamental systems of neighborhood at any point, an immediate consequence of Theorem 2.1.2 is the following important result.

**Theorem 2.1.3.** *Let $\mathcal{S}$ be one of the sheaves $\mathcal{E}$, $\mathcal{A}$, $\mathcal{O}$, $\mathcal{D}'$, $\mathcal{B}$. Then the sequence*

$$0 \longrightarrow \mathcal{S}^P \longrightarrow \mathcal{S}^{r_0} \overset{P(D)}{\longrightarrow} \mathcal{S}^{r_1} \overset{P_1(D)}{\longrightarrow} \mathcal{S}^{r_2} \ldots$$
$$\ldots \longrightarrow \mathcal{S}^{r_{m-1}} \overset{P_{m-1}(D)}{\longrightarrow} \mathcal{S}^{r_m} \longrightarrow 0$$

*is a resolution of $\mathcal{S}^P$.*

This last result becomes particularly interesting when considering elliptic operators by virtue of the following result due to Hörmander, see [88]:

**Theorem 2.1.4.** *If $P(D)$ is an elliptic operator, then $\mathcal{A}^P = \mathcal{D}'^P$, hence any distribution solution $f$ to the equation $P(D)f = g$ is real analytic on the open set on which $g$ is real analytic.*

Theorem 2.1.4 was improved by the Bengel, Harvey and Komatsu, see [108] and the references therein, who showed independently the following generalization:

**Theorem 2.1.5.** *If $P(D)$ is elliptic, then $\mathcal{A}^P = \mathcal{B}^P$.*

As a consequence, if $P(D)$ is an elliptic operator, we have the equalities

$$\mathcal{A}^P = \mathcal{D}'^P = \mathcal{B}^P$$

and the resolution in Theorem 2.1.3, which can be rewritten as

$$0 \longrightarrow \mathcal{A}^P \longrightarrow \mathcal{B}^{r_0} \overset{P(D)}{\longrightarrow} \mathcal{B}^{r_1} \overset{P_1(D)}{\longrightarrow} \ldots \mathcal{B}^{r_{m-1}} \overset{P_{m-1}(D)}{\longrightarrow} \mathcal{B}^{r_m} \longrightarrow 0, \qquad (2.12)$$

provides a flabby resolution for the sheaf $\mathcal{A}^P$.

Now we follow Komatsu [108] to discuss the significance of the Ext–modules defined in (2.8) and their analytical meaning; to this aim, we need to recall some further notations and propositions concerning relative cohomology groups, see Chapter 1. For $\mathcal{S}^P$ the sheaf of solutions defined before, where $\mathcal{S} = \mathcal{E}$, or $\mathcal{D}'$ or $\mathcal{B}$, one can consider a set $K$ relatively compact in an open set $\Omega \subset \mathbb{R}^n$, and the long exact sequence associated to the pair $(\Omega, K)$, with coefficients in $\mathcal{S}^P$, i.e.,

$$\begin{aligned} 0 \to H^0_K(\Omega, \mathcal{S}^P) \to H^0(\Omega, \mathcal{S}^P) \to H^0(\Omega \setminus K, \mathcal{S}^P) \to \\ \to H^1_K(\Omega, \mathcal{S}^P) \to H^1(\Omega, \mathcal{S}^P) \to H^1(\Omega \setminus K, \mathcal{S}^P) \to \dots . \end{aligned} \tag{2.13}$$

An interesting result is the fact the (2.13) can be actually decomposed into short exact sequences.

**Theorem 2.1.6.** *Let $\mathcal{S}$ be one of the sheaves $\mathcal{E}$, $\mathcal{D}'$ or $\mathcal{B}$. Let $\Omega$ be an open set in $\mathbb{R}^n$ and let $K$ be a relatively compact (and locally closed) set in $\Omega$. Then the following sequences are exact:*

$$\begin{aligned} 0 \to H^0_K(\Omega, \mathcal{S}^P) \to H^0(\Omega, \mathcal{S}^P) \to H^0(\Omega \setminus K, \mathcal{S}^P) \to \\ \to H^1_K(\Omega, \mathcal{S}^P) \to 0 \end{aligned}$$

*and*

$$\begin{aligned} 0 \to H^j(\Omega, \mathcal{S}^P) \to H^j(\Omega \setminus K, \mathcal{S}^P) \to \\ \to H^{j+1}_K(\Omega, \mathcal{S}^P) \to 0, \quad j \geq 1. \end{aligned}$$

*Proof.* By virtue of Theorem 1.1.12, it suffices to prove that

$$0 \to H^j(\Omega, \mathcal{S}^P) \to H^j(\Omega \setminus K, \mathcal{S}^P)$$

is exact for $j \geq 1$. Let us begin by considering $\mathcal{S} = \mathcal{B}$. By Theorem 2.1.3, we obtain a flabby resolution of $\mathcal{B}^P$ from which it follows that

$$H^j(\Omega, \mathcal{B}^P) = \Gamma(\Omega, P_{j-1}\mathcal{B}^{r_{j-1}})/P_{j-1}\Gamma(\Omega, \mathcal{B}^{r_{j-1}})$$

and

$$H^j(\Omega \setminus K, \mathcal{B}^P) = \Gamma(\Omega \setminus K, P_{j-1}\mathcal{B}^{r_{j-1}})/P_{j-1}\Gamma(\Omega \setminus K, \mathcal{B}^{r_{j-1}}).$$

Note that the restriction map $H^j(\Omega, \mathcal{B}^P) \to H^j(\Omega \setminus K, \mathcal{B}^P)$ is induced by the restriction at the level of sections.

Let $\phi$ be an element in $\Gamma(\Omega, P_{j-1}\mathcal{B}^{r_{j-1}})$ whose restriction to $\Omega \setminus K$ is cohomologous to zero in $\Gamma(\Omega \setminus K, P_{j-1}\mathcal{B}^{r_{j-1}})$; then $\phi_{|\Omega \setminus K} = P_{j-1}\psi$ for some $\psi \in \Gamma(\Omega \setminus K, \mathcal{B}^{r_{j-1}})$. As $\mathcal{B}$ is a flabby sheaf, $\psi$ has an extension $\psi_1$ to $\Omega$. Then $\phi_1 = \phi - P_{j-1}\psi_1 \in \Gamma(\Omega, \mathcal{B}^{r_{j-1}})$ has support in $K$, therefore $\phi_1$ has a trivial extension, still denoted by $\phi_1$, to $\mathbb{R}^n$. Now we have $P_j\phi_1 = 0$ or $\phi_1 \in \Gamma(\mathbb{R}^n, P_{j-1}\mathcal{B}^{r_{j-1}})$. Since $\mathbb{R}^n$ is convex there exists $\psi_2 \in \Gamma(\mathbb{R}^n, \mathcal{B}^{r_{j-1}})$ such that $\phi_1 = P_{j-1}\psi_2$ and so $\phi = P_{j-1}(\psi_1 + \psi_2)$ belongs to $P_{j-1}\Gamma(\Omega, \mathcal{B}^{r_{j-1}})$

and this completes the proof for $\mathcal{S} = \mathcal{B}$. When $\mathcal{S}$ is either $\mathcal{D}'$ or $\mathcal{E}$, Theorem 2.1.3 gives a soft resolution of $\mathcal{S}^P$. We will prove that if $\phi \in \Gamma(\Omega, P_{j-1}\mathcal{S}^{r_{j-1}})$ has its restriction in $P_{j-1}\Gamma(\Omega \setminus K, P_{j-1}\mathcal{S}^{r_{j-1}})$, then $\phi \in P_{j-1}\Gamma(\Omega, \mathcal{S}^{r_{j-1}})$. Let $\varphi \in C_0^\infty(\Omega)$ such that $\varphi = 1$ in a neighborhood of $\bar{K}$. If the restriction of $\phi$ to $\Omega \setminus K$ is of the form $\phi_{|\Omega\setminus K} = P_{j-1}\psi$ where $\psi \in \Gamma(\Omega \setminus K, \mathcal{S}^{r_{j-1}})$, then $\phi$ can be written as

$$\phi = P_{j-1}((1-\varphi)\psi) + \phi_1,$$

where $(1-\varphi)\psi$ is extended to $\Omega$ by zero. We obviously have that $\phi_1$ belongs to $\Gamma(\Omega, P_{j-1}\mathcal{S}^{r_{j-1}})$ and it has compact support in $\Omega$. The rest of the proof can be carried out as in the case $\mathcal{S} = \mathcal{B}$. □

The most interesting applications of the previous theorem occur when $H^j_K(\Omega, \mathcal{S}^P)$ vanishes. The simplest cases to characterize, from an analytical point of view, are when $j = 0$ or 1.

**Proposition 2.1.2.** *Let $\Omega$ be an open set in $\mathbb{R}^n$ and let $K$ be a relatively compact set in $\Omega$. The condition $H^0_K(\Omega, \mathcal{S}^P) = 0$ is equivalent to the unique continuation property of solutions $f \in \mathcal{S}$ of $P(D)f = 0$ on $\Omega \setminus K$ inside $K$, while $H^1_K(\Omega, \mathcal{S}^P) = 0$ is equivalent to the existence of a continuation inside $K$.*

*Proof.* In view of Theorem 2.1.6 the vanishing of $H^0_K(\Omega, \mathcal{S}^P)$ implies the injectivity of the map $H^0(\Omega, \mathcal{S}^P) \to H^0(\Omega \setminus K, \mathcal{S}^P)$ that gives uniqueness of the continuation. If $H^1_K(\Omega, \mathcal{S}^P) = 0$, then the map $H^0(\Omega, \mathcal{S}^P) \to H^0(\Omega \setminus K, \mathcal{S}^P)$ is surjective and this proves the existence of such a continuation. □

Note that if the vanishing of $H^j_K(\Omega, \mathcal{S}^P)$ occurs for all $j \geq 2$, then there are the isomorphisms

$$H^j(\Omega, \mathcal{S}^P) \cong H^j(\Omega \setminus K, \mathcal{S}^P), \quad \forall j \geq 1.$$

To determine vanishing conditions, we are naturally lead to consider the $R$-modules $\operatorname{Ext}^j(M, R)$, i.e., the cohomology groups of the complex (2.7), whose vanishing has several interesting consequences.

**Theorem 2.1.7.** *The following conditions are all equivalent:*

$$\begin{aligned}&\operatorname{Ext}^0(M,R) = 0 \Leftrightarrow H^0_c(\mathbb{R}^n, \mathcal{B}^P) = 0 \Leftrightarrow H^0_c(\mathbb{R}^n, \mathcal{D}'^P) = 0\\ &\Leftrightarrow H^0_c(\mathbb{R}^n, \mathcal{E}^P) = 0 \Leftrightarrow H^0_{\{0\}}(\mathbb{R}^n, \mathcal{B}^P) = 0\\ &\Leftrightarrow H^0_{\{0\}}(\mathbb{R}^n, \mathcal{D}'^P) = 0.\end{aligned}$$

*Proof.* It is obvious that we have the following implications

$$H^0_c(\mathbb{R}^n, \mathcal{B}^P) = 0 \Rightarrow H^0_c(\mathbb{R}^n, \mathcal{D}'^P) = 0 \Rightarrow H^0_c(\mathbb{R}^n, \mathcal{E}^P) = 0$$

and

$$H^0_c(\mathbb{R}^n, \mathcal{B}^P) = 0 \Rightarrow H^0_{\{0\}}(\mathbb{R}^n, \mathcal{B}^P) = 0 \Rightarrow H^0_{\{0\}}(\mathbb{R}^n, \mathcal{D}'^P) = 0,$$

and finally

$$H^0_c(\mathbb{R}^n, \mathcal{D}'^P) = 0 \Rightarrow H^0_{\{0\}}(\mathbb{R}^n, \mathcal{D}'^P) = 0.$$

Let us now prove that

$$\operatorname{Ext}^0(M, R) = 0 \Rightarrow H^0_c(\mathbb{R}^n, \mathcal{B}^P) = 0.$$

Let $f \in \mathcal{B}(\mathbb{R}^n)^{r_0}$ be a compactly supported solution of $P(D)f(x) = 0$. By taking the Fourier transform we obtain the equation

$$P(\lambda)\hat{f}(\lambda) = 0, \qquad \lambda \in \mathbb{C}^n.$$

Since $\operatorname{Ext}^0(M, R) = 0$, the homomorphism $P(\lambda)$ is injective, i.e., the rank of the matrix $P(\lambda)$ is equal to $r_0$ almost everywhere. This fact implies that $\hat{f}(\lambda) = 0$ almost everywhere and so $f = 0$.

Now we show the implication

$$H^0_c(\mathbb{R}^n, \mathcal{E}^P) = 0 \Rightarrow H^0_c(\mathbb{R}^n, \mathcal{D}'^P) = 0.$$

Let $f \in \mathcal{D}'(\mathbb{R}^n)^{r_0}$ with compact support (i.e., $f \in \mathcal{E}'(\mathbb{R}^n)^{r_0}$) be a solution to the equation $P(D)f = 0$. Its regularization $f_\varepsilon = \omega_\varepsilon * f$, where $\omega_\varepsilon$, for all $\varepsilon > 0$, is a suitable function in $\mathcal{D}(\mathbb{R}^n)$ (see for example [1]), belongs to $\mathcal{E}^P(\mathbb{R}^n)$ and it has compact support. From Theorem 1.2.13, we have

$$P(D)f_\varepsilon = P(D)[\omega_\varepsilon * f] = \omega_\varepsilon * [P(D)f] = 0$$

and so $f_\varepsilon = 0$. Thanks to the fact that $f_\varepsilon$ converges to $f$ in $\mathcal{E}'(\mathbb{R}^n)^{r_0}$ we get $f = 0$.

Finally, we show that $H^0_{\{0\}}(\mathbb{R}^n, \mathcal{D}'^P) = 0$ implies $\operatorname{Ext}^0(M, R) = 0$. Suppose the contrary; then the columns of the matrix $P(\lambda)$ are linearly dependent over $R$. A nontrivial vector $\hat{f}(\lambda)$ with polynomial components such that $P(\lambda)\hat{f}(\lambda) = 0$ is the Fourier transform of a distribution $f$ with support at the origin (see Remark 1.2.9) satisfying $P(D)f(x) = 0$, thus contradicting the hypothesis. □

**Corollary 2.1.1.** *Let $K$ be a relatively compact set in an open set $U \subseteq \mathbb{R}^n$ and let $\mathcal{S}$ be one of the sheaves $\mathcal{E}$, $\mathcal{D}'$ or $\mathcal{B}$. If $\operatorname{Ext}^0(M, R) = 0$, then $H^0_K(U, \mathcal{S}^P) = 0$.*

*Proof.* It is an immediate consequence of the fact that

$$H^0_K(U, \mathcal{S}^P) \subset H^0_c(U, \mathcal{S}^P).$$

□

**Theorem 2.1.8.** *For $j \geq 1$, the following conditions are equivalent:*

*(a)* $\operatorname{Ext}^j(M, R) = 0$;

*(b)* $H^j_K(\mathbb{R}^n, \mathcal{B}^P) = 0$, *where $K$ is a bounded convex set in $\mathbb{R}^n$;*

*(c)* $H^j_{\{0\}}(\mathbb{R}^n, \mathcal{B}^P) = 0$;

*(d)* $H^j_c(U, \mathcal{B}^P) = 0$, *where $U$ is a convex open set in $\mathbb{R}^n$.*

*Proof.* To prove the theorem, it is useful to note that $(b)$ is equivalent to the exactness of the sequence

$$\mathcal{B}[K]^{r_{j-1}} \xrightarrow{P_{j-1}(D)} \mathcal{B}[K]^{r_j} \xrightarrow{P_j(D)} \mathcal{B}[K]^{r_{j+1}}.$$

Let us denote by $b'$) this condition of exactness and let us start by proving the implication $(a) \Rightarrow (b')$. The condition $(a)$ means that $P'_{j-1}(D)$ is the compatibility system for $P'_j(D)$. If $K$ is a compact convex set, Theorem 2.1.2 implies that the sequence

$$\mathcal{A}(K)^{r_{j-1}} \xleftarrow{P'_{j-1}(D)} \mathcal{A}(K)^{r_j} \xleftarrow{P'_j(D)} \mathcal{A}(K)^{r_{j+1}}$$

is exact and the image of the map $P'_{j-1}(D)$ is closed in $\mathcal{A}(K)^{r_{j-1}}$. By Remark 1.3.8, $\mathcal{B}[K]^{r_j}$ is an FS space whose strong dual is $\mathcal{A}(K)^{r_j}$ and $P_j(D)$ and $P'_j(D)$ are continuous linear maps dual to each other, so the sequence $b'$) is exact by the Serre's duality (see [177]).

Let us prove that $(b') \Rightarrow (c)$. Let us consider a bounded convex set $K$ in $\mathbb{R}^n$. An element $f$ in the kernel of $P_j(D)$ has support $S$ in $K$. The convex hull $\widehat{S}$ of $S$ is a compact convex set contained in $K$. We can write the sequence $(b')$ for the compact set $\widehat{S}$, so that $f$ belongs to $P_{j-1}(D)\mathcal{B}[\widehat{S}]^{r_{j-1}}$. The implication $(b) \Rightarrow (c)$ follows immediately.

Let us prove that $(c) \Rightarrow (a)$. We can regard vectors $F \in R^{r_j}$ with polynomial entries as the Fourier transform of distributions supported at the origin (see Remark 1.2.9). If $F$ satisfies $P(\lambda)F(\lambda) = 0$ then, by the sequence $(b')$ with $K = \{0\}$, we can find a vector $u$ with entries hyperfunctions supported at the origin and whose Fourier transform $\hat{u}(\lambda)$ satisfies

$$P_{j-1}(\lambda)\hat{u}(\lambda) = F(\lambda), \quad \lambda \in \mathbb{C}^n.$$

Since $F$ is a vector of polynomials, then, for some positive $\nu \in \mathbb{R}$, there exists a positive constant $C$ such that

$$|F(\lambda)| \leq C(1 + |\lambda|^2)^\nu.$$

From the classical $L^2$-estimates of Hörmander, see [90], we know that there exists a vector $U(\lambda)$ of holomorphic functions and positive constants $C_1$, $N$ such that

$$P_{j-1}(\lambda)U(\lambda) = F(\lambda)$$

and

$$|U(\lambda)| \leq C_1(1 + |\lambda|^2)^{\nu+N}.$$

This last inequality shows that $U(\lambda)$ is a vector with polynomial entries so the sequence

$$R^{r_{j-1}} \xrightarrow{P_{j-1}} R^{r_j} \xrightarrow{P_j} R^{r_{j+1}}$$

is exact and $(a)$ holds. It is immediate to note that $(b) \Rightarrow (d)$, so we now prove that $(d) \Rightarrow (a)$. To this end, let us note that the above considerations show that $(a)$ holds if there is an analytic functional $u$ solution of $P_{j-1}(D)u = f$ for any $f \in \mathcal{D}'[\{0\}]^{r_j}$ with $P_j(D)f = 0$, moreover $(a)$ holds when $(b)$ is true for a nonempty set $K$. In particular, $(a)$ holds when $(d)$ is true for a nonempty set. □

With similar arguments (see [108]), one can prove an analogous statement in the case of distributions:

**Theorem 2.1.9.** *For $j \geq 1$, the following conditions are equivalent:*

*(a)* $\mathrm{Ext}^j(M, R) = 0$;

*(b) the sequence*

$$\mathcal{D}'[K]^{r_{j-1}} \xrightarrow{P_{j-1}(D)} \mathcal{D}'[K]^{r_j} \xrightarrow{P_j(D)} \mathcal{D}'[K]^{r_{j+1}}$$

*is exact;*

*(c)* $H^j_K(\mathbb{R}^n, \mathcal{D}'^P) = 0$, *where $K$ is a bounded convex open set in $\mathbb{R}^n$;*

*(d)* $H^j_c(U, \mathcal{D}'^P) = 0$, *where $U$ is a convex open set in $\mathbb{R}^n$.*

In the case of infinitely differentiable functions, we have the following theorem due to Malgrange (see [124]):

**Theorem 2.1.10.** *For $j \geq 1$, the following conditions are equivalent:*

*(a)* $\mathrm{Ext}^j(M, R) = 0$;

*(b)* $H^j_K(\mathbb{R}^n, \mathcal{E}^P) = 0$, *where $K$ is a bounded convex open set in $\mathbb{R}^n$;*

*(c)* $H^j_c(U, \mathcal{E}^P) = 0$, *where $U$ is a convex open set in $\mathbb{R}^n$.*

**Remark 2.1.3.** In the case of infinitely differentiable functions, the condition (a) of Theorem 2.1.10 does not necessarily follows from the condition

$$H^j_{\{0\}}(\mathbb{R}^n, \mathcal{E}^P) = 0.$$

In fact, if we consider the wave operator

$$P(D) = \frac{\partial^2}{\partial x^2} - \frac{\partial^2}{\partial y^2},$$

then we see that every solution $u$ to the equation $P(D)u = 0$ in $\mathcal{E}(\mathbb{R}^n \setminus \{0\})$ can be extended to an infinitely differentiable solution on $\mathbb{R}^n$. However

$\mathrm{Ext}^1(M, R) \neq 0$; in fact the fundamental solution $E$, solving $P(D)E = \delta$, gives a distribution solution on $\mathbb{R}^n \setminus \{0\}$ which cannot be extended to a solution on $\mathbb{R}^n$.

The vanishing of the Ext-modules allows to prove some interesting duality theorems in the same spirit of Serre's duality theorem see [177].

**Theorem 2.1.11.** *Let $K$ be a compact set in $\mathbb{R}^n$, let $P(D)$ be the matrix associated to a system such that $\mathrm{Ext}^j(M, R) = 0$ for $j = 0, \dots, m-1$, where $m$ is the length of its minimal free resolution (2.6), and let $Q(D) = P'_{m-1}(D)$. Moreover, suppose that either*

$$\dim H^j_K(\mathbb{R}^n, \mathcal{B}^P) < +\infty, \quad j = 1, \dots, m$$

*or*

$$\dim H^{m-j}(K, \mathcal{A}^Q) \leq \aleph_0, \quad j = 0, 1, \dots, m-1.$$

*Then $H^j_K(\mathbb{R}^n, \mathcal{B}^P)$ and $H^{m-j}(K, \mathcal{A}^Q)$ are respectively a FS-space and an DFS-space, and, for $j = 0, 1, \dots, m$, they are strong dual to each other.*

*Proof.* In view of Theorem 2.1.3, we have the following resolutions of $\mathcal{B}^P$ and $\mathcal{A}^Q$, respectively:

$$0 \longrightarrow \mathcal{B}^P \longrightarrow \mathcal{B}^{r_0} \xrightarrow{P(D)} \mathcal{B}^{r_1} \xrightarrow{P_1(D)} \dots \longrightarrow \mathcal{B}^{r_{m-1}} \xrightarrow{P_{m-1}(D)} \mathcal{B}^{r_m} \longrightarrow 0 \tag{2.14}$$

and

$$0 \longrightarrow \mathcal{A}^Q \longrightarrow \mathcal{A}^{r_m} \xrightarrow{Q(D)} \mathcal{A}^{r_{m-1}} \longrightarrow \dots \longrightarrow \mathcal{A}^{r_1} \xrightarrow{Q_{m-1}(D)} \mathcal{A}^{r_0} \longrightarrow 0,$$

where we have set $Q_j(D) = P'_{m-j-1}(D)$. Since (2.14) is a flabby resolution of $\mathcal{B}^P$ we have that the groups

$$H^j_K(\mathbb{R}^n, \mathcal{B}^P)$$

are the cohomology groups of the complex

$$0 \longrightarrow \mathcal{B}_K(\mathbb{R}^n)^{r_0} \xrightarrow{P(D)} \mathcal{B}_K(\mathbb{R}^n)^{r_1} \xrightarrow{P_1(D)} \mathcal{B}_K(\mathbb{R}^n)^{r_2} \longrightarrow \dots$$

$$\dots \longrightarrow \mathcal{B}_K(\mathbb{R}^n)^{r_{m-1}} \xrightarrow{P_{m-1}(D)} \mathcal{B}_K(\mathbb{R}^n)^{r_m} \longrightarrow 0. \tag{2.15}$$

Since $\mathcal{B}_K(\mathbb{R}^n)^{r_\ell}$ are FS-spaces, and dim $H^j_K(\mathbb{R}^n, \mathcal{B}^P) < +\infty$, one can use the Schwartz lemma (see [175]) to show that the maps $P_\ell(D)$ have closed ranges. The groups $H^{m-j}(K, \mathcal{A}^Q)$ are the cohomology groups of the complex

$$0 \longrightarrow \mathcal{A}(K)^{r_m} \xrightarrow{Q(D)} \mathcal{A}(K)^{r_{m-1}} \longrightarrow \dots \longrightarrow \mathcal{A}(K)^{r_1} \xrightarrow{Q_{m-1}(D)} \mathcal{A}(K)^{r_0} \longrightarrow 0 \tag{2.16}$$

and, since $\dim H^{m-j}(K, \mathcal{A}^Q) \leq \aleph_0$ for $j = 0, 1, \dots, m-1$, again by the Schwartz Lemma, we get that the ranges of $Q_\ell(D)$ are closed. Finally, since the sequence (2.15) and (2.16) are dual one another, by the Serre's Theorem [177] we get the desired result. □

**Remark 2.1.4.** If $K$ is a compact convex set, then $H^{m-j}(K, \mathcal{A}^Q) = 0$ for $j = 0, \dots, m-1$, then $H^j_K(\mathbb{R}^n, \mathcal{B}^P) = 0$ for $j = 0, \dots, m-1$ and $H^m_K(\mathbb{R}^n, \mathcal{B}^P) \cong (\mathcal{A}^Q(K))'$.

**Remark 2.1.5.** We point out that Theorem 2.1.11 can be also formulated in the framework of the sheaves $\mathcal{D}'$ and $\mathcal{E}$ (instead of $\mathcal{B}$ and $\mathcal{A}$), where $K$ is a bounded open set, with a proof which follows exactly the lines of the proof of Theorem 2.1.11.

**Remark 2.1.6.** Provided the necessary vanishing of the Ext-modules, from the previous theorem we get that, if $K$ is a convex set in $\mathbb{R}^n$, then

$$H^m_K(\mathbb{R}^n, \mathcal{D}'^P) \cong (H^0(K, \mathcal{E}^Q))'. \tag{2.17}$$

This is indeed an interesting generalization of the classical theory of analytical functionals due to Fantappiè. In particular, when $\mathbb{R}^{2n}$ is identified with $\mathbb{C}^n$, and $P$ is the Cauchy–Riemann system, then $Q$ is (up to a sign) again the Cauchy–Riemann system and $m = n$. We can then deduce that (under the hypothesis of finiteness of cohomology), $H^n_K(\mathbb{R}^{2n}, \mathcal{O}) = \mathcal{B}_K(\mathbb{R}^n)$, which is the fundamental result of Sato on compactly supported hyperfunctions.

Let us discuss an important result which follows from the algebraic analysis described above. It was the first result ever proved in this framework, namely Ehrenpreis' new interpretation of Hartogs' theorem on the removability of compact singularities for holomorphic functions in several variables. The original result, see [85], was established by Hartogs in 1906: it is a major result in the theory of several complex variables.

**Theorem 2.1.12.** (Hartogs). *Let $U \subseteq \mathbb{C}^n$, $n > 1$, be an open set and let $K$ be a compact set in $U$ such that $U \setminus K$ is connected. Then any holomorphic function on $U \setminus K$ can be extended to a unique holomorphic function on $U$.*

By proving that holomorphic functions in $n \geq 2$ variables cannot have compact singularities, Hartogs demonstrated that the theory of several complex variables was indeed very different from the theory of one complex variable. Ehrenpreis, however, shed a totally new light on this question, by proving that the result has really nothing to do with complex analysis.

We now state the theorem for the case of $r_0 = 1$, $r_1 = r$ and we provide the original proof of Ehrenpreis [66] which, due to its clarity, provides an important insight in the methods of algebraic analysis. To give the exact statement and proof of the result we need some notation and terminology.

Let $\Omega_0 \subset \mathbb{R}^n$, $n > 1$, be such that there exists two open relatively compact sets $\Omega_1$, $\Omega_2$ in $\mathbb{R}^n$ such that $\bar{\Omega}_1 \subset \Omega_2$ and $\Omega_0 = \Omega_2 \setminus \Omega_1$. Denote by $\Gamma_j$ the boundary of $\Omega_j$, $j = 1, 2$, by $N(\Gamma_1)$ a small neighborhood of $\Gamma_1$ in $\Omega_2$, by $N'(\Gamma_1)$ a small neighborhood of $\Gamma_1$ in $\Omega_2$ whose closure is contained in $N(\Gamma_1)$ and set $\Omega_3 = \Omega_1 \setminus N'(\Gamma_1)$ (see Figure 2.1). Let $p_1, \dots, p_r$ be $r \geq 2$ polynomials in the ring of polynomials in $n$ variables with complex coefficients. We make the following hypotheses:

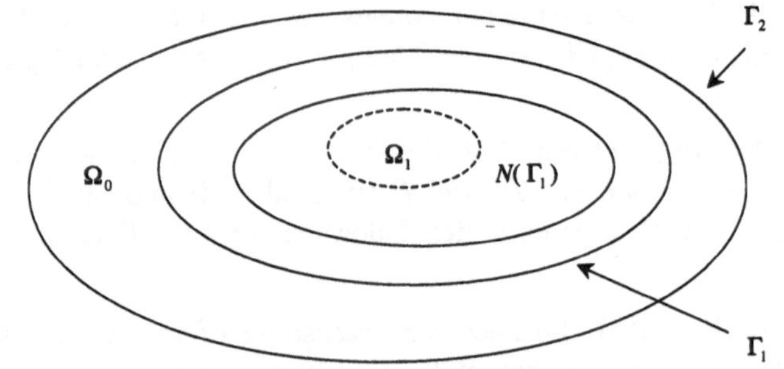


**Figure 2.1.**

($\alpha$) the polynomials $p_1, \ldots, p_r$ have no common factors;

($\beta$) for any $N'(\Gamma_1)$ the following unique continuation property holds: if $g \in \mathcal{E}(\mathbb{R}^n)$, $g = 0$ outside the convex hull $\widehat{\Omega}_1$ of $\Omega_1$ and $P(D)g = (p_1(D), \ldots, p_r(D))g = 0$ outside $\Omega_3$, then $g = 0$ outside some compact subset $\Omega_4$ of $\Omega_1$.

We are now ready to state and proof Ehrenpreis' extension result.

**Theorem 2.1.13.** *Assumptions* ($\alpha$) *and* ($\beta$) *are necessary and sufficient in order that any* $f \in \mathcal{E}(\Omega_0 \cup N(\Gamma_1))$ *satisfying*

$$P(D)f = (p_1(D), \ldots, p_r(D))f = 0,$$

*on* $\Omega_0 \cup N(\Gamma_1)$, *for some* $N(\Gamma_1)$, *has a unique extension* $\tilde{f} \in \mathcal{E}(\Omega_2)$ *with* $P(D)\tilde{f} = 0$ *on* $\Omega_2$ *and* $\tilde{f} = f$ *on the union of* $\Omega_0$ *with some neighborhood of* $\Gamma_1$ *in* $\Omega_1$.

*Proof.* We start by proving the sufficiency of ($\alpha$) and ($\beta$). Consider $g \in \mathcal{E}(\Omega_2)$ such that $g_{|\Omega_0 \cup N'(\Gamma_1)} = f$, where $N'(\Gamma_1)$ is as above, and $\Omega_3 = \Omega_1 \setminus N'(\Gamma_1)$. By hypothesis $P(D)g = 0$ on $\Omega_2 \setminus \Omega_3$. Let $g_j = p_j(D)g$ for $j = 1, \ldots, r$. It is immediate to see that for every $j$, the function $g_j$ is infinitely differentiable and has its support contained in $\bar{\Omega}_3$. Moreover, the commutativity of the operators implies the (cocycle) relations

$$p_j(D)g_i = p_i(D)g_j, \quad i, j = 1, \ldots, r. \tag{2.18}$$

By taking the Fourier transform of both sides in (2.18), we immediately obtain the equalities

$$p_j\hat{g}_i = p_i\hat{g}_j, \quad i, j = 1, \ldots, r.$$

We denote by $G$ the function $\hat{g}_j/p_j = \hat{g}_i/p_i = \ldots$ and we observe that $G$ is holomorphic outside the algebraic variety

$$W = \{z \in \mathbb{C}^n \; : \; p_1(z) = \ldots = p_r(z) = 0\}.$$

In view of the hypothesis ($\alpha$), the codimension of $W$ is at least 2 and so the function $G$ is in fact entire. As to its growth we know from the Paley–Wiener–Schwartz's theorem that $\hat{g}_j$ satisfies

$$|\hat{g}_j(z)| \leq A(1+|z|)^{-s} \exp(H_{\widehat{\Omega}_3}(\mathrm{Im} z)), \quad j = 1, \dots, r. \tag{2.19}$$

The division lemma of Ehrenpreis–Malgrange (see e.g. [68], Lemma 1.2 and Corollary 1.3) immediately shows that $G$ satisfies the same growth conditions given in (2.19). Again by Paley–Wiener–Schwartz's theorem, we have that there exists $h \in \mathcal{E}(\mathbb{R}^n)$ with support in the closure of $\widehat{\Omega}_3$ such that $G = \hat{h}$. In view of hypothesis ($\beta$) it is possible to assume that $h$ vanishes outside $\Omega_4$. If we set

$$\tilde{f} = g - h,$$

then we clearly have $\tilde{f} \in \mathcal{E}(\Omega_2)$ since both $g$ and $h$ belong to $\mathcal{E}(\Omega_2)$; moreover $\tilde{f} = g$ on $\Omega_2 \setminus \Omega_4$ as $h$ is supported in $\Omega_4$ and finally

$$p_j(D)\tilde{f} = p_j(D)g - p_j(D)h = g_j - g_j = 0, \quad \text{on } \Omega_2.$$

Thus $\tilde{f}$ is an extension of $f$. This extension is unique in fact if there were another extension $\tilde{f}'$, then $\tilde{f} - \tilde{f}'$ would be compactly supported as both $\tilde{f}$, $\tilde{f}'$ coincide with $g$ on $\Omega_2 \setminus \Omega_4$ and

$$P(D)(\tilde{f} - \tilde{f}') = 0. \tag{2.20}$$

By taking the Fourier transform of one equation of the system (2.20) we get

$$p_j \widehat{(\tilde{f} - \tilde{f}')} = 0. \tag{2.21}$$

Since $\widehat{\tilde{f} - \tilde{f}'}$ is an entire function, the equation (2.21) implies $\tilde{f} - \tilde{f}' = 0$ which give the uniqueness of the extension.

Let us turn to the proof of the necessity of the two hypotheses.

*Necessity of* ($\alpha$). Suppose the polynomials $p_j$ have a common factor $Q$ whose inverse Fourier transform is denoted by $Q(D)$. By Theorem 2.1.1, $Q(D)$ maps $\mathcal{D}'$ onto itself. In particular, there exists a fundamental solution $E$ of $Q(D)E = \delta_{x_0}$ where $x_0 \in \Omega_1$. Let $m \in \mathcal{D}$ have support so small that the support of $m * \delta_{x_0}$ is still contained in $\Omega_1$. Moreover, we assume that $Q$ does not divide the Fourier transform $\hat{m}$ of $m$. The restriction $f$ of $E * m$ to $\Omega_0 \cup N(\Gamma_1)$ satisfies

$$Q(D)(E * m) = \delta_{x_0} * m = 0, \quad \text{on } \Omega_0 \cup N(\Gamma_1).$$

Hence $P(D)(E * m) = 0$ on $\Omega_0 \cup N(\Gamma_1)$. We claim that there are no functions $\tilde{f} \in \mathcal{E}(\Omega_2)$ such that $P(D)\tilde{f} = 0$ on $\Omega_2$ and $\tilde{f} = f$ on $\Omega_0$. Suppose the contrary; then $\tilde{f}$ would be such that $F = \tilde{f} - E * m = 0$ on $\Omega_0$ (hence it would be in $\mathcal{D}$) and

$$P(D)F = -P(D)(E * m).$$

Nevertheless, $-P(D)(E * m)$ is not of the form $P(D)F$ for $F \in \mathcal{D}$. In fact, let us set

$$p_j(D) = Q(D)Q_j(D)$$

for each $j$. Then we have

$$Q_j(D)Q(D)F = -Q_j(D)m * \delta_{x_0}$$

that is

$$Q_j(D)(Q(D)F + m * \delta_{x_0}) = 0.$$

Since $Q(D)F + m * \delta_{x_0}$ has compact support we have that, as in the proof of the sufficiency, $Q(D)F + m * \delta_{x_0} = 0$. By taking the Fourier transform we obtain that $Q$ divides $\hat{m}$ contrary to our assumptions.
*Necessity of* $(\beta)$. If $(\beta)$ does not hold, then there exists $N'(\Gamma_1)$ and $g \in \mathcal{E}(\mathbb{R}^n)$ such that $g = 0$ outside the convex hull of $\Omega_3$, $P(D)g = 0$ outside $\Omega_3$ and the support of $g$ is not contained in a compact subset of $\Omega_1$. Define the function $g^* \in \mathcal{E}(\Omega_2 \setminus \Omega_3)$ to be the restriction of $g$ to $\Omega_2 \setminus \Omega_3$. Clearly, it is $P(D)g^* = 0$ on $\Omega_2 \setminus \Omega_3$. We claim that for no neighborhood $N(\Gamma_1)$ in $\Omega_2$ there exists a $\tilde{g} \in \mathcal{E}(\Omega_2)$ such that $P(D)\tilde{g} = 0$ on $\Omega_2$ and $\tilde{g} = g^*$ on $\Omega_0 \cup N(\Gamma_1)$. For, we could extend $\tilde{g}$ to a $C^\infty$ function over $\mathbb{R}^n$ by defining $\tilde{g} = g$ outside $\Omega_2$ and we would get $\tilde{g}$ with compact support and $P(D)\tilde{g} = 0$ everywhere. By taking the Fourier transform one gets $\tilde{g} = 0$ on $\mathbb{R}^n$ and $g^* = 0$ on $\Omega_0 \cup N(\Omega_1)$, which prove that $(\beta)$ is necessary. □

One should point out that the $C^\infty$ setting could be replaced by other more "singular" ones. The theorem, for example, holds also in the space of distributions, in view of the flexibility of the Ehrenpreis–Malgrange Lemma that holds also in this setting.

**Remark 2.1.7.** Going back to the statement and proof of the above theorem, one may notice that the unpleasant fact that $\Omega_0 \cup N'(\Gamma_1)$ is not the whole domain of definition of $f$ itself. This fact is clearly unavoidable, as there may not even exist in $C^\infty$ function $g$ which extends to all $\Omega_2$. The obvious reason for this fact is the non-flabbiness of the sheaf $\mathcal{E}$. The situation will improve when Hartogs' phenomenon is studied for hyperfunctions (see Theorem 2.1.15 in the sequel).

**Remark 2.1.8.** As Ehrenpreis himself pointed out, if $P(D)$ is elliptic, then the condition $(\beta)$ holds if and only if the complement of the set $\Omega_1$ is connected. When $P(D)$ is the Cauchy–Riemann system, Theorem 2.1.13 gives exactly Hartogs' theorem 2.1.12.

Let us consider the case in which $P(D)$ is a matrix:

**Theorem 2.1.14.** *Let $P(D)$ be a $r_1 \times r_0$ matrix of linear partial differential operators with constant coefficients, let $K$ be a compact convex set and $\Omega$ a neighborhood of $K$. Then the distributions solutions to the homogeneous system*

*$P(D)f = 0$ on $\Omega \setminus K$ can be uniquely extended to a solution of the the system on $\Omega$ if and only if* $\operatorname{Ext}^0(M,R) = \operatorname{Ext}^1(M,R) = 0$.

**Remark 2.1.9.** Let us add few comments to this last theorem. The system represented by $P(D)$ (and the module $M$) will be said determined if $\operatorname{Hom}(M,R) = \operatorname{Ext}^0(M,R) = 0$ while it will be said overdetermined if it is determined and $\operatorname{Ext}^1(M,R) = 0$. A determined system is characterized by the fact that it has no distribution solution with compact support. An overdetermined system is such that any solution defined in a neighborhood of the boundary of a convex region has a unique extension to the whole region.

The main difficulty for the use of Theorem 2.1.14 is to provide conditions under which a given system is overdetermined. The removability of compact singularities of hyperfunction solutions (in the case of systems represented by a matrix of operators) has been discussed in [134], where a sufficient condition on $[P_{ij}(D)]$ is given. Unfortunately, this condition (Theorems 4.1 and 4.3 in [134]) involves the minors of the matrix $P$ and, since it is quite strong, it does not apply, for example, to the case of the Cauchy–Fueter system that we will study in the next chapter. This extra condition was suggested to the authors of [134] by the need of considering matrices of entire functions but while relevant in that case, it would prevent us from applying the result to a large class of systems. The most suitable condition applicable in our case is the request that the cokernel of the map is torsion free, as proved in the following result.

**Theorem 2.1.15.** *Let $P = [P_{ij}]$ be a $r_1 \times r_0$ matrix of polynomials in $n$ variables with complex coefficients, $1 \le r_0 \le r_1$, $r_1 \ge 2$, such that coker(P) is torsion free. Let $f = (f_1, \dots, f_{r_0})$ be a vector of hyperfunctions on $\mathbb{R}^n$ such that*

$$\sum_{j=1}^{r_0} P_{ij}(D) f_j = 0, \qquad i = 1, \dots, r_1,$$

*on $\mathbb{R}^n \setminus K$, for $K$ a compact convex subset in $\mathbb{R}^n$. Then there exists a vector $f^* = (f_1^*, \dots, f_{r_0}^*)$ of hyperfunctions on $\mathbb{R}^n$ such that*

$$f_j^* = f_j \quad \text{on} \quad \mathbb{R}^n \setminus K, \quad j = 1, \dots, r_0$$

*and*

$$\sum_{j=1}^{r_0} P_{ij}(D) f_j^* = 0, \quad i = 1, \dots, r_1,$$

*on all of $\mathbb{R}^n$.*

*Proof.* Consider an $r_0$-tuple $f = (f_1, \dots, f_{r_0})$ of hyperfunctions on $\mathbb{R}^n \setminus K$, for $K$ a compact convex set. Suppose that $f$ satisfies the $r_1 \times r_0$ system of linear partial differential equations with constant coefficients

$$\sum_{j=1}^{r_0} P_{ij}(D) f_j = 0, \qquad i = 1, \dots, r_1.$$

In view of the flabbiness of the sheaf $\mathcal{B}$, there exist hyperfunctions $\tilde{f}_j \in \mathcal{B}(\mathbb{R}^n)$ such that $\tilde{f}_j = f_j$ on $\mathbb{R}^n \setminus K$. Define now

$$g_i = \sum_{j=1}^{r_0} P_{ij}(D)\tilde{f}_j, \qquad i = 1, \dots, r_1.$$

Clearly, $g_i \in \mathcal{B}_K(\mathbb{R}^n)$, and if $Q = [Q_{jt}]$ is now a matrix which generates the module of relations of the rows of $P$, then

$$\sum_{t=1}^{r_1} Q_{jt}(D)g_t = 0, \qquad j = 1, \dots, r_0.$$

A result of Andreotti–Nacinovich [8] (but mostly due to Ehrenpreis–Malgrange) shows that if coker$(P)$ is torsion free, then there exist entire functions $k_1, \dots, k_{r_0}$ such that

$$\hat{g}_i = \sum_{j=1}^{r_0} P_{ij}k_j, \qquad i = 1, \dots, r_1.$$

But since the $P_{ij}$ are polynomials, and since the growth of $\hat{g}_i$ is given by

$$|\hat{g}_i(z)| \leq A_\varepsilon \exp(H_K(z) + \varepsilon|z|),$$

since $\mathcal{B}_K(\mathbb{R}^n) \cong [\mathcal{O}(K)]'$, where $H_K$ is the supporting function on $K$, one can use the Ehrenpreis–Malgrange division lemma (see e.g. [68], Lemma 1.2 and Corollary 1.3) to conclude the existence of elements $h_j^* \in \mathcal{B}_K(\mathbb{R}^n)$, $j = 1, \dots, r_0$, such that $\hat{h}_j^* \in \widehat{\mathcal{B}_K(\mathbb{R}^n)}$ and

$$\hat{g}_i = \sum_{j=1}^{s} P_{ij}\hat{h}_j^*, \qquad i = 1, \dots, r_1.$$

If we now define

$$f_j^* := \tilde{f}_j - h_j^*, \qquad j = 1, \dots, r_0$$

we have that $f_j^*$ are well defined hyperfunctions on $\mathbb{R}^n$, such that

$$\sum_{j=1}^{r_0} P_{ij}(D)f_j^* = 0$$

everywhere. Moreover, $f_j^* - \tilde{f}_j = f_j$ on $\mathbb{R}^n \setminus K$. □

The relevance of this last theorem lies in the verifiability of the hypothesis on the torsion of coker$(P)$. More specifically, if $R$ is a polynomial ring in $n$ variables over $\mathbb{C}$, we can easily establish conditions under which a map

$$R^s \xrightarrow{A} R^q$$

given by a matrix $A$ has a torsion free cokernel.

Let $B$ be the matrix whose rows form a generating set for the syzygy module of the rows of $A$. It is readily seen that the torsion submodule of coker$(A)$ is $\ker(B)/\langle A\rangle$. Using techniques from Gröbner basis theory, $\ker(B)/\langle A\rangle$ can be explicitly computed in the case in which $R$ is a polynomial ring over a field. But this requires, in addition to computing the syzygies of the rows of $A$, the computation of the syzygies of the columns of $B$ and determining whether this module is equal to the module generated by the columns of $A$ (this is equivalent to checking if the short exact sequence given by the matrices $A$ and $B$ is exact).

However, it is possible to give a very simple condition that guarantees the exactness, under the hypothesis that $A$ is of maximal rank.

Let $A$ be a $q \times s$ matrix over a Unique Factorization Domain (UFD) $R$, $q \geq s$ (obviously the ring $\mathbb{C}[z_1, \dots, z_n]$ is a UFD, see for example [70]).

**Lemma 2.1.1.** *Suppose that $A$ has maximal rank $s$. Then coker$(A)$ is torsion free if and only if the $s \times s$ minors of $A$ are relatively prime.*

*Proof.* We denote the columns of $A$ by $\mathbf{a}_1, \dots, \mathbf{a}_s$. For a vector $\mathbf{a}$ in $R^q$, we denote by gcd$(\mathbf{a})$ a greatest common divisor of the coordinates of $\mathbf{a}$ (if the coordinates are relatively prime, we assume that gcd$(\mathbf{a}) = 1$).
It is convenient to use the notation and some basic tools from exterior algebra (see, for example [9]). We consider the vector $\beta = \mathbf{a}_1 \wedge \dots \wedge \mathbf{a}_s$, let $d = \gcd(\beta)$, and write $\beta = d\alpha$. So we have to prove that coker$(A)$ is torsion free if and only if $d = 1$.

Let us first assume that $d = 1$. Let $\mathbf{a} \in R^q$ and assume that $c \in R$ is such that $c \neq 0$ and $c\mathbf{a} \in \langle A\rangle$. Choose $c_1, \dots, c_s \in R$ such that

$$c\mathbf{a} = c_1\mathbf{a}_1 + \cdots + c_s\mathbf{a}_s.$$

Then for all $j = 1, \dots, s$, we have

$$c\left(\mathbf{a} \wedge \bigwedge_{i\neq j} \mathbf{a}_i\right) = c\mathbf{a} \wedge \bigwedge_{i\neq j} \mathbf{a}_i = \pm c_j\beta = \pm c_j\alpha.$$

Thus $c$ divides $c_j$, since gcd$(\alpha) = 1$. Therefore $\mathbf{a} \in \langle A\rangle$, and so coker$(A)$ is torsion free.

Conversely, assume that $d$ has a prime divisor $p$. We use $\bar{v}$ to denote reduction mod $p$ whenever $v$ is a vector in $R^q$ or an element of $R$. Since $\mathbf{a}_1 \wedge \dots \wedge \mathbf{a}_s = d\alpha$, we have that

$$\bar{\mathbf{a}}_1 \wedge \dots \wedge \bar{\mathbf{a}}_s = \bar{0}.$$

Therefore the vectors $\bar{\mathbf{a}}_1, \dots, \bar{\mathbf{a}}_s$ are linearly dependent over the quotient field of $R/pR$. Thus there exist $c_1, \dots, c_s \in R$, not all divisible by $p$, such that

$$\bar{c}_1\bar{\mathbf{a}}_1 + \cdots + \bar{c}_s\bar{\mathbf{a}}_s = \bar{0}.$$

So there is a vector $\mathbf{a} \in R^q$ such that

$$p\mathbf{a} = c_1\mathbf{a}_1 + \cdots + c_s\mathbf{a}_s \in \langle A \rangle.$$

However $\mathbf{a} \notin \langle A \rangle$, since at least one $c_i$ is not divisible by $p$ and the vectors $\mathbf{a}_1, \dots, \mathbf{a}_s$ are linearly independent over the quotient field of $R$. Therefore $\mathrm{coker}(A)$ has torsion elements. □

## 2.2 The Ehrenpreis–Palamodov Fundamental Principle

In this section we discuss the culminating and most important result of this algebraic theory, independently obtained and proved by Ehrenpreis and Palamodov, the so-called Fundamental Principle of Ehrenpreis–Palamodov. This principle allows one to write the solution of a homogeneous system of partial differential equations with constant coefficients in integral form. It is a very far reaching generalization of Euler's Fundamental Principle.

Consider the ordinary homogeneous linear equation of order $n$ with constant coefficients:

$$a_n \frac{d^n}{dx^n} f(x) + a_{n-1} \frac{d^{n-1}}{dx^{n-1}} f(x) + \cdots + a_0 f(x) = 0, \tag{2.22}$$

where $a_j \in \mathbb{R}$, $a_n \neq 0$. The characteristic polynomial associated to the differential equation is defined by

$$\chi(\lambda) := a_n\lambda^n + a_{n-1}\lambda^{n-1} + \cdots + a_0.$$

Suppose that the characteristic equation $\chi(\lambda) = 0$ has $n$ distinct solutions $\lambda_1, \dots, \lambda_n$. Then a fundamental set of solutions for (2.22) is given by

$$\{e^{\lambda_j x}\}, \quad j = 1, \dots, n$$

and the general integral of the differential equation (2.22) is given by

$$f(x) = \sum_{j=1}^{n} c_j e^{\lambda_j x}$$

where $c_j$, $j = 1, \dots, n$, are arbitrary constants. To verify that the functions $e^{\lambda_j x}$ are linearly independent we construct the Wronskian $W(x)$ of the functions $\{e^{\lambda_j x}\}$

$$W(x) = F(\lambda_1, \dots, \lambda_n) \cdot \exp\left(x \sum_{j=1}^{n} \lambda_j\right)$$

where $F(\lambda_1, \dots, \lambda_n)$ is the Vandermonde determinant, which is not zero if and only if all the $\lambda_j$'s are distinct. In the general case, however, not all the roots of the characteristic equation are distinct, so we have the solutions $\lambda_1, \dots, \lambda_r$, $r \leq n$, with $\lambda_j$ of multiplicity $\mu_j$. Euler's principle establishes that a fundamental set of solutions can still be described explicitly as in the next theorem.

**Theorem 2.2.1.** *Suppose that the characteristic equation $\chi(\lambda) = 0$ has roots $\lambda_j$ with multiplicity $\mu_j$, $j = 1, \dots, r$, with $\sum_{j=1}^{r} \mu_j = n$. Then, a fundamental set of solutions for the differential equation (2.22) is given by*

$$e^{\lambda_1 x}, \dots, x^{\mu_1 - 1} e^{\lambda_1 x}; \quad e^{\lambda_2 x}, \dots, x^{\mu_2 - 1} e^{\lambda_2 x}; \quad \dots; \quad e^{\lambda_r x}, \dots, x^{\mu_r - 1} e^{\lambda_r x}$$

*so the general integral is*

$$f(x) = \sum_{j=1}^{r} c_j(x) e^{\lambda_j x}$$

*with $c_j(x)$, $j = 1, \dots, r$ arbitrary polynomials of degree $\mu_j - 1$.*

For a long time, mathematicians debated the question of whether exponential solutions played the same role in the case of linear partial differential equations with constant coefficients. On a parallel track, they had realized that exponentials were the building blocks for solutions of rather large classes of equations. For example, a periodic function $f$ of period $2\pi$ can be written as

$$f(x) = \sum_{j=-\infty}^{+\infty} c_j e^{j(ix)},$$

and, moreover, it can be thought of as the solution to a very simple convolution equation:

$$f(x + \pi) - f(x - \pi) = (\delta_\pi - \delta_{-\pi}) * f(x) = 0.$$

This consideration led Schwartz to prove, in [174], that if $f$ is a $\mathcal{C}^\infty$ function, solution of a convolution equation

$$\mu * f = 0$$

where $\mu \in \mathcal{E}'$, then $f$ can be given an exponential representation in terms of the exponentials $e^{\lambda x}$, for $\lambda$ a solution of the analytic equation

$$\mathcal{F}\mu(\lambda) = 0.$$

In the 1950s, the best analysts, including Malgrange, Ehrenpreis, Palamodov, and Hörmander, were working on these issues and finally proved [124] that, in fact, exponential solutions were dense in the space of solutions of linear partial differential equations with constant coefficients. But what the Fundamental Principle proves is a much stronger statement: not only exponential solutions are dense in the space of solutions, but an explicit formula that mimics Euler's one can be obtained.

In order to state the Fundamental Principle, we need to introduce some preliminary notations. The crucial point of the theory is that, in order for the exponential representation theorem to hold, one has to consider differential equations in spaces whose dual are isomorphic to spaces of entire functions with suitable growth conditions. Thus, a key ingredient is the space $\mathcal{O}(\mathbb{C}^n)$ of entire functions on $\mathbb{C}^n$. The spaces for which the Fundamental Principle holds, originally introduced by Ehrenpreis, are called Localizable Analytically Uniform spaces (LAU-spaces) and will be presented according to Hansen [84] terminology. We will introduce them after some definitions.

Let $U$ be an open set in $\mathbb{R}^n$. A function $\phi : U \to [-\infty, +\infty)$ is called upper semicontinuous if for every $\varepsilon > 0$ there is a neighborhood $B$ of $x_0$ such that $\phi(x) < \phi(x_0) + \varepsilon$ for all $x \in B$ when $\phi(x_0)$ is finite, while when $\phi(x) = -\infty$ we require that for every positive number $k$ there is a neighborhood $B$ of $x_0$ such that $\phi(x_0) < -k$ for all $x \in B$.

An upper semicontinuous function $\phi : U \to [-\infty, +\infty)$ is called subharmonic if

$$\phi(x_0) \leq \frac{1}{\omega_{n-1}} \int_{\Sigma_{n-1}} \phi(x_0 + rt)\, dt$$

whenever $\{x \in \mathbb{R}^n \ : \ |x - x_0| \leq r\} \subset U$, where $\omega_{n-1}$ is the surface area of the unit sphere $\Sigma_{n-1}$ in $\mathbb{R}^n$ and $dt$ is the element of surface area on $\Sigma_{n-1}$. We recall that a function $\phi : \mathbb{C}^n \to [-\infty, +\infty)$ is said plurisubharmonic if it is upper semicontinuous and if its restriction to every complex line is subharmonic.

Let $\mathcal{K}$ be a nonempty family of positive continuous functions on $\mathbb{C}^n$. We define the space

$$A_{\mathcal{K}} := \{\ u \in \mathcal{O}(\mathbb{C}^n) \mid \frac{|u(z)|}{k(z)} \to 0, \text{ as } |z| \to +\infty, \ \ \forall k \in \mathcal{K}\},$$

with the seminorms

$$p_k(u) := \sup_{z \in \mathbb{C}^n} \frac{|u(z)|}{k(z)}.$$

The pair $(A_{\mathcal{K}}, \mathcal{T}_{p_k})$ (where $\mathcal{T}_{p_k}$ denotes the topology induced by the family $\{p_k\}$) is a Hausdorff locally convex space.

We denote by $\mathcal{M}(\mathcal{K})$ the set of all nonnegative upper semicontinuous functions $m$ on $\mathbb{C}^n$ such that $m(z)/k(z)$ is bounded for any $k \in \mathcal{K}$.

Observe that $\mathcal{M}(\mathcal{K})$ describes the bounded sets in $A_{\mathcal{K}}$ in the sense that, for every bounded set $B$ in $A_{\mathcal{K}}$, there exists a function $m \in \mathcal{M}(\mathcal{K})$ such that

$$B \subseteq \{u \in A_{\mathcal{K}} \ : \ |u(z)| \leq m(z), \ \ z \in \mathbb{C}^n\ \}.$$

Ehrenpreis called the set $\mathcal{M}(\mathcal{K})$ a Bounded Analytic Uniform structure or, briefly, a BAU-structure. To prove the Fundamental Principle, it is necessary to require additional conditions on the family $\mathcal{K}$. In fact, the kernel of a linear differential operator with constant coefficients $P(D)$ is dual, via the Fourier–Laplace transform, to a quotient space of $A_{\mathcal{K}}$. The goal is to find in every coset a representative satisfying good bounds. Such representative can be constructed locally and afterwards globally using the theory of cohomology with bounds, see [90], when $\mathcal{K}$ satisfies the conditions we will introduce below.

**Definition 2.2.1.** *Let $\mathcal{K}$ be a nonempty set of positive continuous functions on $\mathbb{C}^n$. We say that $\mathcal{K}$ is an analytically uniform structure (AU-structure) if for any $k \in \mathcal{K}$ there exists $k' \in \mathcal{K}$ such that*

$$k'(z+z')(2+|z|^2) \leq k(z), \quad \text{for all } z, z' \in \mathbb{C}^n, \ |z'| \leq 1.$$

**Definition 2.2.2.** *Let $\mathcal{K}$ be a nonempty set of positive continuous functions on $\mathbb{C}^n$. We say that $\mathcal{K}$ is a localizable analytically uniform structure (LAU-structure) if it is an AU-structure and if for any $k \in \mathcal{K}$ there exists $k' \in \mathcal{K}$ such that for any $m \in \mathcal{M}(\mathcal{K})$ with $m(z) \leq k'(z)$, $z \in \mathbb{C}^n$, there exists a plurisubharmonic function $\phi$ defined on $\mathbb{C}^n$ with $\exp\phi \in \mathcal{M}$ and*

$$m(z) \leq \exp\phi(z) \leq k(z), \quad \forall z \in \mathbb{C}^n.$$

**Definition 2.2.3.** *Let $W$ be a locally convex space. Assume that there exists an AU- (respectively, LAU-) structure $\mathcal{K}$ and a componentwise continuous bilinear form $\langle\cdot,\cdot\rangle$ on $W \times A_{\mathcal{K}}$ such that the map from $W$ to the strong dual of $A_{\mathcal{K}}$ given by*

$$\omega \to \langle\omega,\cdot\rangle$$

*is a topological isomorphism. Then we call $W$ an AU (respectively, LAU) space.*

Among the linear topological spaces which have the LAU structure there are (see [68], Chapter 5 and [84]):

- the space of polynomials in $n$ variables with complex coefficients,
- the space $\mathcal{O}(\Omega)$ of holomorphic functions on an open convex set $\Omega \subseteq \mathbb{C}^n$,
- the space $\mathrm{Exp}(\mathbb{C}^n)$ of entire functions of exponential type,
- the space $\mathcal{E}(\Omega)$ of $C^\infty$ functions on an open convex set $\Omega$ in $\mathbb{R}^n$,
- the space of distributions $\mathcal{D}'(\Omega)$ where $\Omega$ is an open convex set in $\mathbb{R}^n$.

For other examples of LAU spaces see, e.g., [202].

**Definition 2.2.4.** *Let $V_j$ be algebraic varieties and let $\partial_j$ be differential operators with polynomial coefficients for $j = 1, \ldots, p$. Then we call*

$$V = \{V_1, \partial_1; V_2, \partial_2; \ldots; V_p, \partial_p\}$$

*a multiplicity variety.*

**Theorem 2.2.2.** *Let $I$ be an ideal in the ring $R$ of polynomials in $n$ variables with complex coefficients. Then there exists a multiplicity variety $V$ such that a polynomial $F$ belongs to $I$ if and only if $\partial_j F_{|V_j} = 0$ for every $j = 1, \dots, p$.*

**Remark 2.2.1.** Note that this result is nothing but a version of the famous Nullstellensatz. As is well known, the Nullstellensatz considers the variety $V(I) = \{z \in \mathbb{C}^n \mid P(z) = 0 \text{ for all } P \in I\}$, and shows that if a polynomial $F$ vanishes on $V(I)$, then $F$ must belong to the radical of $I$. The result just quoted is clearly much stronger and provides a nice differential characterization of the ideal $I$.

**Remark 2.2.2.** From a historical point of view, it is interesting to note that when Ehrenpreis was proving his Fundamental Principle, he was convinced that a theorem such as Theorem 2.2.2 had already been established. He realized this was not the case (and he needed to prove the result) after a conversation with Zariski [69].

For the sake of simplicity, we will now state the Fundamental Principle for the case of a single differential equation. This will allow us to provide a reasonably concise sketch of its proof. However, as is well known, the result holds for the more general case of systems of differential equations represented by an $r_1 \times r_0$ matrix $P(D)$. In more recent times, the Fundamental Principle has also been generalized to rectangular systems of a very large class of convolution equations. The statement in all these cases follows the obvious modifications, while the proofs become significantly more involved when convolution equations are concerned.

**Theorem 2.2.3.** (Fundamental Principle). *Let $W$ be a LAU-space with LAU-structure $\mathcal{K}$ and let $P(D) : W \to W$ be a linear differential operator with constant coefficients. Then there exists a multiplicity variety*

$$V = \{V_1, \partial_1; V_2, \partial_2; \dots ; V_p, \partial_p\}$$

*such that an element $f \in W$ is a solution of*

$$P(D)f = 0 \tag{2.23}$$

*if and only if*

$$f(x) = \sum_{j=1}^{p} \int_{V_j} \partial_j \exp(iz, x) \ \ d\mu_j(z)$$

*where $d\mu_j(z)$ are Radon measure supported in $V_j$ which satisfy, for some $k \in \mathcal{K}$,*

$$\int_{V_j} k(z)|d\mu_j|(z) < \infty, \quad j = 1, ..., p. \tag{2.24}$$

The representation above must be correctly interpreted. So, for example, if we are working with distributions the representation must be interpreted in the functional sense.

The proof of the Fundamental Principle is quite complicated and it is based on two theorems (due to Ehrenpreis [68] and Palamodov [142]) to be stated without proof.

**Definition 2.2.5.** *For every plurisubharmonic function $\phi$ defined on $\mathbb{C}^n$ and for a given positive constant $M$ we define the function*

$$\phi_M(z) = \sup\{\ \phi(z+z')\ :\ z' \in \mathbb{C}^n\ \ |z'| \leq M\ \}, \qquad z \in \mathbb{C}^n.$$

**Theorem 2.2.4.** (Division Theorem). *Let $P$ be a polynomial. Then there exist a multiplicity variety $V$ and a positive constant $M$ such that the following condition holds: for every entire function $f : \mathbb{C}^n \to \mathbb{C}$ it is*

$$\partial_j(Pf)\Big|_{V_j} = 0, \quad j = 1, \dots, p;$$

*conversely, if an entire function $g : \mathbb{C}^n \to \mathbb{C}$ satisfies*

$$\partial_j g\Big|_{V_j} = 0, \quad j = 1, \dots, p,$$

*then there exists an entire function $f : \mathbb{C}^n \to \mathbb{C}$ with*

$$g = Pf, \quad in \quad \mathbb{C}^n$$

*and the estimate*

$$\sup_{z\in\mathbb{C}^n} \frac{|f(z)|e^{-\phi_M(z)}}{(2+|z|^2)^M} \leq \sup_{z\in\mathbb{C}^n} |g(z)|e^{-\phi(z)}$$

*holds for every plurisubharmonic function $\phi$ on $\mathbb{C}^n$, where $\phi_M$ is as in Definition 2.2.5.*

Essentially, the division theorem says that if a function belongs to the ideal $\langle P\rangle$ generated by the polynomial $P$, then its "restriction" to the multiplicity variety is zero and, conversely, if the restriction to $V$ of a function $g$ is zero then $g \in \langle P\rangle$. Moreover, if $g \in \mathcal{A}_\mathcal{K}$, then $g = pv$ with $v \in \mathcal{A}_\mathcal{K}$.

**Theorem 2.2.5.** (Extension Theorem). *Let $P$ be a polynomial. Then there exist a multiplicity variety $V$ and a positive constant $M$ such that for any entire function $g : \mathbb{C}^n \to \mathbb{C}$, there exists an entire function $f : \mathbb{C}^n \to \mathbb{C}$ such that*

$$\partial_j(f-g)\Big|_{V_j} = 0, \quad j = 1, \dots, p$$

*and the estimate*

$$\sup_{z\in\mathbb{C}^n} \frac{|f(z)|e^{-\phi_M(z)}}{(2+|z|^2)^M} \leq \max_j \sup_{z\in V_j} |\partial_j g(z)|e^{-\phi(z)}$$

*holds for every plurisubharmonic function $\phi$ on $\mathbb{C}^n$ where $\phi_M$ is as in Definition 2.2.5.*

The extension theorem essentially states that if a given entire function $g$ satisfies assigned growth conditions on $V$, then it can be extended to a function in $\mathcal{A}_{\mathcal{K}}$. We now want to give an idea of how these two theorems can be used to prove the Fundamental Principle.

*Sketch of the proof of the Fundamental Principle.* We first observe that $f \in W$ satisfies (2.23) if and only if it defines, via the pairing $\langle \cdot, \cdot \rangle$, a continuous linear functional on the space $A_{\mathcal{K}} / \overline{PA_{\mathcal{K}}}$. Let $V$ be the multiplicity variety associated to the equation and denote by $\mathcal{C}_{\mathcal{K}}$ the space of the $p$-tuples $h = (h_1, \ldots, h_p)$ of continuous complex valued functions $h_j$ on $V_j$ satisfying

$$\frac{|h_j(z)|}{k(z)} \to 0, \quad \text{as } |z| \to +\infty$$

for $z \in V_j$, for every $k \in \mathcal{K}$. We introduce a topology on the space $\mathcal{C}_{\mathcal{K}}$ via the family of seminorms

$$p_k(h) = \max_j \sup_{z \in V_j} \frac{|h_j(z)|}{k(z)}.$$

Consider now the linear map $N : A_{\mathcal{K}} \to \mathcal{C}_{\mathcal{K}}$ defined by

$$N(f) := (\partial_1 f_{|V_1}, \ldots, \partial_p f_{|V_p}). \tag{2.25}$$

The map $N$ is well defined and continuous since $\mathcal{K}$ is an AU-structure. We have to show that the kernel of $N$ coincides with $\overline{PA_{\mathcal{K}}}$ (which will turn out to be equal to $PA_{\mathcal{K}}$) and that $N$ is a homomorphism, i.e., since the continuity of $N$ is clear, we must show that the inverse of

$$\tilde{N} : \frac{A_{\mathcal{K}}}{\ker N} \to N(A_{\mathcal{K}}) \tag{2.26}$$

is continuous.

Once we have proved these two properties, it will follow that $N$ identifies $A_{\mathcal{K}} / \overline{PA_{\mathcal{K}}}$ with a subspace of $\mathcal{C}_{\mathcal{K}}$ and the statement will follow applying the Hahn–Banach and the Riesz representation theorems.

**Lemma 2.2.1.** *Let $N$ be the operator defined in (2.25). Then we have*

$$\ker N = \overline{PA_{\mathcal{K}}} = PA_{\mathcal{K}}.$$

*Proof.* From the first part of the division theorem it follows that $\overline{PA_{\mathcal{K}}}$ is contained in the kernel of $N$. On the other hand, if $g \in A_{\mathcal{K}}$ is such that $Ng = 0$, we can use the division theorem to find an entire function $v : \mathbb{C}^n \to \mathbb{C}$ such that $g(z) = P(z)v(z)$. But since $g \in A_{\mathcal{K}}$ and $\mathcal{K}$ is localizable, there exists a plurisubharmonic function $\phi$ with $\exp \phi \in \mathcal{M}_{\mathcal{K}}$ and $|g(z)| \leq \exp \phi(z)$, $z \in \mathbb{C}^n$. The estimate of the division theorem together with the fact that $\mathcal{K}$ is an AU-structure, yields that $v \in A_{\mathcal{K}}$, so $PA_{\mathcal{K}} = \ker N$ and, in particular, it is closed. This proves the lemma. □

**Lemma 2.2.2.** *The operator $N$ defined in (2.25) is a homomorphism.*

*Proof.* We have to prove that $\tilde{N}^{-1}$ is a continuous operator, that is, we have to show that for any neighborhood $U$ of the origin in $A_{\mathcal{K}}/PA_{\mathcal{K}}$ it is possible to find a neighborhood $V$ of the origin in $N(A_{\mathcal{K}})$ such that $\tilde{N}^{-1}(V) \subset U$. Since a fundamental system of neighborhoods for the origin in $A_{\mathcal{K}}$ is given by the family of sets

$$N_k = \{\psi \in A_{\mathcal{K}} : \ |\psi(z)| \leq k(z), \quad \text{for all } z \in \mathbb{C}^n\}$$

indexed on $k \in \mathcal{K}$, we have to show that for every $k \in \mathcal{K}$ we can find $k'' \in \mathcal{K}$ such that for any $g \in A_{\mathcal{K}}$ with $|\partial_j g(z)|_{V_j}| \leq k''(z)$ (for $j = 1, \ldots, p$) there exists an $f \in A_{\mathcal{K}}$ with $|f(z)| \leq k(z)$, and $Nf = Ng$. Let $k \in \mathcal{K}$ be given. Choose, for the $M$ given by the extension theorem, $k' \in \mathcal{K}$ such that

$$k'(z + z') \leq \frac{k(z)}{(2 + |z|^2)^M}, \quad z, z' \in \mathbb{C}^n, \quad |z'| \leq M. \tag{2.27}$$

Now we choose $k'' \in \mathcal{K}$, depending on $k'$, as in the definition of LAU-structure, and let $g \in A_{\mathcal{K}}$ be such that

$$\left|\partial_j g(z)|_{V_j}\right| \leq k''(z), \quad j = 1, \ldots, p.$$

By the LAU-structure of $\mathcal{K}$ we can find a plurisubharmonic function $\phi$ with $\exp\phi \in \mathcal{M}$, such that

$$|\partial_j g(z)|_{V_j}| \leq \exp\phi(z), \quad j = 1, \ldots, p,$$

and (thanks to the description of the BAU-structure) we have

$$\exp\phi(z) \leq k'(z), \quad z \in \mathbb{C}^n. \tag{2.28}$$

Now the statement follows immediately by the application of the extension theorem which gives the existence of an entire function $f$ such that $N(f - g) = 0$ and

$$|f(z)| \leq (2 + |z|^2)^M \exp\phi_M(z).$$

Using the AU-structure, we get $f \in A_{\mathcal{K}}$ while (2.27) and (2.28) imply $|f(z)| \leq k(z)$, for all $z \in \mathbb{C}^n$. □

**Example 2.2.1.** Let $n = 1$; then the multiplicity variety is particularly easy to describe. The varieties $V_j$ reduce to points in $\mathbb{C}$ and each operator $\partial_j$ can be identified with an integer representing a differentiation order. Thus, in $\mathbb{C}$, a multiplicity variety is a set of pairs $(z_k, m_k)$, with $z_k \in \mathbb{C}$ and $m_k \in \mathbb{N}$. If $f \in \mathcal{O}(\mathbb{C})$, its multiplicity variety will be defined by

$$V = \{(z_k, m_k) : f(z_k) = 0, \quad m_k = \text{multiplicity of } z_k\}.$$

We can now associate to such a multiplicity variety a closed ideal:

$$I(V) = \{\psi \in \mathcal{O}(\mathbb{C}) : \ z_k \text{ is a zero of } \psi \text{ with multiplicity at least } m_k\}$$

and it is possible to prove that $\mathcal{O}(V) := \mathcal{O}(\mathbb{C})/I(V)$ can be identified with the space of all sequences $\{a_{k,\ell}\}$, $0 \le \ell \le m_k$, $k \in \mathbb{N}$, which hence describe the analytic functions on $I(V)$. Finally the restriction

$$\rho_V : \mathcal{O}(\mathbb{C}) \to \mathcal{O}(V)$$

is simply defined by

$$\rho_V(\psi) = \left\{\frac{d^\ell \psi}{dz^\ell}(z_k)\right\}_{0\le\ell\le m_k,\ k\in\mathbb{N}}.$$

Consider now the ordinary differential equation of order $m$

$$\sum_{j=0}^{m} a_j D^j f(x) = 0, \tag{2.29}$$

where $a_j \in \mathbb{C}$, $a_m \neq 0$ and $D = -i\, d/dx$. In this case

$$V = \{z \in \mathbb{C} : \ P(z) = \sum_{j=0}^{m} a_j z^j = 0\}$$

is clearly discrete. Thus, the $V_j$ and the $\partial_j$ which appear in the Fundamental Principle are simply the points of $V$, and the operators are ordinary derivatives of a suitable order. As a consequence, if $f$ is a $C^\infty$ solution of (2.29), it can always be represented as

$$f(x) = \sum_j c_{\alpha_j}(x) e^{\alpha_j x},$$

where the summation is extended to all the $\alpha_j$ which are roots of the equation

$$\sum_{j=0}^{m} a_j \alpha^j = 0 \tag{2.30}$$

and the $c_{\alpha_j}$ are polynomials of degree less than the multiplicity of $\alpha_j$ as the root of the algebraic equation (2.30).

In order to fully state the Ehrenpreis–Palamodov Fundamental Principle for homogeneous systems of linear partial differential with constant coefficients, we need to introduce the notion of multiplicity variety associated to a system (see [64], [142] for more details).

**Theorem 2.2.6.** *Let $P$ be a $r_1 \times r_0$ matrix of polynomials. There exists a positive integer $s$, algebraic varieties $V_j$, and $r_0$-tuples $\partial_j = (\partial_{j1}, \dots, \partial_{jr_0})$ of linear differential operators with polynomial coefficients, $j = 1, \dots, s$, such that $F \in \langle P^t R^{r_1} \rangle$ if and only if $\partial_j F_{|V_j} = 0$ for all $j = 1, \dots, s$.*

**Definition 2.2.6.** *We will say that $V = \{V_1, \partial_1; \dots ; V_s, \partial_s\}$ is a multiplicity variety associated to the $r_1 \times r_0$ matrix $P$.*

We state, without proof for which we refer the reader to [64], [142], the Ehrenpreis–Palamodov Fundamental Principle for systems.

**Theorem 2.2.7.** *Let $W$ be a LAU-space with a LAU-structure $\mathcal{K}$ and let $P(D) : W^{r_0} \to W^{r_1}$ be a matrix of linear partial differential operators with constant coefficients, let $V$ be its multiplicity variety. Then an element $f \in W^{r_0}$ is a solution to*

$$P(D)f = 0$$

*if and only if the $k$-th component $f_k(x)$ of $f(x)$ can be written as*

$$f_k(x) = \sum_{j=1}^{s} \int_{V_j} \partial_{jk} \exp(iz, x)\ d\mu_j(z), \tag{2.31}$$

*where $d\mu_j(z)$ is a Radon measure with support in $V_j$ such that*

$$\int_{V_j} k(z) |d\mu_j|(z) < \infty, \quad j = 1, \dots, s \tag{2.32}$$

*for some $k \in \mathcal{K}$.*

The differential operators $\partial_{jk}$ introduced in the theorems above are known as the noetherian operators associated to the system of differential equations. In the general case, such operators have polynomial coefficients. In fact, it suffices to consider the matrix

$$P^t = (z_1^2, z_2^2, z_2 - z_1 z_3).$$

One can show that the module $\langle P^t R^3 \rangle$ is primary, but there exists no noetherian operator with constant coefficients for this matrix. This example is due to Palamodov and is clearly explained in [142], page 183.

As in the case of the previous results, the Fundamental Principle is difficult to apply because the multiplicity variety associated to a polynomial may be very difficult to construct explicitly. It is even more difficult to construct concrete examples when one considers several equations, or, in the most general case, when one considers rectangular systems of differential equations.

In some cases, it is possible to explicitly construct their multiplicity varieties, and therefore to write down explicitly the integral representation of the solutions (see for example Palamodov's work [143]).

## 2.3 The Fundamental Principle for hyperfunctions

In this section we show that the Fundamental Principle holds also for hyperfunction solution of homogeneous systems of partial differential equations with constant coefficients. This is an important, and somewhat surprising, fact because the spaces of hyperfunctions (even on convex sets) are not LAU-spaces. The proof was originally given by Kaneko, see [96], [97], and was later on simplified by Oshima in [141]. We will mention here this second approach based on the fact that hyperfunctions are sums of boundary values of holomorphic functions, and that the space of holomorphic functions on convex sets is a LAU-space.

**Theorem 2.3.1.** *Given an $r_1 \times r_0$ matrix $P(D)$ of partial differential operators with constant coefficients and the system $P(D)f = 0$ where $P(D): \mathcal{B}(\Omega)^{r_0} \to \mathcal{B}(\Omega)^{r_1}$ and $\Omega$ is a convex domain in $\mathbb{R}^n$, the functions $f$ in the kernel of $P(D)$ have components that can be represented by the integral (2.31) where the measures $d\mu_j$ satisfy the estimate*

$$\int_{V_j} \exp(-\varepsilon|z| + h_K(z))\, d\mu_j(z) < +\infty, \quad \forall \varepsilon > 0, \ \forall K \subset \Omega, \tag{2.33}$$

*where $h_K(z) := \sup_{x\in K} Re(ix, z)$. The integral (2.31) is considered in the sense of hyperfunctions.*

*Proof.* We will start by proving the theorem in the space of holomorphic functions. Let us set

$$U = \{u \in \mathbb{C}^n \mid Re(u) = (Re(u_1), \dots, Re(u_n)) \in \Omega\}$$

and $U_i = \{u \in U \mid \mathrm{Im}(u_i) \neq 0\}$. The sets $U$ and $U_i$ are Stein open sets in $\mathbb{C}^n$, so we can consider the coverings $\mathcal{U} = \{U, U_1, \dots, U_n\}$ and $\mathcal{U}' = \{U_1, \dots, U_n\}$ to get that

$$H^n(\mathcal{U}, \mathcal{U}', \mathcal{O}) = H^n_\Omega(U, \mathcal{O}) = \mathcal{B}(\Omega).$$

Let us denote by $\mathcal{R}$ the ring of linear partial differential operators with constant coefficients in $n$ variables. We have the free resolution

$$0 \longleftarrow \mathcal{R}^{r_0}/P^t(D)\mathcal{R}^{r_1} \longleftarrow \mathcal{R}^{r_0} \overset{P^t(D)}{\longleftarrow} \mathcal{R}^{r_1} \overset{P^t_1(D)}{\longleftarrow} \mathcal{R}^{r_2} \longleftarrow \dots.$$

We can define the double complex (see [79]):

$$K^{p,q} = C^p(\mathcal{U}, \mathcal{U}', \mathcal{O}^{r_q}), \qquad p, q \geq 0$$

where we use the coboundary operator

$$d' : C^p(\mathcal{U}, \mathcal{U}', \mathcal{O}^{r_q}) \to C^{p+1}(\mathcal{U}, \mathcal{U}', \mathcal{O}^{r_q})$$

and the operator

$$d'' : C^p(\mathcal{U}, \mathcal{U}', \mathcal{O}^{r_q}) \to C^p(\mathcal{U}, \mathcal{U}', \mathcal{O}^{r_{q+1}})$$

defined by

$$d'' = (-1)^p P_q(D),$$

and with $P_0(D) = P(D)$. By standard arguments on spectral sequences, see [79], and by the exactness of the free resolution, we have (using standard notation and recalling that $\mathcal{O}^P$ denotes the solution sheaf)

$$'E_1^{p,q} = \begin{cases} \mathcal{C}^p(\mathcal{U},\mathcal{U}',\mathcal{O}^P) & \text{if } p \leq n \text{ and } q = 0 \\ 0 & \text{if } p > n \text{ or } q \neq 0 \end{cases}$$

which follows from the fact that the system is solvable in a convex domain (see Theorem 2.1.1). Moreover we have

$$''E_1^{p,q} = \begin{cases} H^n(\mathcal{U},\mathcal{U}',\mathcal{O}^{r_q}) = (\mathcal{B}(\Omega))^{r_q} & \text{if } \ p = n \\ 0 & \text{if } \ p \neq n, \end{cases}$$

hence $E^{p,q} = 0$ when $p \neq n$ or $q \neq 0$. Finally,

$$E^{n,0} = \frac{\mathcal{C}^n(\mathcal{U},\mathcal{U}',\mathcal{O}^P)}{d'\mathcal{C}^{n-1}(\mathcal{U},\mathcal{U}',\mathcal{O}^P)} = \ker\{P(D) : (\mathcal{B}(\Omega))^{r_0} \to (\mathcal{B}(\Omega))^{r_1}\}.$$

Now we set $I = \{(\sigma_1, \dots, \sigma_n) \mid \sigma_i = 1 \text{ or } -1\}$ and

$$W_\sigma = \{u \in U \mid \sigma_i \max \operatorname{Im}(u_i) > 0, \quad 1 \leq i \leq n\}$$

for $\sigma \in I$. Recall that an element in $\mathcal{C}^n(\mathcal{U},\mathcal{U}',\mathcal{O}^P)$ is the set of $2^n$ tuples of vectors of holomorphic functions $\{F_\sigma(u)\}_{\sigma \in I}$ where $F_\sigma(u)$ is defined on $W_\sigma$ and satisfies $P(D)F_\sigma(u) = 0$. Therefore any $f \in \mathcal{B}(\Omega)^{r_0}$ satisfying $P(D)f = 0$ is represented by the cohomology class of $\{F_\sigma(u)\}_{\sigma \in I}$. By the Fundamental Principle 2.2.7, $F_\sigma(u)$ can be written in the form

$$(\prod_{i=1}^n \sigma_i) F_\sigma(u) = \sum_{j=1}^s \int_{V_j} \partial_j \exp(iz, x)\, d\mu_j(z)$$

and the measures $d\mu_j$ satisfy

$$\sum_{j=1}^s \int_{V_j} \exp(h_L(z)) |d\mu_j|(z) < \infty \tag{2.34}$$

for all $L \subset W_\sigma$, where given $\varepsilon > 0$ and $K \subset \Omega$, we set

$$L = \{u \in \mathbb{C}^n \mid \operatorname{Re}(u) \in K, \quad \sigma_i \operatorname{Im}(u_i) = \varepsilon, \text{ for } 1 \leq i \leq n\} \subset W_\sigma.$$

Obviously we have

$$\sup_{x \in K} \operatorname{Re}(ix, z) \leq \sup_{\zeta \in L} \operatorname{Re}(i\zeta, z) + \varepsilon |\zeta|,$$

hance the measures $d\mu_j$ satisfy the conditions (2.33). We now have to explain the meaning (2.31) in the sense of hyperfunctions. We can write $d\mu_j = \sum_{\sigma\in I} d\mu_{j\sigma}$ such that each $d\mu_{j\sigma}$ satisfies (2.33) and has its support in

$$\Gamma_{j\sigma} = \{\zeta \in \mathbb{C}^n \; |\mathrm{Re}(\sigma_j\zeta_j) \geq 0 \text{ for } 1 \leq j \leq n\}.$$

We set

$$G_\sigma(u) = (\prod_{i=1}^{n} \sigma_i) \sum_{j=1}^{s} \int_{V_j} \partial_j \exp(iz, x)\; d\mu_{j\sigma}(z)$$

and we obtain that $G_\sigma(u)$ is holomorphic in $W_\sigma$ and (2.31) is a vector $f(x)$ of hyperfunctions solution to $P(D)f = 0$, since $P(D)G_\sigma = 0$. To finish the proof we need to show that $\{F_\sigma(u)\}_{\sigma\in I}$ represents the same cohomology class of $\{G_\sigma(u)\}_{\sigma\in I}$ but this follows from the fact that they are congruent modulo $dC^{n-1}(\mathcal{U},\mathcal{U}',\mathcal{O}^P)$ in $C^n(\mathcal{U},\mathcal{U}',\mathcal{O}^P)$, see [141]. □

Berenstein, Kawai, Struppa in [19] have proved that the Fundamental Principle also holds for infinite order differential operators. In what follows, we will denote by $\mathscr{D}^\infty$ the ring of infinite order differential operators with constant coefficients. The space of symbols of these operators is the space $\mathrm{Expo}(\mathbb{C}^n)$ of functions of infraexponential type. As is well known among people working on the Fundamental Principle, it is impossible, in general, to prove the Principle for convolution equations (and infinite order differential operators are a special case) unless additional conditions are imposed. The reason for this difficulty is algebraic in nature and quite complicated to discuss in detail. Probably the most complete treatment of the problems which arise in this context is offered in [202]. In short, one can show (see [20] and [202]) that a version of the Fundamental Principle can be restored for a special (though large) class of systems which go under the name of slowly decreasing systems. That this is the case for systems of infinite order differential operators was first shown in [102], [202] and it was finally given a complete description in [18].

The condition of slow decrease for systems of infinite order differential operators is quite complicated, and will be given in two different definitions according to the number of operators considered. Let us begin with the general case.

**Definition 2.3.1.** *Let* $P = (P_1, \ldots, P_r)$, $1 \leq r \leq n$, $P_i \in\mathrm{Expo}(\mathbb{C}^n)$, $i = 1, \ldots, r$, *be such that*

$$\{\; z \in \mathbb{C}^n \;:\; P_1(z) = \ldots = P_r(z) = 0 \;\}$$

*is a complete intersection* $(n-r)$*-dimensional variety. We say that* $P$ *is slowly decreasing if there exist a family* $\mathcal{L} = \{L\}$ *of* $r$*-dimensional affine complex spaces with*

$$\bigcup_{L\in\mathcal{L}} L = \mathbb{C}^n,$$

*and there are constants* $C_1 > 0$, $C_2 > 0$ *such that for every* $\delta > 0$ *there exist* $A_\delta > 0$ *(we require* $A_\delta$ *to be a nondecreasing function as* $\delta$ *decreases to zero) such that*

(i) *for each* $L \in \mathcal{L}$, *the set*

$$S_L(P,\delta) = \{z \in L \ : \ |P(z)| < A_\delta \exp(-\delta|z|)\}$$

*has relatively compact connected components;*

(ii) *if* $z_1$, $z_2$ *are two points in the closure of the same component of some* $S_L(P,\delta)$, *then*

$$|z_1| \leq C_1|z_2| + C_2.$$

In the case in which $P$ has as many operators as variables, the previous definition becomes simpler and reduces to the following:

**Definition 2.3.2.** *Let* $P = (P_1, \ldots, P_n)$, $P_i \in Expo(\mathbb{C}^n)$, $i = 1, \ldots, n$, *be such that*

$$\{z \in \mathbb{C}^n \ : \ P_1(z) = \ldots = P_n(z) = 0\}$$

*is a discrete variety. We say that* $P$ *is slowly decreasing if there are positive constants* $C_1$, $C_2$ *such that*

(i) *for any* $\delta > 0$ *there exists* $A_\delta > 0$ *such that all the connected components of*

$$S(P,\delta) = \{z \in \mathbb{C}^n \ : \ |P_1(z)| + \ldots. + |P_n(z)| < A_\delta \exp(-\delta|z|)\}$$

*are relatively compact;*

(ii) *if* $z_1$, $z_2$ *are two points in the closure of the same component of* $S(P,\delta)$, *then*

$$|z_1| \leq C_1|z_2| + C_2;$$

(iii) $A_\delta$ *can be chosen to be nondecreasing as* $\delta$ *decreases to zero.*

The reader interested in constructing examples of such slowly decreasing vectors can see [18]. By using the results in [18] and [105], one can state the following theorem (see [18]) that gives the representation of holomorphic solutions of infinite order differential equation in the case scalar case.

**Theorem 2.3.2.** *Let* $P = (P_1, \ldots, P_r)$, $1 \leq r \leq m$, $P_i \in \mathrm{Expo}(\mathbb{C}^n)$, $i = 1, \ldots, r$, *be slowly decreasing. Let* $V = \{z \in \mathbb{C}^n \mid P_1(z) = \ldots = P_r(z) = 0\}$. *Then there exist a locally finite family* $\{V_j\}_{j \in J}$ *of closed subsets of* $V$, *partial differential operators* $\partial_j$ *on* $\mathbb{C}^n$ *with analytic coefficients on the regular points of* $V$ *and a partition of* $J$ *into finite subsets* $\{J_k\}_{k \in K}$ *such that if* $f \in \mathcal{O}(\mathbb{C}^n)$ *satisfies* $P(D)f = 0$, *then*

$$f(x) = \sum_{k \in K} \sum_{j \in J_k} \int_{V_j} \partial_j \exp(ix, z) \ \ d\nu_j(z),$$

*where the series converges in* $\mathcal{O}(\mathbb{C}^n)$ *and* $d\nu_j$ *are Radon measures with support in* $V_j$.

**Remark 2.3.1.** The proof of Theorem 2.3.2 requires knowledge about the module

$$\mathcal{M} = \mathscr{D}^{\infty}/P(\mathscr{D}^{\infty})^r = \frac{\mathscr{D}^{\infty}}{P_1\mathscr{D}^{\infty} + \ldots + P_r\mathscr{D}^{\infty}};$$

the hypothesis of slow decrease allows one to show that the appropriate Koszul complex provides a free resolution for the module $\mathcal{M}$. With an abuse of notation, we will say that the module

$$\mathcal{M} = \mathscr{D}^{\infty}/P(\mathscr{D}^{\infty})^r$$

is slowly decreasing if the matrix $P$ satisfies the slowly decreasing condition.

**Lemma 2.3.1.** *Let $\mathcal{M}$ be a slowly decreasing $\mathscr{D}^{\infty}$-module. Let $\Omega$ be an open convex set in $\mathbb{R}^n$, and set $U = \{z \in \mathbb{C}^n : Re z \in \Omega\}$, $U_i = \{z \in U : \mathrm{Im} z_i \neq 0\}$. Consider, furthermore, Leray's covering: $\mathcal{U} = \{U, U_1, \ldots, U_m\}$, $\mathcal{U}' = \{U_1, \ldots, U_m\}$. Define the solution sheaves*

$$\mathcal{B}^P = \mathrm{Hom}_{\mathscr{D}^{\infty}}(\mathcal{M}, \mathcal{B})$$

*and*

$$\mathcal{O}^P = \mathrm{Hom}_{\mathscr{D}^{\infty}}(\mathcal{M}, \mathcal{O}).$$

*Then every $u \in \mathcal{B}^P(\Omega)$ can be represented by the cohomology class of a suitable cochain $\{F_\sigma\} \in C^n(\mathcal{U}, \mathcal{U}', \mathcal{O}^P)$.*

*Proof.* Considering the free resolution of $\mathcal{M}$ given by the generalized Koszul complex and mimicking the proof of Theorem 2.3.1 one gets the result. □

We can finally state the Fundamental Principle for hyperfunctions and infinite order differential operators.

**Theorem 2.3.3.** *Let $P = (P_1, \ldots, P_r)$ be a slowly-decreasing $r$-tuple in $Exp_0(\mathbb{C}^n)$. Then there are analytic varieties $V_j$ and differential operators $\partial_j$ as in Theorem 2.3.2 such that if $u$ is a hyperfunction on $\Omega \subseteq \mathbb{R}^n$ satisfying*

$$P_1(D)u = \ldots = P_r(D)u = 0,$$

*then*

$$u(x) = \sum_{k \in K} \sum_{j \in J_k} \int_{V_j} \partial_j(\exp(ix, z)) \; d\nu_j(z),$$

*where $d\nu_j$ are Radon measures supported in $V_j$.*

*Proof.* From Lemma 2.3.1 we see that $U$ can be represented by a cochain $\{f_\sigma\} \in \mathcal{C}^{n-1}(\mathcal{U}, \mathcal{U}', \mathcal{O}^P)$. By Theorem 2.3.2 we now get that each $f_\sigma$ can be given an exponential representation with suitable measures. This immediately yields the corresponding representation for $u$. □

## 2.4 Using computational algebra software: a section in experimental mathematics

The systems we are interested in have an arbitrarily large number of equations and unknowns. This makes it impossible to compute directly the Gröbner bases for the modules involved as well as their syzygies. Given the complexity of the computations one has to carry on, it is usually difficult to make reasonable guesses on what the results should be. To obviate this difficulty, it is necessary to use some computer algebra software to help us make appropriate conjectures on very large systems. We have been experimenting with two computer algebra packages. One of them is Macaulay2, developed by D. R. Grayson and M. E. Stillman, the second one is CoCoA version 4.0 (the acronym stands for Computational Commutative Algebra). Macaulay2 is a software for computations in algebraic geometry and commutative algebra that is extremely rich and allows a large number of manipulations. The program is available over the web at `http://www.math.uiuc.edu`. CoCoA was developed at the University of Genova, by a group of algebraic geometers and is available by anonymous ftp at `http://cocoa.dima.unige.it`. We have worked with both packages, though we have eventually decided to use CoCoA as our default software. We will provide several concrete examples of systems which we have studied with the help of CoCoA, and we include some of the programs we have used and their output.

To give an overview of the software we provide some examples of computations in the setting of quaternionic analysis. We will omit the details of the theory here and, instead, we focus our attention on the computational aspects. Let us start with the Cauchy–Fueter system in two variables (see the Chapter 3):

$$\begin{cases} \partial_{\bar{q}_1} f = g_1 \\ \partial_{\bar{q}_2} f = g_2. \end{cases}$$

From an algebraic point of view the system corresponds to eight equations and can be described by the matrix

$$P = \begin{bmatrix} x_0 & -x_1 & -x_2 & -x_3 \\ x_1 & x_0 & -x_3 & x_2 \\ x_2 & x_3 & x_0 & -x_1 \\ x_3 & -x_2 & x_1 & x_0 \\ y_0 & -y_1 & -y_2 & -y_3 \\ y_1 & y_0 & -y_3 & y_2 \\ y_2 & y_3 & y_0 & -y_1 \\ y_3 & -y_2 & y_1 & y_0 \end{bmatrix} = \begin{bmatrix} P_1 \\ P_2 \end{bmatrix}$$

that is associated to the operator $P(D) = (\partial_{\bar{q}_1}, \partial_{\bar{q}_2})^t$.

The operators $\partial_{\bar{q}_1}$, $\partial_{\bar{q}_2}$ do not commute, as one can easily see by multiplying the corresponding matrices $P_1$, $P_2$. In view of their noncommutativity, the first syzygies cannot be given by the usual Koszul complex, but one cannot exclude (in principle) that they are linear. It is known that the use of tools coming from invariant operator theory (see e.g. [13]) implies that quadratic syzygies suffice; that this is the case was shown in [183], but this result is actually a consequence of [2], [6].

Let us now describe in some more detail the concrete steps which are necessary to work with the two computer algebra softwares we have mentioned above. With both the systems, the first instruction one has to give is the choice of a ring $R$ of polynomials over a field. We recall that the finitely generated module $M$ associated to a system is an $R$-module where $R$ is a ring of polynomials over the field $\mathbb{C}$, while the fields available with CoCoA or Macaulay2 are either $\mathbb{Z}_p$ or $\mathbb{Q}$. But $\mathbb{C}$ is an extension of the field $\mathbb{Q}$ so if we denote by $S$ the ring of polynomials in the same number of variables as in $R$ but with coefficients in $\mathbb{Q}$, we have the following result (see [113]).

**Theorem 2.4.1.** *If $N$ is a finitely generated $S$-module and $M$ is the $R$-module generated by the elements in $N$ then a Gröbner basis for $N$ is a Gröbner basis for $M$.*

Since the computations of resolutions, syzygies and Ext-modules rely on computations of Gröbner bases techniques, one can use the ring $S$ of polynomials over $\mathbb{Q}$ instead of $R$.

We now have to write the module $M$ generated by the rows of the homogeneous matrix $P$, symbol of a system. With Macaulay2 one can assign a module as the cokernel of a suitable map. In this case, the map is induced by the matrix $P^t$ so that we assign first the matrix $P$ by the command `P= matrix` $\{\{\dots\},\dots,\{\dots\}\}$ where $\{\dots\}$ are the rows of the matrix $P$ then `Q= cokernel transpose P`. To compute the resolution, it suffices to give the system the instruction `res Q`. With CoCoA the module can be assigned by writing `M=Module([... ],... ,[... ]);` where [...] are the rows of the matrix $P$. The resolution of the module $M$ associated to two Cauchy-Fueter operators can be computed giving the command `Res(M)`, as in the following example:

```
Use
R::=Q[x[0..3],y[0..3]];
M:=Module([x[0],-x[1], -x[2],-x[3]],
[x[1] , x[0], -x[3], x[2]],
[x[2], x[3], x[0], -x[1]],
[x[3], -x[2], x[1], x[0]],
[y[0],-y[1], -y[2],-y[3]],
[y[1] , y[0], -y[3], y[2]],
[y[2], y[3], y[0], -y[1]],
[y[3], -y[2], y[1], y[0]]);
Res(N);
-------------------------------------
```

The output is

```
0⟶ R^4(-4) ⟶ R^8(-3)⟶ R^8(-1)
-----------------------------------
```

where the rank of the free modules are, as usually, denoted in form of exponents and the shifts into parentheses indicated the degree of the syzygies. Note that the previous module is generated by the rows of two matrices of the same form: the first in the variables $x_i$ and the second in $y_i$ (each of them corresponding to a Cauchy–Fueter operator). If one has to deal with a module generated by the rows of several matrices, all of the same form, one can mimic the following procedure which we show for the Cauchy–Fueter system: first, one defines the matrix associated to the operator in N variables, in a compact form as

```
MatConcat(A,B):=Mat(Concat(List(A),List(B)));
Define FueterMat(X);
N:=Len(X)/4;
Q:=[];
For J:=1 To N Do
K:=4*(J-1);
Q:=MatConcat(Q,Mat[[X[1+K],-X[2+K],-X[3+K],-X[4+K]],
[X[2+K],X[1+K],-X[4+K],X[3+K]],
[X[3+K],X[4+K],X[1+K],-X[2+K]],
[X[4+K],-X[3+K],X[2+K],X[1+K]]]);
EndFor;
Return Q;
EndDefine;
```

then, to display the matrix, it suffices to define the desired ring of indeterminates, e.g. $\mathtt{N} = 2$ in the previous case, and then writing the command

```
P:=FueterMat((Indets()));
```

and finally the module can be written using the command

```
M:=Cast(P,MODULE);
M;
```

With Macaulay2, to compute the syzygies of the matrix $P$ it suffices to type the command `syz P`. With CoCoA, if one wishes to display one of the syzygy matrices in the resolution, for example the first one, it suffices to type

```
Syz(M,1)
```

and the output gives the first syzygy module appearing in the minimal free resolution of $M$ by means of its minimal generators. When we are interested in the algebraic analysis of a system, the module $M$ is the only object of interest, and its resolution is the algebraic object which contains all the relevant information. However, in general, if we denote by $A$ the matrix whose rows are the generators of the syzygy module and by $P$ the matrix whose rows generate $M$ then, in general, it is not true that $AP = 0$. What one can do if interested in finding the matrix $A$ is to use the function `SyzOfGens(M)` that calculates the syzygy module for the given set of generators of $M$ and then transform the result into a matrix. This is the case, for example, if one wants to compute explicitly the conservation laws of a given physical system. In that case one actually needs the syzygies in matrix form for a very specific set of generators.

Throughout the book, we will need to compute the Ext-modules related to the dual of the resolutions computed using CoCoA. They can be calculated directly, but it is also possible to use the script written by A. Damiano [52] giving the command

```
Alias Ext := $contrib/ext;
```

and then computing the Ext–modules $\text{Ext}^j(M, R)$ using

```
Ext.Ext(R∧s/M,R∧1,j);
```

if $M$ is a submodule of $R^s$. More details on how to use the package CoCoA to compute the resolutions for specific systems will be shown in the chapters to follow. Here, we will limit ourselves to write the matrices associated to those systems and to show how to perform the computations.

When dealing with systems of linear partial differential operators with constant coefficients arising in physics, one is faced with operators whose symbol is not necessarily homogeneous. A typical example that we have treated in Chapter 5 is given by the Proca system (see [46]). As we will see, the Proca equations generalize the Maxwell field for massive spin 1 particles and the system can be written in matrix form as follows

$$P_{\mathcal{P}}(D)F = G$$

where $F$, $G$ are the transpose of

$$[E_1, E_2, E_3, B_1, B_2, B_3, A_0, A_1, A_2, A_3],$$

$$[\rho^e, 0, 0, 0, 0, j_1^e, j_2^e, j_3^e, B_1^{ext}, B_2^{ext}, B_3^{ext}, E_1^{ext}, E_2^{ext}, E_3^{ext}]$$

respectively. The matrix $P_{\mathcal{P}}$, up to a multiplication by the imaginary unit, is

$$\begin{bmatrix} x & y & z & 0 & 0 & 0 & -m^2 & 0 & 0 & 0 \\ 0 & 0 & 0 & x & y & z & 0 & 0 & 0 & 0 \\ 0 & z & -y & -t & 0 & 0 & 0 & 0 & 0 & 0 \\ -z & 0 & x & 0 & -t & 0 & 0 & 0 & 0 & 0 \\ y & -x & 0 & 0 & 0 & -t & 0 & 0 & 0 & 0 \\ -t & 0 & 0 & 0 & -z & y & 0 & -m^2 & 0 & 0 \\ 0 & -t & 0 & z & 0 & -x & 0 & 0 & -m^2 & 0 \\ 0 & 0 & -t & -y & x & 0 & 0 & 0 & 0 & -m^2 \\ 0 & 0 & 0 & 1 & 0 & 0 & 0 & 0 & z & -y \\ 0 & 0 & 0 & 0 & 1 & 0 & 0 & -z & 0 & x \\ 0 & 0 & 0 & 0 & 0 & 1 & 0 & y & -x & 0 \\ 1 & 0 & 0 & 0 & 0 & 0 & x & t & 0 & 0 \\ 0 & 1 & 0 & 0 & 0 & 0 & y & 0 & t & 0 \\ 0 & 0 & 1 & 0 & 0 & 0 & z & 0 & 0 & t \end{bmatrix}$$

and has nonzero entries equal to either $\pm x$, $\pm y$, $\pm z$ $\pm m^2$ or 1. Then its symbol is nonhomogeneous since $m^2, 1$ are elements of the field $\mathbb{Q}$. The rows of the matrix $P_{\mathcal{P}}$ are homogeneous only if considered in a module with suitable shifts. Using CoCoA it is not possible to compute the resolution for such modules and a strategy one can apply is the following. First of all, since $m^2$ is a constant we can write $v$ instead of $m^2$. Then, we can introduce a new variable, say $u$, and multiply all the entries of the matrix by suitable powers of $u$ to get homogeneous elements. This corresponds to writing $u$ instead of 1. We now compute the resolution of the module generated by the new matrix and then we come back by setting $u = 1$, $v = m^2$. The computation of the resolution with CoCoA gives the following result:

$$0 \longrightarrow R(-3) \xrightarrow{P_2^t} R^5(-2) \xrightarrow{P_1^t} R^{14}(-1) \xrightarrow{P_{\mathcal{P}}^t} R^{10}$$

where $R = \mathbb{Q}[t, x, y, z, u, v]$. The same technique to add a variable can be used when we need to homogenize the generators of a module multiplying them by suitable powers of a new variable $u$. One can perform the computations and then come back to the original module by setting $u = 1$.

When dealing with systems in a Clifford algebra, the dimension of the matrices grows exponentially since an operator acting on $\mathbb{R}_n$-valued functions is represented by a $2^n \times 2^n$ matrix. Even reducing the size of the matrices involved trough the use of the spinor formalism (see Chapter 4) one can still have problems in completing the computation of resolutions. Since the beginning of our experiments with CoCoA, computers have become faster and the versions of CoCoA more powerful, so that our experiments have become increasingly probing. For example, in the paper [2] we could only compute the resolutions up to three Cauchy–Fueter operators in the quaternionic algebra while in the paper

[164] we were able to compute the resolutions for systems of three operators in the Clifford algebra $\mathbb{C}_8$ and of four operators in the Clifford algebra $\mathbb{C}_4$. Nevertheless, the problem of computing resolutions exceeding the possibility of the system still exists. In this case, one can try to compute a part of the resolution, up to the maximum allowed by the system. In fact, denoting as usual by $M$ the module generated by the rows of the matrix $P$, we can give the following commands:

```
Set Verbose,
GB.Start_Res(M),
GB.Steps(M,n),
GB.GetBettiMatrix(M),
```

where n is the desired number of steps of computations. With all those instructions, CoCoA performs a step-by-step computation and the verbose mode allows one to monitor the execution. After the given number of steps, CoCoA returns the Betti numbers computed so far, from which we can obtain the degree and the number of generators of each free module appearing in the resolution. A similar set of instructions can be used also with Macaulay2:

```
T=res(M,SyzygyLimit=> n)
betti T
```

We wish to display the results that can be obtained in the case of the matrix $P$ associated to the Cauchy–Fueter system in two variables. Although in this case the resolution can be easily computed, we have chosen to treat this example for sake of simplicity. Let us begin by computing the first ten steps of the procedure illustrated above.

```
GB.Steps(M,15);
GB.GetBettiMatrix(M);
..........15
-------------------------------------
IPs   IVs  Gens   GBases   MinGens  MinDeg
-------------------------------------
6     0     8      12       8        3
0     1     1       1       1       3
-------------------------------------
Betti numbers: 1 8
15 steps of computation
-------------------------------
---------
---------
0    8
0    0
1    0
---------
```

Now we will proceed with the next 100 steps

```
GB.Steps(M,100);
GB.GetBettiMatrix(M);
.....................21
-----------------------------------
IPs   IVs  Gens   GBases   MinGens  MinDeg
-----------------------------------
0     0     8      12        8       -1
0     0     8      8        8       -1
0     0     4      4        4       -1
-----------------------------------
Betti numbers: 4 8 8
21 steps of computation
------------------------------
---------
---------
0   0   8
0   0   0
0   8   0
4   0   0
---------
------------------------------
```

The computations, in this case, stopped after 21 steps and the results in the table, the so-called Betti matrix, can be read as follows: the Betti numbers 4, 8, 8 are written in the order they appear in the resolution, while the order of the syzygies are denoted by the vertical shift. We have 8 linear syzygies (first row of the table), then 8 quadratic (third row) and finally 4 linear. The result is the resolution already obtained:

$$0 \longrightarrow R^4(-4) \longrightarrow R^8(-3) \longrightarrow R^8(-1).$$

An important further step in this experimental mathematics is given by the possibility to compute resolutions and syzygies in the case of systems with polynomial coefficients. The script to perform the computations in the Weyl algebra is available in Macaulay2 (it will also be available in the newest versions of CoCoA). To give an example, let us consider the so-called Hypergeometric system (see [167]). Let $f = f(x, y)$, $f : \mathbb{R}^2 \to \mathbb{R}$ and consider the system $D_x f = D_y f = 0$ where

$$D_x = x(1-x)\frac{\partial^2}{\partial x^2} + (1-2x)\frac{\partial}{\partial x}$$

$$D_y = y(1-y)\frac{\partial^2}{\partial y^2} + (1-2y)\frac{\partial}{\partial y}.$$

These two equations are hypergeometric equations with parameters $(1, 0, 1)$. The matrix representing the system is $[D_x,\ D_y]^t$. With Macaulay2 one has to declare the Weyl algebra in which to work as follows:

```
R=QQ[x,dx,y,dy, WeylAlgebra =>{x=>dx,y=>dy}];
```

then we can write in the usual way the matrix $P$ associated to the system and with the command `res P` we get the resolution. Since we are working in the real setting, we can get the syzygies of the system using the command `syz P` and we obtain $[-D_y,\ D_x]$. The immediate consequence of this fact is that the inhomogeneous system $[D_x,\ D_y]^t f = [g_1,\ g_2]^t$ can be solved only if the datum $[g_1\ g_2]$ satisfies the system $D_y g_1 = D_x g_2 = 0$.

# 3

# The Cauchy–Fueter System and Its Variations

## 3.1 Regular functions of one quaternionic variable

In this section we introduce the main results for regular functions of one quaternionic variable. This material is purely instrumental since it is needed as background to develop the theory of quaternionic hyperfunctions in one variable. Therefore we give an overview of the theory without proofs, for which we give references, pointing out the main differences and the similarities with the theory of holomorphic functions in one complex variable.
Let $\mathbb{H}$ be the real associative algebra of quaternions with the standard basis 1, $i$, $j$, $k$ such that

$$i^2 = j^2 = k^2 = -1, \quad ij = -ji = k, \quad jk = -kj = i, \quad ki = -ik = j.$$

Note that $\mathbb{H}$ is an example of a Clifford algebra (see Chapter 4): it is the Clifford algebra over two units $e_1$, $e_2$ that, for historical reasons, are denoted by $i$, $j$ while their product $e_1e_2$ is denoted $k$.

We write an element $q \in \mathbb{H}$ in the form

$$q = x_0 + ix_1 + jx_2 + kx_3,$$

where $x_\ell \in \mathbb{R}$, for $\ell = 0, 1, 2, 3$ and we set

$$\text{Re } q = x_0, \qquad \text{Im } q = ix_1 + jx_2 + kx_3, \qquad |q| = \sqrt{x_0^2 + x_1^2 + x_2^2 + x_3^2};$$

Re $q$, Im $q$ and $|q|$ are called the real part, the imaginary part and the module of $q$, respectively. The quaternion $\bar{q} = \text{Re } q - \text{Im } q = x_0 - ix_1 - jx_2 - kx_3$ is called the conjugate of $q$ and satisfies

$$|q| = \sqrt{q\bar{q}} = \sqrt{\bar{q}q}.$$

Sometimes it will be useful to write a quaternion in a more compact way as $q = \sum_{\ell=0}^{3} i_\ell x_\ell$ where $x_\ell \in \mathbb{R}$ and $i_0 = 1$, $i_1 = i$, $i_2 = j$, $i_3 = k$. The multiplication rules of the units $i$, $j$, $k$ show that $\mathbb{H}$ is noncommutative: it is a skew field, in which the inverse of any element $q \neq 0$ is given by

$$q^{-1} = \frac{\bar{q}}{|q|^2}.$$

Let us consider a function $f : \mathbb{H} \to \mathbb{H}$. One can extend the notion of holomorphicity to functions of one quaternionic variable, but while in the complex case there are several equivalent definitions, in the quaternionic case there is only one definition which is meaningful, and it consists in defining a regular function (or quaternionic holomorphic function) as a function defined on an open set of the space of quaternions which is in the kernel of the so-called Cauchy–Fueter operator (a natural generalization of the Cauchy–Riemann operator). Note that the notion of regularity in one variable was designed to avoid triviality. Indeed, functions with (left) quaternionic derivatives are only the linear functions, since the limit

$$\lim_{h\to 0} h^{-1}(f(q+h) - f(q)), \qquad q, h \in \mathbb{H}$$

exists and it is finite if and only if $f$ is of the form $f(q) = qa + b$, with $a, b \in \mathbb{H}$. Also the theory of functions with quaternionic power series expressions is trivial, because it coincides with the theory of all real analytic functions in the real variables $x_\ell$ (see [206]). In fact, the generalization of the complex term $a_n z^n$ to the quaternionic case is given by a monomial of the type $a_0 q a_1 q \ldots a_n q$, with $a_i \in \mathbb{H}$, but if we require that a function in the variable $q$ is a sum of such monomials we do not have any restriction. In fact, $x_\ell$ can be expressed in terms of $q$ for every $\ell = 0, \ldots, 3$, therefore every polynomial in $x_\ell$ is a polynomial in $q$.

Let us now introduce the two differential operators which generalize the Cauchy–Riemann operator to the quaternionic case:

$$\frac{\partial_l}{\partial \bar{q}} = \frac{\partial}{\partial x_0} + i\frac{\partial}{\partial x_1} + j\frac{\partial}{\partial x_2} + k\frac{\partial}{\partial x_3},$$

$$\frac{\partial_r}{\partial \bar{q}} = \frac{\partial}{\partial x_0} + \frac{\partial}{\partial x_1}i + \frac{\partial}{\partial x_2}j + \frac{\partial}{\partial x_3}k.$$

The two operators are called the left and right Cauchy–Fueter operators, respectively. We also define their conjugate operators

$$\frac{\partial_l}{\partial q} = \frac{\partial}{\partial x_0} - i\frac{\partial}{\partial x_1} - j\frac{\partial}{\partial x_2} - k\frac{\partial}{\partial x_3},$$

$$\frac{\partial_r}{\partial q} = \frac{\partial}{\partial x_0} - \frac{\partial}{\partial x_1}i - \frac{\partial}{\partial x_2}j - \frac{\partial}{\partial x_3}k.$$

**Definition 3.1.1.** *Let $U \subseteq \mathbb{H}$ be an open set and let $f: U \to \mathbb{H}$ be a real differentiable function. We say that $f$ is left regular on $U$ if*

$$\frac{\partial_l f}{\partial \bar{q}} = \frac{\partial f}{\partial x_0} + i\frac{\partial f}{\partial x_1} + j\frac{\partial f}{\partial x_2} + k\frac{\partial f}{\partial x_3} = 0. \tag{3.1}$$

*We say that $f$ is right regular on $U$ if*

$$\frac{\partial_r f}{\partial \bar{q}} = \frac{\partial f}{\partial x_0} + \frac{\partial f}{\partial x_1}i + \frac{\partial f}{\partial x_2}j + \frac{\partial f}{\partial x_3}k = 0. \tag{3.2}$$

*We denote by $\mathcal{R}_l(U)$ the set of left regular functions on $U$ and by $\mathcal{R}_r(U)$ the set of right regular functions on $U$.*

**Remark 3.1.1.** Let $\Delta$ be the Laplace operator in $\mathbb{R}^4$. It is immediate to verify that

$$\Delta = \frac{\partial_l}{\partial q}\frac{\partial_l}{\partial \bar{q}} = \frac{\partial_l}{\partial \bar{q}}\frac{\partial_l}{\partial q} = \frac{\partial_r}{\partial q}\frac{\partial_r}{\partial \bar{q}} = \frac{\partial_r}{\partial \bar{q}}\frac{\partial_r}{\partial q}. \tag{3.3}$$

The theory of left regular functions is completely equivalent to the theory of right regular functions so, classically, the theory is usually developed for the case of left regular functions. In this section, for sake of clarity, we will keep the subscripts "$l$" and "$r$" referring to the left and right case respectively. For all the statements which are not proved in this section, we refer the reader to the classical reference [206].

A first result, that can be proved with direct computations, is the following.

**Proposition 3.1.1.** *Let $f, g : U \subseteq \mathbb{H} \to \mathbb{H}$ be regular functions on an open set $U$ and let $a \in \mathbb{H}$. Then the following equalities hold:*

*(i)* $\dfrac{\partial_l(f+g)}{\partial \bar{q}} = \dfrac{\partial_l f}{\partial \bar{q}} + \dfrac{\partial_l g}{\partial \bar{q}}$,

*(ii)* $\dfrac{\partial_l(fa)}{\partial \bar{q}} = \dfrac{\partial_l f}{\partial \bar{q}}a$,

*(iii)* $\dfrac{\partial_l(fg)}{\partial \bar{q}} = \dfrac{\partial_l f}{\partial \bar{q}}g + \sum_{\ell=0}^{3} i_\ell f \dfrac{\partial_l g}{\partial x_\ell}$.

**Remark 3.1.2.** Equalities (*i*) and (*ii*) prove that $\mathcal{R}_l(U)$ is a right vector space on $\mathbb{H}$, while (*iii*) shows that $\mathcal{R}_l(U)$ is not an algebra. Analogously, we can verify that $\mathcal{R}_r(U)$ is a left vector space on $\mathbb{H}$.

We now introduce the notion of $\mathbb{H}$–differential forms. We point out that such forms have important applications in physics (see e.g. [10]).

**Definition 3.1.2.** *Let $\mathcal{E}^p(U)$ be the vector space of the $C^\infty$ $p$-forms on an open set $U$ in $\mathbb{H} \cong \mathbb{R}^4$. An element $\omega \in \mathcal{E}^p_{\mathbb{H}}(U) = \mathcal{E}^p(U) \otimes_{\mathbb{R}} \mathbb{H}$ is called a $p$-form with* *values in* $\mathbb{H}$ *(of class* $C^\infty$*). A p–form* $\omega$ *is of the type*

$$\omega = \sum_{0\le i_1\le\ldots\le i_p\le 3} a_{i_1\ldots i_p} dx_{i_1}\wedge\ldots\wedge dx_{i_p},$$

*where* $a_{i_1\ldots i_p}$ *are* $\mathbb{H}$*-valued functions of class* $C^\infty$ *on* $U$.

**Remark 3.1.3.** The following forms will be useful in the theory of regular functions:

$$\begin{aligned}
&dq = dx_0 + idx_1 + jdx_2 + kdx_3 \in \mathcal{E}^1_{\mathbb{H}}(\mathbb{H}),\\
&\overline{dq} = dx_0 - idx_1 - jdx_2 - kdx_3 \in \mathcal{E}^1_{\mathbb{H}}(\mathbb{H}),\\
&Dq = dx_1\wedge dx_2\wedge dx_3 - idx_0\wedge dx_2\wedge dx_3 + jdx_0\wedge dx_1\wedge\\
&\wedge dx_3 - kdx_0\wedge dx_1\wedge dx_2 \in \mathcal{E}^3_{\mathbb{H}}(\mathbb{H}),\\
&\upsilon = dx_0\wedge dx_1\wedge dx_2\wedge dx_3 \in \mathcal{E}^4_{\mathbb{H}}(\mathbb{H}).
\end{aligned}$$

The form $\upsilon$ is called canonical volume form in $\mathbb{H}$, and it is such that

$$4\upsilon = \overline{dq}\wedge Dq = -Dq\wedge\overline{dq}.$$

**Theorem 3.1.1.** *Let* $f$, $g$ *be two continuously differentiable functions on an open set* $U$*. Then for any four dimensional compact, oriented manifold* $S\subset U$ *with boundary* $\partial S$*, it is*

$$\int_{\partial S} gDqf = \int_S \left\{\left(\frac{\partial_r g}{\partial\bar q}\right) f + g\left(\frac{\partial_l f}{\partial\bar q}\right)\right\}\upsilon.$$

*Proof.* Standard computations show the equality:

$$d(gDqf) = dg\wedge Dqf - gDq\wedge df = \left\{\left(\frac{\partial_r g}{\partial\bar q}\right) f + g\left(\frac{\partial_l f}{\partial\bar q}\right)\right\}\upsilon. \tag{3.4}$$

Using Stokes' theorem for real-valued functions we get the statement. □

Substituting $g = 1$ into equality (3.4) we obtain

$$d(Dqf) = -Dq\wedge df = \frac{\partial_l f}{\partial\bar q}\upsilon. \tag{3.5}$$

From (3.4) and (3.5) one deduces the analogue of the Cauchy formula in the case of regular functions of one quaternionic variable. The proofs of this and the following theorems are similar to the proofs for the corresponding results in the complex case. For this reason we omit the proofs and we refer the reader to [206] for the details.

**Theorem 3.1.2.** (Cauchy–Fueter I). *Let* $f, g : U\subseteq\mathbb{H}\to\mathbb{H}$ *be functions such that* $f\in\mathcal{R}_l(U)$ *and* $g\in\mathcal{R}_r(U)$*. If* $S$ *is a four-dimensional compact, oriented manifold* $S\subset U$ *with boundary* $\partial S$*, then*

$$\int_{\partial S} g(q)Dqf(q) = 0. \tag{3.6}$$

*In particular, if $f \in \mathcal{R}_l(U)$, then*

$$\int_{\partial S} Dqf(q) = 0,$$

*for every such manifold $S$.*

Let us now introduce a function, denoted by $G(q)$, that generalizes the Cauchy kernel $1/z$:

**Definition 3.1.3.** *The function $G(q)$ defined by*

$$G(q) = \frac{q^{-1}}{|q|^2} = \frac{\bar{q}}{|q|^4} \tag{3.7}$$

*is called the Cauchy–Fueter kernel.*

Very simple computations prove the following.

**Proposition 3.1.2.** *The function $G(q)$ is left and right regular on $\mathbb{H}\setminus\{0\}$.*

**Theorem 3.1.3.** (Cauchy–Fueter II). *Let $f : U \subseteq \mathbb{H} \to \mathbb{H}$ be a left regular function on $U$. If $S$ is a four dimensional compact, oriented manifold $S \subset U$ with boundary $\partial S$ and $q_0$ belongs to the interior of $S$, then*

$$f(q_0) = \frac{1}{2\pi^2} \int_{\partial S} G(q - q_0) Dqf(q). \tag{3.8}$$

Theorem 3.1.3 shows that any left (or right) regular function has derivatives of any order. Equality (3.3) shows that any left (or right) regular function is harmonic and so it is real analytic. From this last property we deduce the following results.

**Theorem 3.1.4.** (Liouville). *Let $f : \mathbb{H} \to \mathbb{H}$ be a left (or right) regular function bounded on $\mathbb{H}$. Then $f$ is constant on $\mathbb{H}$.*

**Theorem 3.1.5.** (Identity Principle). *Let $U \subseteq \mathbb{H}$ be an open connected set and let $f, g \in \mathcal{R}_l(U)$ be such that $f = g$ on an open set contained in $U$. Then $f = g$ on $U$.*

The following theorem is the quaternionic version of the well-known Morera theorem:

**Theorem 3.1.6.** (Morera). *Let $f : U \subseteq \mathbb{H} \to \mathbb{H}$ be a continuous function such that*

$$\int_{\partial V} Dqf(q) = 0,$$

*on every open set $V$ with differentiable boundary, relatively compact in $U$. Then $f$ is left regular on $U$.*

The proof given in the case of holomorphic functions cannot immediately be adapted to regular functions: it is based on a dissection argument as in [173].

We point out that in the case of continuously differentiable functions which are not necessarily regular, it is still possible to give integral representation formulas which will be used later on. For their proofs we refer the reader to [145].

**Theorem 3.1.7.** (Green I). *Let $U$ be an open bounded set in $\mathbb{H}$ and let $f \in C^1(\bar{U})$. Let $q_0$ be a point in $U$. Then we have*

$$f(q_0) = \frac{1}{2\pi^2}\int_{\partial U} G(q-q_0)Dqf(q) - \frac{1}{2\pi^2}\int_U G(q-q_0)Dq\frac{\partial f}{\partial \bar{q}}(q). \tag{3.9}$$

**Theorem 3.1.8.** (Green II). *Let $U$ be an open bounded set in $\mathbb{H}$ and let $f \in C^1(\bar{U})$. Let $q_0$ be a point in $U$. Then we have*

$$f(q_0) = \frac{1}{2\pi^2}\int_{\partial U} DqG(q-q_0)f(q) - \frac{1}{2\pi^2}\int_U \frac{\partial_l}{\partial \bar{q}}(G(q-q_0)f(q)). \tag{3.10}$$

We finally mention the generalization of the Poincaré Lemma for the operator $\partial_l/\partial\bar{q}$ (see [29]). Note that if $U$ is an open convex set, then the result follows from Theorem 2.1.1 and the fact that the module of the first syzygies for the system of four real equations corresponding to $\partial_l f/\partial\bar{q} = g$ is trivial.

**Theorem 3.1.9.** *Let $U \subseteq \mathbb{H}$ be an open set and let $g : U \to \mathbb{H}$ be a function of class $C^k$, $k \geq 1$. Then there exists $f \in C^k(U)$ such that $\dfrac{\partial_l f}{\partial \bar{q}} = g$ in $U$.*

Despite the difficulties due to the noncommutativity it is possible to expand regular functions in term of series of suitable homogeneous functions. Let us consider a set $\nu$ of $n$ integers $\nu = \{\lambda_1, \dots, \lambda_n\}$, $i = 1, \dots, n$, with $1 \leq \lambda_i \leq 3$. We can characterize $\nu$ also giving three integers $n_1$, $n_2$, $n_3$, such that the numbers of 1 which appear in $\nu$ is $n_1$, the number of 2 is $n_2$, the number of 3 is $n_3$, and $n_1 + n_2 + n_3 = n$. Let us denote by $\sigma_n$ the set of triples $\nu = [n_1, n_2, n_3]$; then for any $n > 0$ the set $\sigma_n$ contains $\frac{1}{2}(n+1)(n+2)$ triples. When $n = 0$, we set $\nu = \emptyset$. For any $\nu \in \sigma_n$, let

$$\partial_\nu = \frac{\partial^n}{\partial x_1^{n_1}\partial x_2^{n_2}\partial x_3^{n_3}} \qquad \text{and} \qquad G_\nu(q) = \partial_\nu G(q),$$

where $G(q)$ is the Cauchy–Fueter kernel. We define

$$p_\nu(q) = \frac{1}{n!}\sum_{1\leq\lambda_1,\dots,\lambda_n\leq 3} (x_0 i_{\lambda_1} - x_{\lambda_1})\dots(x_0 i_{\lambda_n} - x_{\lambda_n}),$$

where the sum is taken over the $\dfrac{n!}{n_1!n_2!n_3!}$ different alignments of $n_i$ elements equal to $i$, with $i = 1, 2, 3$. The polynomials $p_\nu(q)$ play the same role of the powers $z^n$ in the Taylor expansion of a function $\sum a_n z^n$ holomorphic in the origin.

Let $\mathcal{U}_n$ be the quaternionic right vector space of the functions $f : \mathbb{H} \to \mathbb{H}$ left regular and homogeneous of degree $n \geq 0$ over $\mathbb{R}$, i.e., such that $f(\alpha q) = \alpha^n f(q)$, for any $\alpha \in \mathbb{R}$. We have the following result (see [206]).

**Proposition 3.1.3.** *The polynomials $p_\nu$, $\nu \in \sigma_n$, are left regular and form a basis for $\mathcal{U}_n$. Moreover, if $f \in \mathcal{U}_n$, then*

$$f(q) = \sum_{\nu \in \sigma_n} (-1)^n p_\nu(q) \partial_\nu f(0). \tag{3.11}$$

In an analogous way, we can prove that the polynomials $p_\nu$ are right regular and if $f$ is right regular and homogeneous of degree $n$, we have

$$f(q) = \sum_{\nu \in \sigma_n} (-1)^n \partial_\nu f(0) p_\nu(q).$$

The introduction of the polynomials $p_\nu$ and of the derivatives $G_\nu$ allows one to prove two results (see [206]) which generalize the Taylor and the Laurent expansion series.

**Theorem 3.1.10.** *Let $f : U \subseteq \mathbb{H} \to \mathbb{H}$, $f \in \mathcal{R}_l(U)$, $q_0 \in U$. Then there exists a ball $|q - q_0| < \delta$ with radius $\delta <$ dist $(q_0, \partial U)$ in which $f$ can be represented by a uniformly convergent series of the form*

$$f(q) = \sum_{n=0}^{+\infty} \sum_{\nu \in \sigma_n} p_\nu(q - q_0) a_\nu,$$

*where*

$$a_\nu = (-1)^n \partial_\nu f(q_0) = \frac{1}{2\pi^2} \int_{|q-q_0|=\delta} G_\nu(q - q_0) Dq f(q).$$

**Theorem 3.1.11.** *Let $q_0$ be a point in an open set $U$ in $\mathbb{H}$ and let $f : U \setminus \{q_0\} \to \mathbb{H}$ be a left regular function. There exists a set of the form*

$$\{\, q \in \mathbb{H} \ : \ r_1 < |q - q_0| < r_2 \,\} \quad \textit{with} \quad 0 < r_1 < r_2 < \operatorname{dist}(q_0, \partial U)$$

*on which $f$ can be represented by a uniformly convergent series of the form*

$$f(q) = \sum_{n=0}^{+\infty} \sum_{\nu \in \sigma_n} \left[ p_\nu(q - q_0) a_\nu + G_\nu(q - q_0) b_\nu \right],$$

*where*

$$a_\nu = \frac{1}{2\pi^2} \int_{|q-q_0|=r_2} G_\nu(q - q_0) Dq f(q),$$

$$b_\nu = \frac{1}{2\pi^2} \int_{|q-q_0|=r_1} p_\nu(q - q_0) Dq f(q).$$

We finish this section with some historical remarks on the theory of functions of quaternionic variables which is, undoubtedly, an interesting one. When Hamilton created quaternions, he was not envisioning other than a method to study and model rotations in three-dimensional space. However, the new ring he introduced became rapidly a *cause célèbre*, because it was the first significant example of a noncommutative structure.

But a theory of functions of quaternionic variables did not begin until the Swiss mathematician Fueter decided, in the first half of the $20^{\text{th}}$ century, to use quaternions as a way to find new properties of functions of two complex variables. Actually the Romanian mathematicians Moisil and Theodorescu had also independently developed a similar theory, although the relevance of their work was not fully appreciated until recently (see [112]). Fueter noticed that every quaternion $q = x + iy + jz + kw$ can be thought of as a pair of complex variables $a = x+jz$ and $b = y+jw$, so that $q = a+ib$. He therefore thought that, by studying functions of one quaternionic variable, he could infer properties for functions of two complex variables. His first interest was to find a new proof of Hartogs' theorem on the removability of compact singularities for holomorphic functions of several variables. Such a theorem, which Hartogs had proved in 1906, [85], with techniques essentially based on a clever use of the Cauchy integral formula for polydiscs, was at that time the most important and striking result for the theory of several complex variables, and thus it was natural that mathematicians would want to find more direct proofs for it. In his attempt to do just that, Fueter worked to find the analogue of complex analyticity (holomorphicity) for functions of a quaternionic variable.

Fueter was able to reconstruct most of the theory of holomorphic functions for these new objects, and as a biproduct he obtained in [75], [76] and [77] a new proof of Hartogs' theorem. Even more important, he built a Cauchy kernel for regular functions and used this kernel to construct a new kernel for holomorphic functions of two complex variables. The work of Fueter is very interesting, and historically played a very significant role, since it inspired (independently, and almost simultaneously) two great mathematicians, S. Bochner and E. Martinelli, who on the basis of Fueter's work went on to construct what is now known as the Bochner–Martinelli kernel.

It is interesting the fact that both these authors used the formula to prove their version of Hartogs' theorem (see [26], [131], [146], [147], [176]). From an historical point of view, it seems that, for the first time, the formula appears in 1937 in a paper of Martinelli which surprisingly was not known to Bochner even though he seemed to know the work of Italian school quite well.

**Remark 3.1.4.** The reader interested in the fascinating history of these ideas may want to read [203] or the more recent [150] and the references therein.

Martinelli gave two different proofs of Hartogs' theorem based on completely different integral formulas he discovered in 1937–1938. Let us describe the following two formulas that can be considered as two fundamental steps in the development of complex analysis in several variables. The first Martinelli formula (see [130]), which is the same Bochner discovered in 1941, is introduced as follows.

**Theorem 3.1.12.** *Let $D$ be a bounded domain in $\mathbb{C}^n$ whose boundary $\partial D$ is a closed hypersurface of real dimension $2n-1$. Let $f$ be a holomorphic function on the closure $\bar{D}$ of $D$. Then for every $(\xi_1,\dots,\xi_n)$ in the interior of $D$, the value of $f$ is given by the formula (Bochner–Martinelli)*

$$f(\xi_1,\dots,\xi_n) = \frac{(n-1)!}{(2\pi i)^n}\int_{\partial D} f(z_1,\dots,z_n)\frac{\sum_{\alpha=1}^{n}(-1)^{\alpha-1}(\bar{z}_\alpha-\bar{\xi}_\alpha)}{(\sum_{\alpha=1}^{n}(z_\alpha-\xi_\alpha)(\bar{z}_\alpha-\bar{\xi}_\alpha))^n}$$

$$\cdot dz_1\wedge\dots\wedge dz_n\wedge d\bar{z}_1\wedge\dots\wedge\widehat{d\bar{z}_\alpha}\wedge\dots\wedge d\bar{z}_n. \tag{3.12}$$

As usual, the symbol $\widehat{d\bar{z}_\alpha}$ means that the term $d\bar{z}_\alpha$ is missing. Based on this formula, Martinelli gave a direct proof of Hartogs' theorem different from the proof found by Bochner. In a second paper, Martinelli [131] tried to understand more deeply the reasons for which Hartogs' theorem could not hold in one dimension. Nowadays we know that this peculiar difference is due to the different behavior of the Cauchy–Riemann system of one and several variables. This interpretation was not understood in the 1930s, but Martinelli was led to some interesting topological considerations from which he got a second integral formula, totally different from the previous one and that hints to a possible cohomological treatment of this issue. The result of Martinelli (see [130]) is the following.

**Theorem 3.1.13.** *Let $f$ be a holomorphic function in a domain $D\subseteq\mathbb{C}^n$. Choose a point $\xi=(\xi_1,\dots,\xi_n)$ in $D$ and set*

$$S_n=\{z\in\mathbb{C}^n\ :\ \prod_{i=1}^{n}(z_i-\xi_i)=0\}.$$

*Let $\Gamma_n\subset\mathbb{C}^n\setminus S_n$ be an $n$-cycle such that $\Gamma_n$ is homologous to zero in $(\mathbb{C}^n\setminus S_n)\cup\{0\}$. Then there exists an integer $N$ such that*

$$f(\xi)=\frac{1}{N(2\pi i)^n}\int_{\Gamma_n}\frac{f(z_1,\dots,z_n)}{(z_1-\xi_1)\dots(z_n-\xi_n)}dz_1\wedge\dots\wedge z_n.$$

Let us point out the topological nature of the theorem, in fact the number $N$ is a topological invariant which describes the position of $\Gamma_n$ with respect to $S_n$. It is also evident how, in this case, the integrand is an $n$ differential form integrated on an $n$-dimensional variety. The situation is now very complicated because the $n$-cycle $\Gamma_n$ must be chosen in $\mathbb{C}^n\setminus S_n$ whose homology is nontrivial. This fact corresponds to the presence of the integer $N$ in the integral formula. Martinelli attempts to explain the different nature of those two formulas using the Alexander duality theorem. This fact opens an interesting link between complex analysis in several variables and topology. Martinelli was probably the first to study, in [130], the relations between integral formulas in $\mathbb{C}^n$ and duality theorems of a topological-algebraic nature. For more details on this fascinating aspect the reader can consult the sixth chapter of Ehrenpreis book [64] in which he poses interesting questions on the topic as well as the book of Aizenberg and Dautov [7] or the more recent [101], written by Kashiwara and Schapira.

## 3.2 Quaternionic hyperfunctions in one variable

The theory of regular functions described in Section 3.1 is sufficient to provide the foundation for a quaternionic theory of hyperfunctions (see [72] and [165]), at least for one quaternionic variable. We will denote by $\mathcal{R}_l$ the presheaf $\{U, \mathcal{R}_l(U)\}$, with the restriction maps:

$$\rho_V^U : \mathcal{R}_l(U) \to \mathcal{R}_l(V), \qquad V \subseteq U.$$

It is immediate to verify that the presheaf $\mathcal{R}_l$ satisfies the conditions defining a sheaf, so we have

**Proposition 3.2.1.** *The presheaf $\mathcal{R}_l$ is a sheaf.*

To construct the sheaf of $\mathbb{H}$-hyperfunctions as the sheaf of the boundary values of regular functions it is necessary to prove a vanishing theorem generalizing the Mittag–Leffler theorem, so we begin by proving the following result.

**Theorem 3.2.1.** *Let $U \subseteq \mathbb{H}$ be an open set. Then*

$$H^1(U, \mathcal{R}_l) = 0.$$

*Proof.* Let $\mathcal{U} = \{U_i\}_{i\in I}$ be a locally finite open covering of $U$ and let $g = \{g_{ij}\}_{i,j\in I}$ be a 1-cocycle with values in $\mathcal{R}_l$. Then for any $g_{ij} \in \mathcal{R}_l(U_i \cap U_j)$, with $U_i \cap U_j \neq \emptyset$ we have

$$g_{ij} + g_{ji} = 0,$$

and if $U_i \cap U_j \cap U_k \neq \emptyset$ it is

$$g_{ij} + g_{jk} + g_{ki} = 0.$$

Since infinitely differentiable functions form a fine sheaf, we can consider a partition of the unity $\{\varphi_i\}$ associated to the covering $\mathcal{U}$, i.e., $\varphi_i \in C^\infty(U_i)$ with $\sum_i \varphi_i = 1$. Let us set

$$h_j = \sum_i \varphi_i g_{ij}.$$

We have that $h_j \in \mathcal{C}^\infty(U_j)$. Moreover, on $U_j \cap U_k$, it is

$$\begin{aligned}\frac{\partial_l}{\partial \bar{q}}(h_k - h_j) &= \frac{\partial_l}{\partial \bar{q}}(\sum_i \varphi_i g_{ik} - \sum_i \varphi_i g_{ij}) \\ &= \frac{\partial_l}{\partial \bar{q}}(\sum_i \varphi_i (g_{ik} + g_{ji})) = \frac{\partial_l}{\partial \bar{q}}(\sum_i \varphi_i g_{jk}) = \frac{\partial_l g_{jk}}{\partial \bar{q}} = 0.\end{aligned}$$

The function $h = \left\{\frac{\partial_l h_j}{\partial \bar{q}}\right\}_{j \in I}$ belongs to $\mathcal{C}^\infty(U)$ and, by the Poincaré lemma, there is a function $u \in \mathcal{C}^\infty(U)$ such that $\frac{\partial_l u}{\partial \bar{q}} = h$. Let us set

$$g_j = h_j - u \qquad \text{on} \quad U_j,$$

so that

$$\frac{\partial_l g_j}{\partial \bar{q}} = \frac{\partial_l h_j}{\partial \bar{q}} - \frac{\partial_l u}{\partial \bar{q}} = 0,$$

and

$$\begin{aligned}g_j - g_k = h_j - h_k = \sum_i \varphi_i g_{ij} - \sum_i \varphi_i g_{ik} \\ = \sum_i \varphi_i g_{kj} = g_{kj} \quad \text{on} \quad U_j \cap U_j, \quad \forall j, k \in I.\end{aligned}$$

This shows that $H^1(\mathcal{U}, \mathcal{R}_l) = 0$ and taking the inductive limit we conclude the proof. □

**Remark 3.2.1.** Note that the preceding theorem is, in general, false for open sets in $\mathbb{H}^n$ as we will see in the next sections.

Let us introduce the following sets

$$\widetilde{\mathbb{H}} = \left\{ q = \sum_{\ell=0}^{3} i_\ell x_\ell \in \mathbb{H} \text{ such that } x_0 = 0 \right\}, \tag{3.13}$$

$$\mathbb{H}^+ = \left\{ q = \sum_{\ell=0}^{3} i_\ell x_\ell \in \mathbb{H} \text{ such that } x_0 > 0 \right\},$$

$$\mathbb{H}^- = \left\{ q = \sum_{\ell=0}^{3} i_\ell x_\ell \in \mathbb{H} \text{ such that } x_0 < 0 \right\}.$$

Since we wish to introduce quaternionic hyperfunctions as boundary values of regular functions, we state the analogue of the Painlevé theorem for functions of one quaternionic variable.

**Theorem 3.2.2.** *Let $U \subseteq \mathbb{H}$ be an open set and let $U^+ = U \cap \mathbb{H}^+$ and $U^- = U \cap \mathbb{H}^-$. Let $F \in \mathcal{R}_l(U^+ \cup U^-)$ and $F \in \mathcal{C}^0(\bar{U})$. Then $F \in \mathcal{R}_l(U)$.*

*Proof.* By Morera's theorem it suffices to show that for any 3-dimensional closed, smooth and oriented variety $\Gamma$ contained in $U$ one has

$$\int_\Gamma DqF = 0. \tag{3.14}$$

Obviously, if $\Gamma \subset U^+$ or $\Gamma \subset U^-$, then (3.14) is an immediate consequence of the fact that $F \in \mathcal{R}_l(U^+ \cup U^-)$ and of the Cauchy's theorem. Let us now suppose that $\Gamma$ is not contained in $U^+$ or in $U^-$. Let $V$ be the interior of $\Gamma$, set $V^+ = V \cap U^+$ and $V^- = V \cap U^-$. Then

$$\int_\Gamma DqF = \int_{\partial V^+} DqF + \int_{\partial V^-} DqF, \tag{3.15}$$

in fact on $\partial V^+$ and $\partial V^-$ the orientation is such that the integrals on $V \cap \widetilde{\mathbb{H}}$ cancel. We can consider a sequence $\{\Gamma_n\}$ of 3-dimensional closed and smooth variety, such that $\Gamma_n \Subset V^+$, for every $n$. Since $F \in \mathcal{R}_l(V^+)$ it is

$$\int_{\Gamma_n} DqF = 0, \qquad \text{for every } n.$$

If the sequence $\{\Gamma_n\}$ is convergent to $\partial V^+$, as $F$ is continuous on $\bar{U}$, we obtain

$$\int_{\partial V^+} DqF = 0.$$

Analogously, $\int_{\partial V^-} DqF = 0$. From (3.15) and Morera's theorem, the statement follows. □

We can now give the definition of $\mathbb{H}$-hyperfunction.

**Definition 3.2.1.** *Let $U$ be an open set in $\widetilde{\mathbb{H}}$ and let $V$ be an open set in $\mathbb{H}$ such that $U$ is relatively closed in $V$. The right vector space on $\mathbb{H}$ defined by*

$$\mathcal{F}(U) := \frac{\mathcal{R}_l(V \setminus U)}{\mathcal{R}_l(V)} \tag{3.16}$$

*is said to be the vector space of (left) $\mathbb{H}$–hyperfunctions.*

The definition of $\mathcal{F}(U)$ does not depend on the choice of the open set $V$. In fact, we have the following proposition whose proof, which mimics the proof of Proposition 1.3.1, is a consequence of the Mittag–Leffler theorem.

**Proposition 3.2.2.** *Let $V_1$, $V_2$ be two open sets in $\mathbb{H}$ such that $\bar{U} \subset V_i$, $i = 1, 2$. Then there is an isomorphism of right vector spaces over $\mathbb{H}$*

$$\frac{\mathcal{R}_l(V_2 \setminus U)}{\mathcal{R}_l(V_2)} \cong \frac{\mathcal{R}_l(V_1 \setminus U)}{\mathcal{R}_l(V_1)}.$$

The most important result on the sheaf of hyperfunctions introduced in Chapter 1 is that it is a flabby sheaf. The same result (essentially with the same proof) holds also in the quaternionic case.

**Theorem 3.2.3.** *The assignment*

$$U \to \mathcal{F}(U),$$

*for every open set $U$ in $\widetilde{\mathbb{H}}$, defines a flabby sheaf $\mathcal{F}$ on $\widetilde{\mathbb{H}}$.*

**Example 3.2.1.** Let $F \in \mathcal{R}_l(\mathbb{H}^+)$. The function

$$\tilde{F}^+ = \begin{cases} F & \text{on } \mathbb{H}^+ \\ 0 & \text{on } \mathbb{H}^- \end{cases}$$

belongs to $\mathcal{R}_l(\mathbb{H} \setminus \widetilde{\mathbb{H}})$ and defines an element $[\tilde{F}^+]$ in $\mathcal{F}(\widetilde{\mathbb{H}})$ which (in view of Painlevé's theorem) represents the boundary value of $F$. Analogously, if $F \in \mathcal{R}_l(\mathbb{H}^-)$ and

$$\tilde{F}^- = \begin{cases} 0 & \text{on } \mathbb{H}^+ \\ F & \text{on } \mathbb{H}^-, \end{cases}$$

then $[\tilde{F}^-] \in \mathcal{F}(\widetilde{\mathbb{H}})$. If $F \in \mathcal{R}_l(\mathbb{H} \setminus \widetilde{\mathbb{H}})$, then we can write (with obvious meaning of the symbols)

$$[F] = [\tilde{F}^+] + [\tilde{F}^-].$$

As for any element in a sheaf, we can define the support of a $\mathbb{H}$–hyperfunction.

**Definition 3.2.2.** *We will denote by $\mathcal{F}[K]$ the space of $\mathbb{H}$-hyperfunctions with support contained in the compact set $K \subset \widetilde{\mathbb{H}}$.*

An $\mathbb{H}$-hyperfunction $f \in \mathcal{F}(U)$, with $U \subseteq \tilde{\mathbb{H}}$ has its support in $K \subset U$ if and only if it is defined by a function $F \in \mathcal{R}_l(V \setminus K)$, for $V$ such that $U$ is relatively closed in $V$.

**Remark 3.2.2.** As in the complex case we have the following characterization of the space $\mathcal{F}[K]$:

$$\mathcal{F}[K] \cong \frac{\mathcal{R}_l(U \setminus K)}{\mathcal{R}_l(U)} \cong H^1_K(U, \mathcal{R}_l).$$

We now give the notion of integral of a hyperfunction $f = [F]$ with compact support $K$: its integral over an open set $U$ is given by

$$\int_U Dqf := -\int_\Gamma DqF,$$

where $\Gamma \subset V$ is a 3-dimensional closed and smooth variety such that $\Gamma$ is the boundary of a set $\Sigma$ of dimension 4, diffeomorphic to a sphere containing $K$. The Cauchy–Fueter theorem assures that the previous definition of integral does not depend on the choice of $F$ and $\Gamma$. This notion will be used to extend the classical duality theorem 1.3.21 to the quaternionic case, but the novelty here is that, by duality, left and right regular functions interchange their roles, so we need the following definition.

**Definition 3.2.3.** *Let $K$ be a noncompact set in $\mathbb{H}$. The space $\mathcal{G}(K)$ of germs of right $\mathbb{H}$-analytic functions on $K$ is defined by*

$$\mathcal{G}(K) := \varinjlim_{U \text{ open} \supset K} \mathcal{R}_r(U).$$

*The space $\mathcal{G}(K)$ is endowed with the inductive limit topology.*

**Theorem 3.2.4.** *Let $K \subset \mathbb{H}$ be a compact set and let $V \supset K$ be an open set. Then there is an algebraic isomorphism*

$$(\mathcal{G}(K))' \cong \frac{\mathcal{R}_l(V \backslash K)}{\mathcal{R}_l(V)},$$

*where $(\mathcal{G}(K))'$ is the space of left $\mathbb{H}$-linear continuous functionals on $\mathcal{G}(K)$.*

*Proof.* Let us show that it is possible to associate to each function $f$ belonging to $\mathcal{R}_l(V\backslash K)/\mathcal{R}_l(V)$ a functional $\mu_f \in (\mathcal{G}(K))'$. Let $F \in \mathcal{R}_l(V \setminus K)$ be a representative of $f$ and let $\varphi \in \mathcal{G}(K)$. With an abuse of notation, we will denote by $\varphi$ also a regular extension of $\varphi$ to a neighborhood $U \subseteq V$ of $K$. If we set

$$\langle \mu_f, \varphi \rangle := \int_{\partial U} \varphi Dq F, \tag{3.17}$$

the functional $\mu_f$ is well defined by virtue of Cauchy–Fueter's theorem and it turns out to be left $\mathbb{H}$-linear and continuous on $\mathcal{G}(K)$. Let us define the map:

$$T : \frac{\mathcal{R}_l(V\backslash K)}{\mathcal{R}_l(V)} \to (\mathcal{G}(K))', \qquad T(f) = \mu_f.$$

Then $T$ is right $\mathbb{H}$-linear. To show that $T$ is an isomorphism, it suffices to find a map

$$S : (\mathcal{G}(K))' \longrightarrow \frac{\mathcal{R}_l(V\backslash K)}{\mathcal{R}_l(V)}$$

that is the inverse of $T$. To this end, consider $\mu \in (\mathcal{G}(K))'$ and the function $F$ defined by

$$F(q) := \langle \mu_p, G(p-q) \rangle, \qquad p, q \in \mathbb{H}. \tag{3.18}$$

The notation $\mu_p$ indicates that the functional $\mu$ acts with respect to the variable $p$. Let us verify that $F$ is regular on $\mathbb{H} \setminus K$:

$$\frac{\partial F}{\partial \bar{q}} = \frac{\partial}{\partial \bar{q}} \langle \mu_p, G(p-q) \rangle = \langle \mu_p, \frac{\partial}{\partial \bar{q}} G(p-q) \rangle.$$

Note that $\frac{\partial}{\partial \bar{q}} G(p-q) = 0$, for $q \notin K$, so $F \in \mathcal{R}_l(\mathbb{H} \setminus K)$. Then $F$ defines an element $f$ in the quotient $\mathcal{R}_l(V \setminus K)/\mathcal{R}_l(V)$, so that we have a right $\mathbb{H}$-linear map $S$ such that $S(\mu) = f$. Let us verify that $S \cdot T = T \cdot S = \mathrm{id}$.
Let $\mu \in (\mathcal{G}(K))'$; then $S(\mu)$ is defined by $F$ as in (3.18). $T(S(\mu))$ is the functional $\nu_F$ such that

$$\langle \nu_F, \varphi \rangle = \int_{\partial U} \varphi D q F = \int_{\partial U} \varphi D q \langle \mu, G(p-q) \rangle$$

$$= \langle \mu, \int_{\partial U} \varphi D q G(p-q) \rangle = \langle \mu, \varphi \rangle,$$

so $T(S(\mu)) = \mu$. Conversely, let $F \in \mathcal{R}_l(V \setminus K)$. Then $T(F)$ is the functional $\mu_F$ such that

$$\langle \mu_F, \varphi \rangle = \int_{\partial U} \varphi D q F,$$

for any function $\varphi$ $\mathbb{H}$–analytic and right regular. Moreover

$$S(\mu_F) = S(T(F))(q) = \langle \mu_F, G(p-q) \rangle = \int_{\partial U} G(p-q) D q F = F,$$

so that $S(T(F)) = F$, and this completes the proof. □

As an immediate corollary, one obtains the following.

**Corollary 3.2.1.** *If $K$ is a compact set in $\widetilde{\mathbb{H}}$, then*

$$\mathcal{G}(K)' \cong \mathcal{F}[K].$$

The isomorphism in Theorem 3.2.4 can be made into a topological one, once that a topology on the space of left (and right) regular functions on an open set in $\mathbb{H}$ is given.

**Theorem 3.2.5.** *For any open set $U \subseteq \mathbb{H}$, $\mathcal{R}_l(U)$ is a Fréchet space, whose topology is given by the countable family of seminorms:*

$$\|f\|_j = \max_{K_j} |f|, \qquad f \in \mathcal{R}_l(U),$$

*where $\{K_j\}$ is an increasing and exhaustive sequence of compact subsets of $U$, i.e., $K_j$ is contained in the interior of $K_{j+1}$, $j = 1, 2, \ldots$ and $\cup_j K_j = U$.*

The proof of this theorem relies on the fact that the $\mathbb{H}$-vector space of functions $\phi: \ U \subseteq \mathbb{R}^4 \to \mathbb{H}$, $k$-times differentiable is a Fréchet space for any $k \geq 0$, with respect to the countable family of seminorms given above. Moreover, $\mathcal{R}_l(U)$ is a closed subspace of $\mathcal{C}^\infty(U)$. Indeed, if $f_j \to f$ in $\mathcal{R}_l(U)$, then $f$ is infinitely differentiable and satisfies the Cauchy–Fueter equation, because its derivatives $\partial f / \partial x_\ell$ are limit of derivatives satisfying the Cauchy–Fueter equation.
For the readers acquainted with Montel spaces (see for example [137]), we mention also the following result:

**Corollary 3.2.2.** *For any open set $U \subseteq \mathbb{H}$, $\mathcal{R}_l(U)$ is a Montel space.*

**Remark 3.2.3.** The space $\mathcal{G}(K)$ is a limit of Fréchet spaces, and it is naturally endowed with a LF-topology: the continuous seminorms on $\mathcal{G}(K)$ are those which are continuous on all $\mathcal{R}_r(U)$, $U$ neighborhood of $K$. Unfortunately $\mathcal{G}(K)$ turns out not to be a Fréchet space, so we cannot characterize it in terms of convergence of sequences; however the definition of inductive limit topology of $\mathcal{G}(K)$ allows us to say that a sequence $\{\varphi_j\}$ of germs in $\mathcal{G}(K)$ converges to a germ $\varphi \in \mathcal{G}(K)$, if $\varphi_j(q)$ converges uniformly to $\varphi(q)$ in a neighborhood $U \subset \mathbb{H}$ of $K$.

To give a characterization of the convergence in $\mathcal{G}(K)$ involving the compact $K$ only, it is necessary to introduce a suitable class of infinite order differential operators.

**Definition 3.2.4.** *Let*

$$p_\nu(q) = \frac{1}{m!} \sum_{1 \leq \lambda_1, \ldots, \lambda_m \leq 3} (i_{\lambda_1} x_0 - i_0 x_{\lambda_1}) \ldots (i_{\lambda_m} x_0 - i_0 x_{\lambda_m}).$$

*We define the operator*

$$p_\nu(D) = \frac{1}{m!} \sum_{1 \leq \lambda_1, \ldots, \lambda_m \leq 3} \left( \frac{\partial}{\partial x_{\lambda_1}} i_0 - \frac{\partial}{\partial x_0} i_{\lambda_1} \right) \cdots \left( \frac{\partial}{\partial x_{\lambda_m}} i_0 - \frac{\partial}{\partial x_0} i_{\lambda_m} \right), \tag{3.19}$$

*and the formal infinite order differential operator*

$$F(D) = \sum_{m=0}^{+\infty} \sum_{\nu \in \sigma_m} p_\nu(D) a_\nu, \qquad a_\nu \in \mathbb{H}. \tag{3.20}$$

The operators $p_\nu(D)$ have been obtained (formally) by replacing $i_\ell$ by $\partial/\partial x_\ell$ and $x_\lambda$ by $i_\lambda$. Another way to introduce the operators $p_\nu(D)$ is to consider them as the Cauchy–Kowalewski extension (C–K extension) of suitable operators. In fact, every analytic function $f : \widetilde{\mathbb{H}} \to \mathbb{H}$ admits a unique left regular extension $\tilde{f} : \mathbb{H} \to \mathbb{H}$ called (left) C–K extension (see [29] for the details). The functions $p_\ell : \widetilde{\mathbb{H}} \to \mathbb{H}$ defined by $p_\ell(q) = x_\ell$, $\ell = 1, 2, 3$ admit $\tilde{p}_\ell(q) = x_\ell i_0 - x_0 i_\ell$ as a C–K extension, while a function of the type $x_1^{m_1} x_2^{m_2} x_3^{m_3}$ has $p_\nu(q)$, $\nu = [m_1, m_2, m_3]$, as a C–K extension. It is then natural to consider the operator

$$\frac{\partial}{\partial x_\ell} i_0 - \frac{\partial}{\partial x_0} i_\ell$$

as the C–K extension to $\mathbb{H}$ of the operator $\partial/\partial x_\ell$ acting on $\widetilde{\mathbb{H}}$, and $p_\nu(D)$ as the C–K extension of

$$\frac{\partial^m}{\partial x_1^{m_1} \partial x_2^{m_2} \partial x_3^{m_3}}.$$

The differential operator $F(D)$ acts formally on left regular functions, although it is not always the case that $F(D)f$ is convergent for every regular function $f$ (see Theorem 3.2.6 below). An analogous operator acting on right regular functions, can be defined by $F(D)$ as $\sum a_\nu p_\nu(D)$, where $a_\nu \in \mathbb{H}$ and $p_\nu(D)$ contains factors of the type

$$\left(i_0\frac{\partial}{\partial x_\ell} - i_\ell\frac{\partial}{\partial x_0}\right).$$

**Lemma 3.2.1.** *The following commutation relation holds:*

$$p_\nu(D)\frac{\partial}{\partial \bar{q}} = \frac{\partial}{\partial \bar{q}}p_\nu(D).$$

*As a consequence, $p_\nu(D) : \mathcal{R}_l \to \mathcal{R}_l$ is a sheaf homomorphism.*

*Proof.* It suffices to prove that $\dfrac{\partial}{\partial \bar{q}}$ commutes with

$$\left(\frac{\partial}{\partial x_\lambda}i_0 - \frac{\partial}{\partial x_0}i_\lambda\right).$$

Computing

$$\left(\sum_{\ell=0}^{3} i_\ell\frac{\partial}{\partial x_\ell}\right)\left(\frac{\partial}{\partial x_\lambda}i_0 - \frac{\partial}{\partial x_0}i_\lambda\right) = \sum_{\ell=0}^{3} i_\ell\frac{\partial^2}{\partial x_\ell \partial x_\lambda}i_0 - \sum_{\ell=0}^{3} i_\ell\frac{\partial^2}{\partial x_\ell \partial x_0}i_\lambda,$$

and

$$\left(\frac{\partial}{\partial x_\lambda}i_0 - \frac{\partial}{\partial x_0}i_\lambda\right)\left(\sum_{\ell=0}^{3} i_\ell\frac{\partial}{\partial x_\ell}\right) = \sum_{\ell=0}^{3} i_\ell\frac{\partial^2}{\partial x_\lambda \partial x_\ell}i_0 - \sum_{\ell=0}^{3} i_\ell\frac{\partial^2}{\partial x_0 \partial x_\ell}i_\lambda,$$

the statement easily follows. □

Under suitable conditions an infinite order differential operator of the type (3.20) acts as a sheaf homomorphism on $\mathcal{R}_l$.

**Theorem 3.2.6.** *Let $F(D)$ be an operator as in (3.20) and let $f(q)$ be a germ of a left regular function in an open set $U \subseteq \mathbb{H}$. If*

$$\lim_{|\nu|\to+\infty} \sqrt[|\nu|]{|a_\nu|} = 0, \tag{3.21}$$

*then $F(D)f(q)$ is a germ of a left regular function on $U$. Thus, $F(D)$ is a sheaf homomorphism from $\mathcal{R}_l$ to itself.*

*Proof.* It suffices to show that if $f(q)$ is regular in $q_0 \in U$, then also $F(D)f(q)$ is regular $q_0$. Let us suppose, without loss of generality, that $q_0 = 0$. The operators $p_\nu(D)$ defined by (3.19) are linear combinations of partial derivatives of order $m$. If $f(q)$ is regular at the origin, then in a neighborhood $\mathcal{U}(0,\delta) = \{q \in \mathbb{H} \mid |q| \leq \delta\}$ we have

$$\begin{aligned}
|p_\nu(D)f(q)| &\leq \frac{1}{m!}\left(\sum_{0\leq\lambda_i\leq 3} |\partial_\nu f(q)|\right) \\
&\leq \frac{1}{m!}\left(\sum_{0\leq\lambda_i\leq 3}\left|\frac{1}{2\pi^2}\int_{|p-q|=\delta'}(\partial_\nu G(p))Dpf(p)\right|\right) \\
&\leq \frac{1}{m!}\frac{1}{2\pi^2}\max_{|p|\leq\delta}|f(p)|\left(\sum_{0\leq\lambda_i\leq 3}\int_{|p-q|=\delta}|(\partial_\nu G(p))Dp|\right),
\end{aligned}$$

where we have chosen $\delta'$ such that the ball with center in $q$ and radius $\delta'$ is contained in $\mathcal{U}(0,\delta)$. Now we estimate the function $\partial_\nu G(p)$ on the ball $\mathcal{U}(0,\delta)$: note that $\partial_\nu G(p)$ is homogeneous of degree $(-m-3)$ and its denominator is the function $|p|^{2^{m+2}}$, so its numerator has to be a homogeneous polynomial of degree $(2^{m+2}-m-3)$ in the variables $x_0,\ldots,x_3$. We have the following inequalities

$$\begin{aligned}
|\partial_\nu G(p)| &\leq 4\binom{2^{m+2}-m}{2^{m+2}-m-3}\frac{|p|^{2^{m+2}-m-3}}{|p|^{2^{m+2}}} \\
&\leq 4\frac{2^{m+2}-m}{2}\frac{1}{|p|^{m+3}}.
\end{aligned}$$

Setting $M = \max_{|p|\leq\delta}|f(p)|$, we can write

$$\begin{aligned}
|p_\nu(D)f(q)| &\leq \frac{1}{m!}\frac{M}{2\pi^2}\frac{m!2^m}{m_1!m_2!m_3!}\frac{2^{m+2}-m}{2}\frac{4}{\delta^{m+3}}\int_{|p-q|=\delta'}|Dp| \\
&\leq \frac{M}{2\pi^2}\frac{2^{2m+2}-m}{2}\frac{4}{\delta^{m+3}}\int_{|p-q|=\delta'}|Dp| \\
&\leq MC\frac{2^m}{\delta^{m+3}},
\end{aligned}$$

where $C$ is a constant. By hypothesis, for all $\varepsilon > 0$ there exists $C_\varepsilon > 0$ such that $|a_\nu| \leq C_\varepsilon \varepsilon^m$ for all $a_\nu$ with $|\nu| = m$, so that for $q \in \mathcal{U}(0,\delta)$ we have

$$|F(D)f(q)| \leq \sum_{m=0}^{+\infty}\frac{(m+1)(m+2)}{2}\frac{MC2^{m+2}}{\delta^{m+3}}C_\varepsilon\varepsilon^m. \tag{3.22}$$

Putting $\varepsilon = \delta/4$ we get

$$|F(D)f(q)| \leq \sum_{m=0}^{+\infty} \frac{C'(m+2)^2}{\delta^3}\frac{1}{2^m}$$

so that the series converges uniformly in $\mathcal{U}(0,\delta)$, and so we get a regular function. □

We will see later (see Lemma 3.2.2 and Remark 3.2.6) that the necessary condition in the theorem is indeed sufficient.

**Definition 3.2.5.** *We denote by $\mathcal{I}$ the set of operators $F(D)$ defined in (3.20) and satisfying (3.21).*

**Proposition 3.2.3.** *Every $F(D) \in \mathcal{I}$ acts as a sheaf endomorphism on the sheaves $\mathcal{R}_l$ and $\mathcal{F}$. Moreover $F(D)$ is continuous on $\mathcal{R}_l(U)$ for any open set $U \subset \mathbb{H}$.*

*Proof.* The first part of the statement follows from the fact that a sheaf endomorphism on $\mathcal{R}_l$ is a set of $\mathbb{H}$–linear maps $F_U : \mathcal{R}_l(U) \to \mathcal{R}_l(U)$, for any open set $U \subseteq \mathbb{H}$, commuting with any inclusion $V \hookrightarrow U$. The case of $\mathcal{F}$ is analogous. From the proof of Theorem 3.2.6 it also follows that the estimate (3.22) only depends on the maximum of $|f|$, so that $F(D)$ is continuous on $\mathcal{R}_l(U)$. □

**Corollary 3.2.3.** *Every operator $F(D) \in \mathcal{I}$ is continuous on $\mathcal{G}(K)$.*

*Proof.* It is obvious that Theorem 3.2.6 holds for operators $F(D)$ acting on $\mathcal{R}_r(U)$. Then Proposition 3.2.3 is valid for $\mathcal{R}_r(U)$, so $F(D)$ is continuous on $\mathcal{G}(K)$. □

Given a hyperfunction $f \in \mathcal{F}[K]$, it is possible to find a representative in a canonical way. This representative will be said to be the standard defining function and it will be defined outside $K$, more precisely, in the complement of $K$ taken in the one point compactification of $\mathbb{H}$, denoted by $\mathbb{HP}^1 := \mathbb{H} \cup \{\infty\}$.

**Theorem 3.2.7.** *Let $f \in \mathcal{F}[K]$ and define*

$$E(q) = \frac{1}{2\pi^2}\int_{\widetilde{\mathbb{H}}} G(p-q)Dpf(p); \tag{3.23}$$

*then $E \in \mathcal{R}_l(\mathbb{HP}^1 \backslash K)$, $[E] = f$ and $E(\infty) = 0$. Moreover, the function $E$ with those properties is unique.*

*Proof.* Let $F \in \mathcal{R}_l(V \setminus K)$ be a representative of $f$. We can choose an open set $U$ such that $q \in \mathbb{H} \setminus K$, and $q \notin U$. Then

$$E(q) = -\frac{1}{2\pi^2}\int_{\partial U} G(p-q)DpF(p).$$

By definition, $E(q)$ is regular in $\mathbb{H} \setminus K$, so $[E(q)]$ is an $\mathbb{H}$-hyperfunction with support in $K$. We now show that $[E(q)] = [F(q)] = f$, by proving that the function $E(q) - F(q)$ can be extended to a regular function in a neighborhood of $K$. Let $U'$ be an open set containing $q$, such that $U$ and $U'$ have no intersection along the boundary. From Theorem 3.1.3 we get

$$F(q) = \frac{1}{2\pi^2} \int_{\partial U'} G(p-q)DpF(p),$$

so that

$$\begin{aligned} F(q) - E(q) &= \frac{1}{2\pi^2} \int_{\partial U'} G(p-q)DpF(p) + \frac{1}{2\pi^2} \int_{\partial U} G(p-q)DpF(p) \\ &= \frac{1}{2\pi^2} \int_{\partial(U \cup U')} G(p-q)DpF(p). \end{aligned}$$

The interior of $U \cup U'$ contains $q$, therefore $F(q) - E(q)$ can be extended to a regular function on $U \cup U'$. Moreover, by its very definition, $E(q)$ is regular at infinity and $E(\infty) = 0$. The uniqueness of the function $E$ follows from Liouville's theorem. □

**Definition 3.2.6.** *We will denote by* $\mathcal{R}_l^\infty(\mathbb{HP}^1 \setminus K)$ *the space of left regular functions on* $\mathbb{HP}^1 \setminus K$ *vanishing at infinity.*

As a consequence of Theorem 3.2.7 and the uniqueness of the canonical representative of a hyperfunction, we have the following corollary.

**Corollary 3.2.4.** *Let $K$ be a compact set in* $\widetilde{\mathbb{H}}$. *Then* $\mathcal{F}[K] \cong \mathcal{R}_l^\infty(\mathbb{HP}^1 \setminus K)$.

**Theorem 3.2.8.** *Let $K$ be a compact set in* $\widetilde{\mathbb{H}}$ *and* $f \in \mathcal{F}[K]$. *The map*

$$A : \mathcal{F}[K] \to \mathcal{R}_l(\mathbb{H} \setminus K)$$

*defined by*

$$A(f) = \frac{1}{2\pi^2} \int_{\widetilde{\mathbb{H}}} G(p-q)Dpf(p),$$

*is injective and its image is a closed subspace of the Fréchet space* $\mathcal{R}_l(\mathbb{H} \setminus K)$.

*Proof.* The map $A$ associates to any $\mathbb{H}$-hyperfunction $f \in \mathcal{F}[K]$ its standard defining function $A(f)$, so $A$ is injective. Let $\{f_j\}$ be a sequence in $\mathcal{F}[K]$ such that $A(f_j)$ converges to $F \in \mathbb{R}_l(\mathbb{H} \setminus K)$, with the topology that makes it a Fréchet space. Let us prove that the $\mathbb{H}$-hyperfunction $f$ defined by $F$ and supported by $K$ is such that $A(f) = F$. We can compute $A(f_j)$ using any defining function for $f_j$, and if we choose $f_j = [F_j]$, then we have

$$A(f_j) = \frac{1}{2\pi^2} \int_{\widetilde{\mathbb{H}}} G(p-q)DpF_j(p) = F_j(q).$$

By hypothesis, $F_j \to F$ uniformly on any compact set that does not meet $\widetilde{\mathbb{H}}$, and this concludes the proof. □

**Corollary 3.2.5.** *$\mathcal{F}[K]$ is a Fréchet space.*

*Proof.* $\mathcal{F}[K]$ is isomorphic to a closed subspace of $\mathcal{R}_l(\mathbb{H}\backslash K)$. □

We are now in position to prove the "dual" version of Theorem 3.2.4:

**Theorem 3.2.9.** *Let $K \subset \tilde{\mathbb{H}}$ be a compact set. Then*

$$(\mathcal{F}[K])' \cong \mathcal{G}(K)$$

*where $(\mathcal{F}[K])'$ denotes the space of right $\mathbb{H}$-linear continuous functionals.*

*Proof.* Consider, for any $f \in \mathcal{F}[K]$ and for any $\varphi \in \mathcal{G}(K)$, the duality bracket:

$$< f, \varphi >:= \int_{\tilde{\mathbb{H}}} \varphi Dqf = \int_{\partial U} \varphi DqF \tag{3.24}$$

where, as usual, $f = [F]$ and $U$ is any open set containing $K$. If we fix $\varphi \in \mathcal{G}(K)$, (3.24) defines a continuous right $\mathbb{H}$-functional $\phi$ on $\mathcal{F}[K]$ and if $\varphi_1 \neq \varphi_2$, then the functionals $\phi_1$ and $\phi_2$ are different. Indeed, let $q_0 \in K$, we consider $\delta(q_0) = [G(q - q_0)]$. We have

$$\phi_1(\delta) = \int_{\partial U} \varphi_1(q) DqG(q - q_0) = \varphi_1(q_0),$$

and

$$\phi_2(\delta) = \int_{\partial U} \varphi_2(q) DqG(q - q_0) = \varphi_2(q_0),$$

hence we have a one to one mapping:

$$T : \mathcal{G}(K) \longrightarrow (\mathcal{F}[K])'.$$

Conversely: let $\phi : \mathcal{F}[K] \longrightarrow \mathbb{H}$ be a continuous right $\mathbb{H}$-linear map. Then $\phi$ acts continuously on the closed subspace of $\mathbb{R}_l(\mathbb{H} \setminus K)$ defined in Theorem 3.2.8, with the topology induced. According to the Hahn–Banach theorem (see [21]), $\phi$ extends to act continuously on all $\mathbb{R}_l(\mathbb{H} \setminus K)$, and allows us to define

$$\varphi(q) = \phi_p(G(p - q)). \tag{3.25}$$

In particular, there exists a compact set $K' \subset \mathbb{H} \setminus K$ such that

$$\max_{q \in K'} |E(q)| \leq \eta \Longrightarrow |\phi(f)| \leq \varepsilon.$$

The extension of $\phi$ preserves this condition hence $\varphi \in \mathcal{G}(K)$. We will call $S : (\mathcal{F}[K])' \to \mathcal{G}(K)$ the map that we have just built. Now we have to show that $T \cdot S = 1$. However,

$$\begin{aligned} T \cdot S(\phi)(f) &= T[S(\phi)](f) = T[\phi_p(G(p - q))](f) \\ &= \int \phi_p(G(p - q)) Dqf = \phi_p \left[ \int G(p - q) Dqf(q) \right] = \phi[f(p)], \end{aligned}$$

which concludes the proof. □

**Corollary 3.2.6.** *The dual space of* $\mathbb{R}_l^\infty(\mathbb{HP}^1 \setminus \{|q| \leq r\})$ *is the space of all left-regular functions defined in a neighborhood of* $|q| \leq r$.

*Proof.* It suffices to repeat the proof of the previous theorem, where the duality bracket is defined by $< F(q), \varphi(q) >= \int \varphi(q) Dq F(q)$. □

Now we can rephrase Remark 3.2.3 that describes the convergence in $\mathcal{G}(K)$, with a result involving only the compact set $K$, following an argument originally given in [98].

**Theorem 3.2.10.** *Let* $\{\varphi_k\}$ *be a sequence in* $\mathcal{G}(K)$ *and let* $\varphi \in \mathcal{G}(K)$. *Then* $\{\varphi_k\} \longrightarrow \varphi$ *in* $\mathcal{G}(K)$ *if and only if the sequence* $\{F(D)\varphi_k(q)\}$ *converges pointwise on* $K$.

*Proof.* The necessity follows from the continuity of $F(D)$, so we have only to prove the sufficiency. We can assume that $K = \{0\}$ and we prove that if $F(D)\varphi_k(0)$ is convergent for any $F(D)$, then $\varphi_k(q)$ converges uniformly in some neighborhood of the origin. We can define an operator $\tilde{F}_D \in (\mathcal{G}(\{0\}))'$ as follows:

$$\tilde{F}_D(\varphi) = F(D)\varphi(0), \quad \forall \varphi \in \mathcal{G}(\{0\}).$$

By Corollary 3.2.1 we have that there exists a function $\tilde{F} \in \mathcal{F}[\{0\}]$ such that $\tilde{F}_D(\varphi) = \int_{\partial U} \varphi Dq \tilde{F}$, where $U$ is a neighborhood of the origin. By Corollary 3.2.4, we can identify $\mathcal{F}[\{0\}]$ with $\mathbb{R}_l^\infty(\mathbb{HP}^1 \setminus \{0\})$ so that we can think of $\tilde{F}$ as an element of $\mathbb{R}_l^\infty(\mathbb{HP}^1 \setminus \{0\})$. We set

$$< \tilde{F}, \varphi_k >= \int_{\partial U} \varphi_k Dq \tilde{F} \tag{3.26}$$

and, in the sequel, we will use the notation $< \tilde{F}, \varphi_k >$ instead of $F(D)\varphi_k(0)$. We divide the proof in four steps.

**Step 1.** There is a suitable $\varepsilon > 0$ such that for any $\tilde{F} \in \mathcal{R}_l^\infty(\mathbb{HP}^1 \backslash \{0\})$ we have

$$\sup_{q \geq \varepsilon} |\tilde{F}(q)| \leq \varepsilon \Longrightarrow | < \tilde{F}(q), \varphi_k(q) > | \leq 1, \quad k = 1, 2, \ldots \tag{3.27}$$

In fact, if we negate the claim, we can inductively choose elements $\tilde{F}_k(q)$ in $\mathcal{R}_l^\infty(\mathbb{HP}^1 \backslash \{0\})$ and an increasing sequence $\{n_k\}$, $k = 1, 2, \ldots$, such that

$$\sup_{|q| \geq 1/2^k} |\tilde{F}_k(q)| \leq \frac{1}{2^k} \tag{3.28}$$

$$| < \tilde{F}_k(q), \varphi_j(q) > | \leq \frac{1}{2^k} \quad j = 1, 2, \ldots, n_{k-1} \tag{3.29}$$

$$| < \tilde{F}_k(q), \varphi_{n_k}(q) > | \geq 2^k + \sum_{j=1}^{k-1} | < F_j(q), \varphi_{n_k}(q) > |. \tag{3.30}$$

In fact, suppose that $n_k$ is the least integer for which (3.29) does not hold. We know that the inner product $< \tilde{F}, \varphi >$ is continuous, so that (3.29) follows for finitely many $\varphi_j$, possibly replacing $1/2^k$ on the right side of (3.28) by a smaller $\delta_k$. Moreover, $< \tilde{F}(q), \varphi_k(q) >$ is convergent for any $\tilde{F}$ by hypothesis, therefore $< \tilde{F}_j(q), \varphi_k(q) >$, $k \in \mathbf{N}$, are bounded for finitely many elements $\tilde{F}_j(q)$, with $j = 1, \dots, k-1$. So the right–hand side of (3.30) is bounded by the number $c_k$ obtained by adding $2^k$ to $(k-1)$–times the least upper bound of the above sequences. Since we have negated (3.27), for $\varepsilon = \min\{1/2^k, \delta_k\}/c_k$, we can find $\tilde{F}_k(q)$ and $\varphi_{n_k}(q)$ such that

$$\sup_{|q|\geq 1/2^k} |\tilde{F}_k(q)| \leq \frac{\min\{1/2^k, \delta_k\}}{c_k}, \quad | < \tilde{F}_k(q), \varphi_{n_k}(q) > | \geq 1$$

The functions $c_k\tilde{F}_k(q)$ satisfy (3.28), (3.29), (3.30), the series

$$\tilde{F}(q) = \sum_{k=1}^{\infty} c_k \tilde{F}_k(q)$$

converges locally uniformly on $\mathbb{HP}^1 \setminus \{0\}$, and its sum is in $\mathbb{R}_l^\infty(\mathbb{HP}^1 \setminus \{0\})$. It follows that

$$| < \tilde{F}(q), \varphi_{n_k}(q) > | \geq | < c_k\tilde{F}_k(q) > | - \sum_{j=1}^{k-1} | < c_j\tilde{F}_j(q), \varphi_{n_k}(q) > |$$

$$- \sum_{j=k+1}^{\infty} | < c_j\tilde{F}_j(q), \varphi_{n_k}(q) > | \geq 2^k - \sum_{j=k+1}^{\infty} \frac{1}{2^j} \geq 2^k - 1$$

but this inequality is absurd, because by hypothesis $< \tilde{F}(q), \varphi_k(q) >$ is convergent.

**Step 2.** Now we want to prove that $\varphi_k(q)$ are right-regular in a neighborhood $\{q \in \mathbb{H}\ |q| \leq r\}$ of the origin, for all $k$. According to the Corollary 3.2.6, it suffices to prove that $\varphi_k$ defines a continuous linear functional on $\mathbb{R}_l^\infty(\mathbb{HP}^1 \setminus \{|q| \leq r\})$. Choose a positive number $\varepsilon$ satisfying (3.27), take $r = \varepsilon/2$ and set $\rho = \varepsilon$. Up to a constant factor, for every $\varepsilon > 0$ there exists $\delta$ such that the implication

$$\sup_{|q|\geq\rho} |\tilde{F}(q)| \leq \delta \Longrightarrow | < \tilde{F}(q), \varphi_k(q) > | \leq \varepsilon \tag{3.31}$$

holds for every $\tilde{F}(q) \in \mathcal{R}_l^\infty(\mathbb{HP}^1 \setminus \{0\})$. Now we can consider the power series expansion for $\tilde{F}(q)$ at infinity. It is easy to see that we can approximate $\tilde{F}(q)$ by a sequence $\{\tilde{F}_h(q)\}$ in $\mathcal{R}_l^\infty(\mathbb{HP}^1 \setminus \{0\})$ uniformly on $\mathbb{HP}^1 \setminus \{|q| \leq \rho\}$. So we can extend the inner product (3.26) to the elements of $\mathbb{R}_l^\infty(\mathbb{HP}^1 \setminus \{|q| \leq r\})$ as follows:

$$< \tilde{F}(q), \varphi_k(q) > = \lim_{h\to\infty} < \tilde{F}_h(q), \varphi_k(q) > .$$

This limit exists because (3.31) expresses the continuity and it satisfies a similar inequality. Now we have obtained a continuous linear functional that can be expressed, by Corollary 3.2.6, as

$$< \tilde{F}(q), \varphi_k(q) > = \int_{|q|=\rho'} \varphi_k(q) Dq \tilde{F}(q) \tag{3.32}$$

where $\varphi_k(q)$ is left-regular in a neighborhood of $\{|q| \leq r\}$. Now we take as $\tilde{F}(q)$ the function $1/(2\pi^2)G(q-p)$, with $p$ in the original domain of regularity of $\varphi_k(q)$. Let $\{G_h\}$ be an approximating sequence of $G$. We have

$$\begin{aligned} < \frac{1}{2\pi^2} G(q), \varphi_k(q) > &= \lim_{h\to\infty} \int_{\partial U} \varphi_k(q) Dq G_h(q-p) \\ &= \int_{\partial U} \frac{1}{2\pi^2} \varphi_k(q) Dq G(q-p) = \varphi_k(p) \end{aligned} \tag{3.33}$$

By (3.32) we have that $< 1/(2\pi^2)G(q-p), \varphi_k(q) >$ is $\varphi_k(p)$, therefore $\varphi_k(p)$ is left-regular on $\{|q| \leq r\}$.

**Step 3**. We want to prove that, chosen $\tilde{F}(q) \in \mathbb{R}_l^\infty(\mathbb{HP}^1 \setminus \{|q| \leq r\})$, the sequence $\{< \tilde{F}(q), \varphi_k(q) >\}$ is convergent. In fact, let $H(q)$ be a function in $\mathcal{R}_l^\infty(\mathbb{HP}^1 \setminus \{0\})$ such that

$$\sup_{|q|\geq\rho>0} |\tilde{F}(q) - H(q)| \leq \delta.$$

Then for sufficiently large $k_0$ we have that the condition $k, h \geq k_0$ implies

$$| < \tilde{F}(q), \varphi_k(q) > - < \tilde{F}(q), \varphi_h(q) > | \leq | < \tilde{F}(q) - H(q), \varphi_k(q) > | +$$

$$+ | < \tilde{F}(q) - H(q), \varphi_h(q) > | + | < H(q), \varphi_k(q) > - < H(q), \varphi_h(q) > | \leq 3\varepsilon$$

by virtue of (3.31). So the previous numerical sequence converges.

**Step 4**. We will prove that the sequence $\varphi_k(q)$ converges uniformly in a neighborhood of the origin. This will conclude the proof.

From (3.33) we have

$$\varphi_k(p) =< 1/(2\pi^2)G(q-p), \varphi_k(q) > . \tag{3.34}$$

Since (3.31) holds also for elements in $\mathbb{R}_l^\infty(\mathbb{HP}^1 \setminus \{|q| \leq r\})$, (3.34) is bounded on $|p| \leq r$. Hence $\{\varphi_k(q)\}$ comprises a normal family so that, by Montel's theorem, a subsequence of it contains a subsequence that converges locally uniformly on $|p| \leq r$. We also know, from (3.33), that (3.34) is convergent for each fixed $p$ in this disc, so the sequence $\varphi_k(q)$ converges pointwise to a limit function $\varphi(p)$ that must coincide with the limit of any subsequence. Then the sequence $\varphi_k(p)$ converges to $\varphi(p)$ locally uniformly on $|p| \leq r$. □

**Remark 3.2.4.** The space $(\mathcal{F}[K])'$ has the topology of uniform convergence on bounded sets.

Now we can topologize the isomorphism described in Theorem 3.2.9.

**Theorem 3.2.11.** *The isomorphism $(\mathcal{F}[K])' \cong \mathcal{G}(K)$ is topological.*

*Proof.* We have to show that $\varphi_k \longrightarrow \varphi$ in $\mathcal{G}(K)$ if and only if $\phi_k \longrightarrow \phi$ in $(\mathcal{F}[K])'$, where $\phi_k$ is related to $\varphi_k$ by (3.25), i.e.,

$$\varphi_k(q) = \phi_k[G(p-q)] = \phi_{k,p}[G(p-q)].$$

If $\varphi_k \in \mathcal{G}(K)$ converges to $\varphi$, this means that $\varphi_k \longrightarrow \varphi$ uniformly in a neighborhood of $K$. With respect to the duality defined by (3.24), we have $< f, \varphi_k > \longrightarrow < f, \varphi >$ uniformly when $f$ varies in a bounded subset $B$ of $\mathcal{F}[K]$. Moreover if $f$ varies in a bounded set $B$, their defining function $E(q)$, defined by (3.23), are uniformly bounded on the region in which $< f, \varphi_k >$ is computed.

Conversely, suppose that $\phi_k \longrightarrow \phi$ in $(\mathcal{F}[K])'$, with its topology. Then the function $\varphi_k = \phi_{k,p}[G(p-q)]$ is right-regular in a neighborhood $U$ of $K$. Now we have to show that the sequence $\{\varphi_k\}$ converges uniformly in some neighborhood of $K$. By Theorem 3.2.10 it is enough to prove that $\{F(D)\varphi_k\}$ converges pointwise for all infinite order differential operators $F(D)$ satisfying the condition of Theorem 3.2.6. From the continuity of $\phi_k$, fixed a $q \in K$, we have

$$F(D)(\varphi_k)(q) = \phi_{k,p}(F(D)G(p-q)) \longrightarrow \phi_p((F(D)G(p-q)).$$

The statement follows. □

**Corollary 3.2.7.** *The following is a topological isomorphism*

$$(\mathcal{G}(K))' \cong \mathcal{F}[K].$$

*Proof.* We know that $\mathcal{R}_l(U)$ is a Montel space (see Corollary 3.2.2). This fact implies that $\mathcal{F}[K]$ is a Montel space, so $\mathcal{F}[K]$ is reflexive, i.e., the dual of $\mathcal{G}(K)$ is $\mathcal{F}[K]$ itself. □

**Corollary 3.2.8.** *The isomorphism*

$$\mathcal{F}[K] \cong \mathcal{R}_l^{\infty}(\mathbb{HP}^1 \setminus K)$$

*is topological as well.*

We now wish to establish the analogue of Theorem 1.3.7 that states that every hyperfunction supported at the origin can be written as a series of derivatives of the Dirac delta. Let us consider a function $f$ regular outside the origin. By Theorem 3.1.11 $f$ admits a Laurent expansion of the type

$$f(q) = \sum_{m=0}^{+\infty} \sum_{\nu \in \sigma_m} \Big( p_\nu(q) a_\nu + G_\nu(q) b_\nu \Big)$$

in $r_1 < |q| < r_2$. The following lemma gives a condition on the coefficients $b_\nu$, in order that the series $\sum_{m=0}^{+\infty} \sum_{\nu \in \sigma_m} G_\nu(q) b_\nu$ be regular on $\mathbb{H} \setminus \{0\}$.

**Lemma 3.2.2.** *The series*

$$\sum_{m=0}^{+\infty} \sum_{\nu \in \sigma_m} G_\nu(q) b_\nu$$

*converges in* $\mathbb{H} \backslash \{0\}$ *if and only if*

$$\lim_{|\nu| \to +\infty} \sqrt[|\nu|]{|b_\nu|} = 0. \tag{3.35}$$

*Proof.* If (3.35) holds, then for all $\varepsilon > 0$ there is a constant $C_\varepsilon > 0$ such that $|b_\nu| \le C_\varepsilon \varepsilon^m$. Following the proof of Theorem 3.2.6 we get

$$\begin{aligned}
&\left| \sum_{m=0}^{+\infty} \sum_{\nu \in \sigma_m} G_\nu(q) b_\nu \right| \\
&\qquad \le \sum_{m=0}^{+\infty} \sum_{\nu \in \sigma_m} |G_\nu(q) b_\nu| \\
&\qquad \le \sum_{m=0}^{+\infty} \sum_{\nu \in \sigma_m} 2(2^{m+2} - m) \frac{1}{|q|^{m+3}} |b_\nu| \\
&\qquad \le \sum_{m=0}^{+\infty} \frac{(m+1)(m+2)}{2} 2^{m+3} \frac{1}{|q|^{m+3}} C_\varepsilon \varepsilon^m \\
&\qquad \le \sum_{m=0}^{+\infty} (m+2)^2 2^{m+2} \frac{1}{|q|^{m+3}} C_\varepsilon \varepsilon^m .
\end{aligned}$$

Setting $\varepsilon = \dfrac{|q|}{4}$, we obtain

$$\left| \sum_{m=0}^{+\infty} \sum_{\nu \in \sigma_m} G_\nu(q) b_\nu \right| \le \sum_{m=0}^{+\infty} 4(m+2)^2 \frac{C_\varepsilon}{2^m |q|^3},$$

which implies that the series converges outside the origin.

Conversely, let us suppose that (3.35) does not hold. Then for any $\varepsilon > 0$ there exists a sequence $\{\mu_j\}$ such that $|b_{\mu_j}| > \varepsilon^{|\mu_j|}$. By assumption, the series converges for all $q \in \mathbb{H} \backslash \{0\}$ and, in particular for $q = \varepsilon$. Each term of the type $G_\nu(\varepsilon) b_\nu$ eventually goes to zero, so that also $|G_\nu(\varepsilon) b_\nu|$ converges to zero. Nevertheless, for $\nu \in \{\mu_j\}$, we have

$$|G_\nu(\varepsilon) b_\nu| \ge \frac{H(\varepsilon)}{\varepsilon^{2^{m+2}}} |b_\nu|$$

where $H(\varepsilon)$ is a homogeneous polynomial of degree $(2^{m+2} - m - 3)$. Then we have:

$$|G_\nu(\varepsilon) b_\nu| \ge \frac{c_\nu}{\varepsilon^{m+3}} |b_\nu| > \frac{c_\nu}{\varepsilon^3}$$

where $c_\nu$ is a positive constant. This shows that $G_\nu(\varepsilon)b_\nu$ cannot converge to zero which is a contradiction. □

**Theorem 3.2.12.** *Let $f \in \mathcal{F}[\{0\}]$. Then $f$ can be uniquely written as*

$$f(q) = \sum_{m=0}^{+\infty} \sum_{\nu\in\sigma_m} \Big(\partial_\nu \delta(q)\Big) b_\nu,$$

*where the coefficients $b_\nu$ satisfy*

$$\lim_{|\nu|\to+\infty} \sqrt[|\nu|]{|b_\nu|} = 0.$$

*Proof.* Since it is

$$\mathcal{F}[\{0\}] \cong \frac{\mathcal{R}_l(\mathbb{H}\backslash\{0\})}{\mathcal{R}_l(\mathbb{H})},$$

$f(q)$ is represented by a function $F(q)$ regular on $\mathbb{H}\backslash\{0\}$. Let us write the Laurent expansion of $F(q)$ in the origin

$$F(q) = \sum_{m=0}^{+\infty} \sum_{\nu\in\sigma_m} (p_\nu(q)a_\nu + G_\nu(q)b_\nu).$$

We can choose the following representative of $f$:

$$\tilde{F}(q) = \sum_{m=0}^{+\infty} \sum_{\nu\in\sigma_m} G_\nu(q)b_\nu,$$

where $\{b_\nu\}$ satisfies (3.35). The Cauchy kernel, neglecting the factor $1/2\pi^2$, defines the Dirac delta function. We have then obtained a bijective correspondence between $\mathbb{H}$–hyperfunctions supported at the origin and series of delta derivatives of the type

$$\sum_{m=0}^{+\infty} \sum_{\nu\in\sigma_m} \Big(\partial_\nu \delta\Big) b_\nu$$

where $b_\nu$ satisfies the growth condition in (3.35). □

We can introduce a second set of infinite order differential operators acting on a function $f$ as follows:

$$J(D)f = \sum_{m=0}^{+\infty} \sum_{\nu\in\sigma_m} (\partial_\nu f)\, b_\nu. \tag{3.36}$$

It is natural to ask under which conditions those kind of operators act as sheaf homomorphism on $\mathcal{R}_l$. The answer is contained in the following result.

**Proposition 3.2.4.** *Let $J(D)$ be the operator defined in (3.36). Then $J(D)$ is a sheaf homomorphism on $\mathcal{R}_l$ if and only if (3.35) holds.*

*Proof.* It is immediate to show that $J(D) : \mathcal{R}_l(U) \longrightarrow \mathcal{R}_l(U)$ commutes with $\partial_l/\partial\bar{q}$ and with respect to every inclusion $V \hookrightarrow U$, so it suffices to prove that $f(q)$ is a germ of a regular function in $q_0$ if and only if so is $J(D)f(q)$. The statement follows from the estimates in the proof of Theorem 3.2.6. □

**Definition 3.2.7.** *We denote by $\mathcal{J}$ the set of operators $J(D)$ defined by (3.36) and satisfying (3.35).*

The operators in the two classes $\mathcal{I}$ and $\mathcal{J}$ can be related when they have the same domain, for example the sheaf $\mathcal{R}_l$.

**Proposition 3.2.5.** *Any element $F(D) \in \mathcal{I}$ belongs to $\mathcal{J}$.*

*Proof.* Let $F(D) \in \mathcal{I}$. Since we consider its action on regular functions, we can substitute $\partial/\partial x_0$ by $-\sum_{\ell=1}^{3} i_\ell \partial/\partial x_\ell$, and we get an operator of the form (3.36). The module of the coefficient of each term $\partial^m/\partial x_1^{n_1}\partial x_2^{n_2}\partial x_3^{n_3}$ is at most $4m^3|a_\nu|$, so $F(D)$ in this new form satisfies (3.35). □

**Remark 3.2.5.** The converse of Proposition 3.2.5 does not hold: it suffices to observe that an operator of the type $\partial/\partial x_i$, $i = 1, 2, 3$, does not belong to $\mathcal{I}$.

**Remark 3.2.6.** Since $\mathcal{I} \subset \mathcal{J}$, condition (3.21) in Theorem 3.2.6 becomes sufficient.

## 3.3 Several quaternionic variables: an analytic approach

This section summarizes the only work of which we are aware, in which a theory of regular functions of several quaternionic variables was developed prior to our own work in [21]. The paper which we refer to is [145], which contains the major results in Pertici's doctoral dissertation. As the reader will see, Pertici's approach consists in mimicking what is usually done in several complex variables, and in so doing he succeeds in proving several interesting and important results, such as a quaternionic analogue of the Bochner–Martinelli formula, and a Hartogs' theorem for regular functions in $\mathbb{H}^n$.

We will begin by giving the definition of regular functions in several quaternionic variables. Let $U$ be an open set in $\mathbb{H}^n$, $n > 1$ and let $f : U \to \mathbb{H}$ be a $C^1$ function. Let $q = (q_1, \dots, q_n)$ be the variable in $\mathbb{H}^n$, i.e., $q_t \in \mathbb{H}$

$$q_t = x_{t0} + ix_{t1} + jx_{t2} + kx_{t3}, \quad t = 1, \dots, n.$$

We say that $f$ is (left) regular in $q = (q_1, \dots, q_n)$ if it satisfies

$$\frac{\partial_l f}{\partial \bar{q}_t} = 0, \qquad t = 1, \dots, n,$$

where

$$\frac{\partial_l}{\partial \bar{q}_t} = \frac{\partial}{\partial x_{t0}} + i\frac{\partial}{\partial x_{t1}} + j\frac{\partial}{\partial x_{t2}} + k\frac{\partial}{\partial x_{t3}}.$$

**Remark 3.3.1.** As in the case of one quaternionic variable, it is possible to define the notion of right regular function by requiring that $f$ be right regular with respect to each variable $q_t$. The two notions give completely analogous theories of regular functions therefore we will limit ourselves to the case of left regularity and, from now on, we will drop the subscript "$l$".

Let us fix some notation that we will use in the sequel. Let $p = (p_1, \dots, p_n)$ be an element in $\mathbb{H}^n$ and let $\Omega_p(q)$ be the $(4n-1)$-form in $\mathbb{H}^n \setminus \{p\}$ defined by

$$\Omega_p(q) = \frac{(2n-1)!}{2\pi^{2n}} \sum_{\ell=1}^{n} \frac{(\bar{q}_\ell - \bar{p}_\ell)}{|q-p|^{4n}} v_1 \wedge \ldots \wedge v_{\ell-1} \wedge Dq_\ell \wedge v_{\ell+1} \wedge \ldots \wedge v_n,$$

where

$$Dq_\ell = dx_{\ell 1} \wedge dx_{\ell 2} \wedge dx_{\ell 3} - i dx_{\ell 0} \wedge dx_{\ell 2} \wedge dx_{\ell 3}$$

$$+ j dx_{\ell 0} \wedge dx_{\ell 1} \wedge dx_{\ell 3} - k dx_{\ell 0} \wedge dx_{\ell 1} \wedge dx_{\ell 2}$$

and

$$v_\ell = dx_{\ell 0} \wedge dx_{\ell 1} \wedge dx_{\ell 2} \wedge dx_{\ell 3}.$$

It can be shown that $d\Omega_p = 0$ in $\mathbb{H}^n \setminus \{p\}$. As a consequence of this fact and of the Stokes' theorem we have

**Theorem 3.3.1.** *Let $U$ be a bounded open set in $\mathbb{H}^n$ with boundary of class $C^1$ and let $p \in \mathbb{H}^n \setminus \bar{U}$. Then*

$$\int_{\partial U} \Omega_p(q) = 0.$$

**Theorem 3.3.2.** *Let $p \in U$ and let $\partial U$ be diffeomorphic to the sphere $S^{4n-1}$ in $\mathbb{R}^{4n}$ and assume that $\partial U$ is of class $C^1$. Then*

$$\int_{\partial U} \Omega_p(q) = 1,$$

*where the orientation on $\partial U$ is induced by the orientation of $\mathbb{R}^{4n}$.*

*Proof.* Let $\varepsilon$ be a positive real number such that

$$\begin{aligned}\int_{\partial U} \Omega_p(q) &= \int_{|q-p|=\varepsilon} \Omega_p(q) \\ &= \frac{(2n-1)!}{2\pi^{2n}\varepsilon^{4n}} \sum_{\ell=1}^{n} \int_{|q-p|=\varepsilon} (\bar{q}_\ell - \bar{p}_\ell) v_1 \wedge \ldots \wedge Dq_\ell \wedge \ldots \wedge v_n.\end{aligned}$$

By Stokes' theorem, the right-hand side of the previous equality turns out to be

$$\frac{(2n-1)!}{2\pi^{2n}\varepsilon^{4n}}\sum_{\ell=1}^{n}\int_{|q-p|\leq\varepsilon} \upsilon_1\wedge\ldots\wedge\overline{dq_\ell}\wedge Dq_\ell\wedge\ldots\wedge\upsilon_n, \qquad \square$$

and recalling that $\overline{dq_\ell}\wedge Dq_\ell = 4\upsilon_\ell$ (see Remark 3.1.3), denoting by $V_\varepsilon$ the volume of the sphere $|q-p|\leq\varepsilon$, we finally get

$$4n\frac{(2n-1)!}{2\pi^{2n}\varepsilon^{4n}}V_\varepsilon = 1.$$

□

One of the main results in [145] is the important integral representation formula known as the quaternionic Bochner–Martinelli formula.

**Theorem 3.3.3.** *Let $f$ be a regular function in an open bounded set $U$ in $\mathbb{H}^n$ with differentiable boundary $\partial U$ and let $p\in U$. Then we have*

$$f(p) = \int_{\partial U}\Omega_p(q)f(q).$$

*Proof.* The differential form $\Omega_p f$ is closed in $U\setminus\{p\}$. In fact by formula (3.5) we have $d\Omega_p = 0$ and $Dq\wedge df = -\dfrac{\partial f}{\partial\bar q}\upsilon$, so that $\Omega_p\wedge df = 0$. Stokes' theorem then gives

$$\int_{\partial U}\Omega_p f = 0.$$

We may now apply this last formula to the open set $U\setminus\{|q-p|\leq\varepsilon\}$ to get

$$\int_{\partial U}\Omega_p f = \int_{|q-p|=\varepsilon}\Omega_p f.$$

For $|q-p|=\varepsilon\to 0$ the right-hand sides converges to $f(p)$. □

**Remark 3.3.2.** The Bochner–Martinelli formula in the case $n=1$ reduces to the usual Cauchy–Fueter formula. Note also that the formula implies that regular functions are infinitely differentiable and harmonic since the Cauchy–Fueter operator factors the Laplacian operator.

We are now ready to discuss the compatibility conditions on the data of the nonhomogeneous Cauchy–Fueter system, although under the additional hypothesis that all the functions considered have compact support. We need a preliminary lemma whose proof, rather technical, is in [145].

**Lemma 3.3.1.** *Let $U$ be an open set in $\mathbb{H}$ and let $f \in C^1(U)$. Assume $D$ is a domain with differentiable boundary, $D$ relatively compact in $U$ such that $0 \notin \partial D$. Then, for $\ell = 0, \dots, 3$, we have*

$$\int_D \left[ i_\ell G(q) \frac{\partial g}{\partial \bar{q}}(q) - \frac{\partial}{\partial \bar{q}}(G(q) i_\ell g(q)) \right] = \int_{\partial D} i_\ell G(q) Dq\, g(q) - Dq\, G(q) i_\ell g(q),$$

*where $G(q)$ is the Cauchy–Fueter kernel.*

**Theorem 3.3.4.** *Let $g_1, \dots, g_n \in C^k(\mathbb{H}^n)$ have compact support and let $n > 1$. If $k \geq 2$ the system*

$$\begin{cases} \dfrac{\partial f}{\partial \bar{q}_1} = g_1 \\ \cdots \\ \dfrac{\partial f}{\partial \bar{q}_n} = g_n \end{cases} \tag{3.37}$$

*admits a compactly supported solution $f \in C^k(\mathbb{H}^n)$ if and only if*

$$\int_{\mathbb{H}} \sum_{\ell=0}^{3} i_\ell G(p) \frac{\partial g_h}{\partial x_{1\ell}}(q_1 + p, q_2, \dots, q_n)$$

$$= \int_{\mathbb{H}} \sum_{\ell=0}^{3} i_\ell G(p) \frac{\partial g_1}{\partial x_{h\ell}}(q_1 + p, q_2, \dots, q_n) \tag{3.38}$$

*for any $h = 1, \dots, n$ and $(q_1, \dots, q_n) \in \mathbb{H}^n$.*

*Proof.* Suppose that system (3.37) admits a solution $f$. We have to prove that the integral conditions (3.38) hold. Since the Cauchy–Fueter kernel $G$ is a regular function, we easily get that

$$\sum_{\ell=0}^{3} i_\ell G(p) \frac{\partial g_h}{\partial x_{1\ell}}(q_1 + p, q_2, \dots, q_n)$$

$$= \sum_{r=0}^{3} \frac{\partial}{\partial \bar{p}} \left[ G(p) i_r \frac{\partial f}{\partial x_{hr}}(q_1 + p, q_2, \dots, q_n) \right].$$

Then we have

$$\sum_{\ell=0}^{3} i_\ell G(p) \frac{\partial g_1}{\partial x_{h\ell}}(q_1 + p, q_2, \dots, q_n)$$

$$= \sum_{\ell=0}^{3} i_\ell G(p) \frac{\partial^2 f}{\partial \bar{p} \partial x_{hr}}(q_1 + p, q_2, \dots, q_n).$$

Since $\partial f/\partial x_{rh}$ have compact support, from Lemma 3.3.1 one can deduce the following:

$$\int_{\mathbb{H}} \sum_{\ell=0}^{3} i_\ell G(p) \frac{\partial^2 f}{\partial \bar{p} \partial x_{hr}}(q_1 + p, q_2, \dots, q_n)$$
$$= \int_{\mathbb{H}} \sum_{r=0}^{3} \frac{\partial}{\partial \bar{p}} \left( G(p) i_r \frac{\partial f}{\partial x_{hr}}(q_1 + p, q_2, \dots, q_n) \right)$$

so we have proved that the integral conditions hold.

Conversely, let us suppose that the integral conditions hold and let $f \in C^k(\mathbb{H}^n)$ be the function defined by

$$\begin{aligned} f(q_1, \dots, q_n) &= -\frac{1}{2\pi^2} \int_{\mathbb{H}} G(q - q_1) g_1(q, q_2, \dots, q_n) \\ &= -\frac{1}{2\pi^2} \int_{\mathbb{H}} G(p) g_1(q_1 + p, q_2, \dots, q_n). \end{aligned}$$

Now, for $\ell = 1, \dots, n$, we have

$$\begin{aligned} \frac{\partial f}{\partial \bar{q}_\ell}(q_1, \dots, q_n) &= -\frac{1}{2\pi^2} \int_{\mathbb{H}} \frac{\partial}{\partial \bar{q}_\ell} \left[ G(p) g_1(q_1 + p, q_2, \dots, q_n) \right] \\ &= -\frac{1}{2\pi^2} \int_{\mathbb{H}} \sum_{h=0}^{3} i_\ell G(p) \frac{\partial g_1}{\partial x_{h\ell}}(q_1 + p, q_2, \dots, q_n). \end{aligned}$$

The integral conditions (3.38) imply that the right-hand side is equal to

$$-\frac{1}{2\pi^2} \int_{\mathbb{H}} \sum_{h=0}^{3} i_\ell G(p) \frac{\partial g_h}{\partial x_{1h}}(q_1 + p, q_2, \dots, q_n) \tag{3.39}$$

and, by the second Green's theorem, (3.39) becomes equal to

$$-\frac{1}{2\pi^2} \int_{\mathbb{H}} \frac{\partial}{\partial \bar{q}} (G(q - q_1) g_h(q, q_2, \dots, q_n))$$
$$= g_h(q_1, q_2, \dots, q_n).$$

Therefore we have that

$$\frac{\partial f}{\partial \bar{q}_\ell} = g_\ell, \quad \ell = 1, \dots, n.$$

It remains to prove that $f$ has compact support. Let

$$\text{supp } g_1 \subseteq \left\{ (q_1, \dots, q_n) \in \mathbb{H}^n \mid \sqrt{\sum_{\ell=1}^{n} |q_\ell|^2} \leq C \right\}$$

where $C$ is a positive real constant. If we choose $q_2, \ldots, q_n$ such that

$$\sum_{\ell=2}^{n} |q_\ell|^2 > C$$

we obviously have that $f(q_1, \ldots, q_n) = 0$ for every $q_1 \in \mathbb{H}$, so $f$ vanishes identically on an unbounded open set in $\mathbb{H}^n$. Since the functions $g_\ell$ are compactly supported, $f$ is regular in the complement of a compact disc in $\mathbb{H}^n$ and also harmonic by the identity principle, and so we have that $f$ vanishes identically on the complement of that compact disc. This completes the proof. □

An important result that we will prove in the next section with algebraic tools is Hartogs' theorem that was originally proved by Pertici [145] with methods similar to those used in the theory of holomorphic functions of several complex variables.

**Theorem 3.3.5.** *Let $U$ be a connected open set in $\mathbb{H}^n$, $n > 1$. Let $K \subset U$ be a compact set such that $U \setminus K$ is connected. Then every regular function in $U \setminus K$ can be extended to a regular function on $U$.*

*Proof.* Let $f$ be a regular function in $U \setminus K$, which implies that $f$ is also harmonic in $U \setminus K$. Let $\varphi \in C^\infty(U)$ be a function with compact support and such that $\varphi = 1$ in a neighborhood of $K$ and set $f_0 = (1 - \varphi)f$. Obviously, $f_0 \in C^\infty(U)$ and it is regular in $U \setminus \{\text{supp } \varphi\}$. Let us set

$$h_\ell = \begin{cases} \dfrac{\partial f_0}{\partial \bar{q}_\ell} & \text{in } U \\ 0 & \text{in } \mathbb{H}^n \setminus U \end{cases}$$

for $\ell = 1, \ldots, n$. We have that $h_\ell \in C^\infty(\mathbb{H}^n)$ and $h_\ell$ is compactly supported since it identically vanishes on $\mathbb{H}^n \setminus \{\text{supp } \varphi\}$. Let us now prove that the functions $h_1, \ldots, h_n$ satisfy the compatibility conditions (3.38), i.e., that the functions

$$\Lambda_s(q_1, \ldots, q_n) = \int_{\mathbb{H}} \sum_{\ell=0}^{3} i_\ell G(p) \left[\frac{\partial h_s}{\partial x_{1\ell}} - \frac{\partial h_1}{\partial x_{s\ell}}\right] (q_1 + p, q_2, \ldots, q_n),$$

for $2 \le s \le n$, are identically zero. Let us begin by proving that for any $s$ the function $\Lambda_s(q_1, \ldots, q_n)$ is harmonic in $\mathbb{H}^n$. To this purpose let us set

$$U(q_2, \ldots, q_n) = \{\, q \in \mathbb{H} \mid (q, q_2, \ldots, q_n) \in U \,\},$$

$$S(q_2, \ldots, q_n) = \{\, q \in \mathbb{H} \mid (q, q_2, \ldots, q_n) \in \text{supp } \varphi \,\}.$$

Note that $S(q_2, \ldots, q_n) \subseteq U(q_2, \ldots, q_n)$. Let now $B(q_2, \ldots, q_n)$ be an open bounded set in $\mathbb{H}$ with differentiable boundary $\partial B$, containing $S(q_2, \ldots, q_n)$

and such that its closure is contained in $U(q_2,\dots,q_n)$. We also need to define the following set

$$B_{q_1}(q_2,\dots,q_n)=\{p\in\mathbb{H}\mid p+q_1\in B(q_2,\dots,q_n)\}.$$

We have

$$\begin{aligned}\Lambda_s(q_1,\dots,q_n)&=\int_{B(q_2,\dots,q_n)}\sum_{\ell=0}^{3}i_\ell G(q-q_1)\left[\frac{\partial h_s}{\partial x_{1\ell}}-\frac{\partial h_1}{\partial x_{s\ell}}\right](q,q_2,\dots,q_n)\\&=\int_{B_{q_1}(q_2,\dots,q_n)}\sum_{\ell=0}^{3}i_\ell G(p)\left[\frac{\partial h_s}{\partial x_{1\ell}}-\frac{\partial h_1}{\partial x_{s\ell}}\right](q_1+p,q_2,\dots,q_n).\end{aligned}$$

Note that when $p\in B_{q_1}(q_2,\dots,q_n)$ we have

$$h_s(q_1+p,\dots,q_n)=\frac{\partial f_0}{\partial\bar q_s}(q_1+p,\dots,q_n),$$

so that

$$\begin{aligned}&\sum_{\ell=0}^{3}i_\ell G(p)\frac{\partial h_s}{\partial x_{1\ell}}(q_1+p,q_2,\dots,q_n)\\&=\sum_{r=0}^{3}\frac{\partial}{\partial\bar p}\left(G(p)i_r\frac{\partial f_0}{\partial x_{sr}}(q_1+p,q_2,\dots,q_n)\right),\end{aligned}\tag{3.40}$$

and we obtain

$$\begin{aligned}&\sum_{\ell=0}^{3}i_\ell G(p)\frac{\partial h_1}{\partial x_{1\ell}}(q_1+p,q_2,\dots,q_n)\\&=\sum_{r=0}^{3}i_rG(p)\frac{\partial}{\partial\bar p}\left(\frac{\partial f_0}{\partial x_{sr}}(q_1+p,q_2,\dots,q_n)\right).\end{aligned}\tag{3.41}$$

By (3.40), (3.41) and by Lemma 3.3.1 we get

$$\begin{aligned}\Lambda_s(q_1,\dots,q_n)=&\int_{\partial B_{q_1}(q_2,\dots,q_n)}\sum_{r=0}^{3}\left[DpG(p)i_r\frac{\partial f_0}{\partial x_{sr}}(q_1+p,q_2,\dots,q_n)\right.\\&\left.-i_rG(p)Dp\frac{\partial f_0}{\partial x_{sr}}(q_1+p,q_2,\dots,q_n)\right]\\=&\int_{\partial B_{q_1}(q_2,\dots,q_n)}\left[DpG(p)\frac{\partial f_0}{\partial\bar q_s}(q_1+p,q_2,\dots,q_n)\right.\\&\left.-\sum_{r=0}^{3}i_rG(p)Dp\frac{\partial f_0}{\partial x_{sr}}(q_1+p,q_2,\dots,q_n)\right].\end{aligned}$$

Since the functions $f_0$ and $f$ coincide on $\partial B_{q_1}(q_2, \dots, q_n)$ and since $f$ is regular, we get that the right-hand side is equal to

$$-\int_{\partial B_{q_1}(q_2,\dots,q_n)} \sum_{r=0}^{3} i_r G(p) Dp \frac{\partial f}{\partial x_{sr}}(q_1 + p, q_2, \dots, q_n).$$

Note that $\Lambda_s$ does not depend on the integration set $\partial B_{q_1}(q_2, \dots, q_n)$ since $B(q_2, \dots, q_n)$ can be any bounded set with differentiable boundary satisfying the hypotheses above. Since $f$ is an harmonic function we have that

$$\Delta\Lambda_s(q_1, \dots, q_n)$$

$$= -\int_{\partial B_{q_1}(q_2,\dots,q_n)} \sum_{r=0}^{3} i_r G(p) Dp \frac{\partial}{\partial x_{sr}}(\Delta f)(q_1 + p, q_2, \dots, q_n) = 0,$$

and so $\Lambda_s$ is harmonic in $\mathbb{H}^n$. Moreover, $\Lambda_s$ vanishes on an open nonempty set in $\mathbb{H}^n$ since the functions $h_\ell$ vanish on $\mathbb{H}^n \setminus \operatorname{supp} \varphi$ and so $\Lambda_s$ vanishes identically on $\mathbb{H}^n$.

By the previous theorem there exists a function $g \in C^\infty(\mathbb{H}^n)$ with compact support which is solution to the system

$$\frac{\partial g}{\partial \bar{q}_\ell} = h_\ell, \quad \ell = 1, \dots, n.$$

The function $g$ is regular on $\mathbb{H}^n \setminus \operatorname{supp} \varphi$ and since it has compact support, by the identity principle it follows that $g$ vanishes identically on the unbounded connected component $C$ of $\mathbb{H}^n \setminus \operatorname{supp} \varphi$. Setting

$$\tilde{f} = f_0 - g,$$

we have that $\tilde{f}$ is regular in $U$ and coincides with $f$ on $C \cap U$. Since $U \setminus K$ is connected, by the identity principle $\tilde{f}$ coincides with $f$ on $U \setminus K$, therefore, it is the regular extension of $f$. □

## 3.4 Several quaternionic variables: an algebraic approach

An algebraic approach to the theory of regular functions of several quaternionic variables was never considered until one of us begun to study Hartogs' phenomenon as part of her doctoral dissertation [157]. As we have seen in the previous section, Hartogs' theorem for regular functions in $\mathbb{H}^n$ was known after the results of Pertici, but we were intrigued by the possibility of using Ehrenpreis' approach to the removability of compact singularities of regular functions in $\mathbb{H}^n$. It was obvious that a direct application of Ehrenpreis' results in [66] would not work because if $q_i$ and $q_j$ are two distinct quaternionic variables, the noncommutative nature of $\mathbb{H}$ implied that

$$\frac{\partial}{\partial \bar{q}_j}\frac{\partial}{\partial \bar{q}_i} \neq \frac{\partial}{\partial \bar{q}_i}\frac{\partial}{\partial \bar{q}_j}.$$

This lack of commutativity was, by itself, an obstacle. On the other hand, [134] provided some explicit conditions for the case of matrix systems and this provided the first inkling on how to reformulate the theory of regular functions in algebraic terms. Let $q = (q_1, \dots, q_n)$ be a point in $\mathbb{H}^n$, i.e., $q_t \in \mathbb{H}$

$$q_t = x_{t0} + ix_{t1} + jx_{t2} + kx_{t3}, \quad t = 1, \dots, n.$$

Let $f : \mathbb{H}^n \to \mathbb{H}$ be a function. We can always write $f = f_0 + if_1 + jf_2 + kf_3$ with $f_0, f_1, f_2, f_3 \in \mathcal{C}^\infty(\mathbb{R}^{4n})$ and real valued. In other words, a function $f : \mathbb{H}^n \to \mathbb{H}$ can be represented via a vector $(f_0, f_1, f_2, f_3)$ of functions from $\mathbb{R}^{4n}$ to $\mathbb{R}$. As before, we say that $f$ is left regular if it satisfies

$$\frac{\partial f}{\partial \bar{q}_t} = 0, \qquad t = 1, \dots, n.$$

But in view of the representation we have just introduced, a function $f$ is left regular if and only if its four real components $f_0, f_1, f_2, f_3$ satisfy the following $4n \times 4$ system of linear partial differential equations with constant coefficients

$$\begin{cases} \dots \\ \dfrac{\partial f_0}{\partial x_{t0}} - \dfrac{\partial f_1}{\partial x_{t1}} - \dfrac{\partial f_2}{\partial x_{t2}} - \dfrac{\partial f_3}{\partial x_{t3}} = 0 \\ \dfrac{\partial f_0}{\partial x_{t1}} + \dfrac{\partial f_1}{\partial x_{t0}} - \dfrac{\partial f_2}{\partial x_{t3}} + \dfrac{\partial f_3}{\partial x_{t2}} = 0 \\ \dfrac{\partial f_0}{\partial x_{t2}} + \dfrac{\partial f_1}{\partial x_{t3}} + \dfrac{\partial f_2}{\partial x_{t0}} - \dfrac{\partial f_3}{\partial x_{t1}} = 0 \\ \dfrac{\partial f_0}{\partial x_{t3}} - \dfrac{\partial f_1}{\partial x_{t2}} + \dfrac{\partial f_2}{\partial x_{t1}} + \dfrac{\partial f_3}{\partial x_{t0}} = 0 \\ \dots . \end{cases} \qquad t = 1, \dots, n.$$

Using the notations introduced in Chapter 2, we see that regularity for functions on $\mathbb{H}^n$ can be fully described in terms of a system

$$[P_{ij}(D)] \;:\; [\mathcal{C}^\infty(\mathbb{R}^{4n})]^4 \to [\mathcal{C}^\infty(\mathbb{R}^{4n})]^{4n}.$$

This system fulfills all the conditions that make it amenable to algebraic treatment, since it is linear and with constant coefficients. Once things have been reformulated in this way, the path ahead is clear, although somewhat complex in view of the computational difficulties. Our treatment in this section is based on our original papers, mostly [2], [5], and [6]. Before we state and prove our most general result, let us deal with two specific low dimensional cases, because their understanding provides the key to the general situation.

We begin with two quaternionic variables, so that the Cauchy–Fueter system we deal with is

$$\frac{\partial f}{\partial \bar{q}_1} = \frac{\partial f}{\partial \bar{q}_2} = 0 \tag{3.42}$$

with $f = f(q_1, q_2)$. If the $q_t$ were complex variables, let us call them $z_1$, $z_2$, the way to analyze algebraically system (3.42) would consist in replacing $\partial/\partial \bar{z}_t$ by its "formal Fourier transform" $\bar{z}_t$ and look at the multiplication by $\bar{z}_1$ and $\bar{z}_2$ on the ring $R$ of polynomials:

$$\begin{array}{ccccc} 0 \longleftarrow \dfrac{R}{I(\bar{z}_1, \bar{z}_2)} & \longleftarrow & R & \longleftarrow & R^2 \\ & & f\bar{z}_1 + g\bar{z}_2 & \longleftarrow & (f, g). \end{array}$$

In the complex case this would immediately yield that, by commutativity, the first syzygy would be given by the multiplication with the vector $(\bar{z}_2, -\bar{z}_1)$. However, in the quaternionic case, it is not true that $\bar{q}_1\bar{q}_2 - \bar{q}_2\bar{q}_1 = 0$, and so, if we want to imitate the complex case, we still need to find "polynomials" $\mathcal{P}_1, \mathcal{P}_2$ in $q_1$, $q_2$, $\bar{q}_1$, $\bar{q}_2$ such that

$$\mathcal{P}_1\bar{q}_1 + \mathcal{P}_2\bar{q}_2 = 0. \tag{3.43}$$

Bear in mind that we are still operating from a purely heuristic point of view: in fact (as we shall soon discover) such an approach has serious flaws. After a bit of experimentation, it becomes clear that $\mathcal{P}_1$ and $\mathcal{P}_2$ cannot be linear in their variables and one is lead to consider second degree expressions. The key for solving (3.43) lies in the realization that in order for some commutativity to be restored, one needs "real" expressions, which will therefore commute with quaternions. The first such real expression is the symbol for the Laplacian $\Delta$ which, in each variable, can be written as $q_t\bar{q}_t = \bar{q}_t q_t$ and will be denoted, with an abuse of notation, by $\Delta_t$. Thus, two natural solutions for (3.43) are $(\mathcal{P}_1, \mathcal{P}_2) = (\bar{q}_2 q_1, -\Delta_1)$ and $(\mathcal{P}_1, \mathcal{P}_2) = (-\Delta_2, \bar{q}_1 q_2)$ since

$$\bar{q}_2\Delta_1 - \Delta_1\bar{q}_2 = 0, \qquad -\Delta_2\bar{q}_1 + \bar{q}_1\Delta_2 = 0.$$

This presents the first set of anomalies when compared with the complex case. First, in the complex case, all syzygies are of degree one; here we have already found a syzygy of degree two. Second, in the complex case the complex ends immediately after the first step as follows:

$$\begin{array}{ccccccc} 0 \longleftarrow \dfrac{R}{I(\bar{z}_1, \bar{z}_2)} & \longleftarrow & R & \longleftarrow & R^2 & \longleftarrow & R \longleftarrow 0 \\ & & f\bar{z}_1 + g\bar{z}_2 & \longleftarrow & (f, g) & & \\ & & & & (h\bar{z}_2 - h\bar{z}_1) & \longleftarrow & h. \end{array}$$

In this case we clearly need at least one more step, since the first syzygy is a $2\times 2$ matrix. Let us point out that there is a second, alternative way to construct this $2 \times 2$ matrix, as described in [158]. To do so, we recall that, as shown in Chapter 2, this matrix has an analytic meaning. Namely the first syzygies of a system yield a matrix which is the symbol of the compatibility system for the corresponding non homogeneous system. To be more explicit, the matrix of the first syzygies is the symbol of two operators $Q_1(D)$ and $Q_2(D)$ such that the system

$$\begin{cases} \dfrac{\partial f}{\partial \bar{q}_1} = g_1 \\[2ex] \dfrac{\partial f}{\partial \bar{q}_2} = g_2 \end{cases} \tag{3.44}$$

has a solution if and only if

$$Q_1(D)g_1 + Q_2(D)g_2 = 0.$$

We can easily show how to construct these operators in an analytic fashion. Let $T_1$, $T_2$ denote the right inverses of $\dfrac{\partial f}{\partial \bar{q}_1}$ and $\dfrac{\partial f}{\partial \bar{q}_2}$, respectively. If $f$ is a solution of the nonhomogeneous system (3.44), then there are functions $h_1$ and $h_2$, $C^\infty$ on $\mathbb{R}^8$, and such that $h_1$ is regular in $q_1$, $h_2$ is regular in $q_2$ and

$$f = T_1 g_1 + h_1 = T_2 g_2 + h_2,$$

i.e.,

$$T_1 g_1 - T_2 g_2 = h_2 - h_1$$

is the sum of a regular function in $q_1$ and a regular function in $q_2$. However the Laplacians

$$\Delta_i = \frac{\partial}{\partial q_i}\frac{\partial}{\partial \bar{q}_i} = \frac{\partial}{\partial \bar{q}_i}\frac{\partial}{\partial q_i}, \qquad i = 1,2$$

are real operators, and therefore $h_1 - h_2$ is in the annihilator of the module generated by $\Delta_1 \dfrac{\partial}{\partial \bar{q}_2}$ and $\Delta_2 \dfrac{\partial}{\partial \bar{q}_1}$. Thus we see that the compatibility conditions for the system (3.44) must include the equations

$$\begin{cases} \dfrac{\partial}{\partial \bar{q}_2}\dfrac{\partial}{\partial q_1} g_1 - \Delta_1 g_2 = 0 \\[2ex] \dfrac{\partial}{\partial \bar{q}_1}\dfrac{\partial}{\partial q_2} g_2 - \Delta_2 g_1 = 0. \end{cases}$$

These conditions are exactly those ones we had found algebraically. Note, however, that both methods only give a set of syzygies (the same set) but do not guarantee the completeness of this set. In other words we do not know, a priori, whether or not other independent syzygies exist. The simplest way we have to confirm that these are indeed all the syzygies is computational. In [2], for example, we abandoned the quaternionic notation, we expressed the various quantities in terms of real coordinates, and we used CoCoA to find a matrix of syzygies. The CoCoA method ensures that this matrix contains all the independent syzygies. In our case we obtained an unwieldy $8 \times 8$ real matrix (see [2] for its explicit expression). At first sight it is hard to imagine what this matrix might be, but in view of our previous heuristic arguments, one can easily recognize that the matrix is exactly the translation of

$$\begin{bmatrix} q_1\bar{q}_2 & -q_2\bar{q}_2 \\ -q_1\bar{q}_1 & q_2\bar{q}_1 \end{bmatrix}$$

into real coordinates. It is then easy to compute the next step in the syzygies, a real $(8 \times 4)$ matrix which is just the translation in real coordinates of

$$\begin{bmatrix} q_2 \\ q_1 \end{bmatrix}$$

so that one has the following exact sequence

$$0 \longleftarrow R \xleftarrow{\begin{bmatrix} \bar{q}_1 & \bar{q}_2 \end{bmatrix}} R^2 \xleftarrow{\begin{bmatrix} q_1\bar{q}_2 & -q_2\bar{q}_2 \\ -q_1\bar{q}_1 & q_2\bar{q}_1 \end{bmatrix}} R^2 \xleftarrow{\begin{bmatrix} q_2 \\ q_1 \end{bmatrix}} R.$$

One is tempted to try this same heuristic approach for the case of three quaternionic variables. Inspired by what we saw in the case of two variables, we expect six syzygies to arise from the use of the three Laplacians, used two at a time:

$$(-q_s\bar{q}_t,\ q_t\bar{q}_t)$$
$$(-q_s\bar{q}_s,\ q_t\bar{q}_s)$$

for $s, t = 1, 2, 3$, $s \neq t$. More generally, given $n$ quaternionic variables, we should always expect $2\binom{n}{2}$ syzygies of this form. We found those syzygies by looking at the ways in which one could construct real expressions starting with two quaternions $q_1$ and $q_2$. One may expect that additional real expressions might be available if we consider three variables. This is, in fact, the case since one can verify that the expressions $q_t\bar{q}_s + q_s\bar{q}_t$ are real whenever we consider quaternions $q_\ell$, $q_s$, $q_t$. Immediately, one sees that there are at least eight first syzygies in the case of three quaternionic variables. More generally, for $n$ quaternionic variables, the system

$$\begin{cases} \dfrac{\partial f}{\partial \bar{q}_1} = g_1 \\ \qquad \cdots \\ \dfrac{\partial f}{\partial \bar{q}_n} = g_n \end{cases}$$

will have at least $2\binom{n}{2}+2\binom{n}{3}$ first syzygies of the following form:

(1) for each of the $2\binom{n}{2}$ ordered pairs of indices $r$, $s$, $1 \le r, s \le n$

$$\partial_{\bar{q}_r}\partial_{q_s}g_s - \partial_{\bar{q}_s}\partial_{q_s}g_r = 0$$

(2) for each of the $\binom{n}{3}$ triples of indices $h$, $r$, $s$, $1 \le h, r, s \le n$

$$\partial_{q_h}\partial_{\bar{q}_r}g_s + \partial_{q_r}\partial_{\bar{q}_h}g_s - \partial_{\bar{q}_s}\partial_{q_r}g_h - \partial_{\bar{q}_s}\partial_{q_h}g_r = 0$$

and

$$\partial_{q_r}\partial_{\bar{q}_s}g_h + \partial_{q_s}\partial_{\bar{q}_r}g_h - \partial_{\bar{q}_h}\partial_{q_r}g_s - \partial_{\bar{q}_h}\partial_{q_s}g_r = 0.$$

This computation would also be in concordance with a first statement of the so called Fischer decomposition theorem (see Chapter 4). It was therefore quite surprising to realize that the application of CoCoA to the case of three operators provided ten syzygies; eight of them are those we expected, but there are two exceptional cases. Those exceptional cases show up even using a higher number of operators and can be described as follows (see also Theorem 3.4.3):

(3) for each of the $\binom{n}{3}$ triples of indices $h$, $r$, $s$, $1 \le h, r, s \le n$

$$(D_{q_r}\partial_{\bar{q}_s} - D_{q_s}\partial_{\bar{q}_r})g_h + (D_{q_s}\partial_{\bar{q}_h} - D_{q_h}\partial_{\bar{q}_s})g_r + (D_{q_h}\partial_{\bar{q}_r} - D_{q_r}\partial_{\bar{q}_h})g_s = 0,$$

$$(D'_{q_r}\partial_{\bar{q}_s} - D'_{q_s}\partial_{\bar{q}_r})g_h + (D'_{q_s}\partial_{\bar{q}_h} - D'_{q_h}\partial_{\bar{q}_s})g_r + (D'_{q_h}\partial_{\bar{q}_r} - D'_{q_r}\partial_{\bar{q}_h})g_s = 0,$$

where

$$D_{q_i} = -j\frac{\partial}{\partial x_{i2}} + k\frac{\partial}{\partial x_{i3}}, \qquad D'_{q_i} = -i\frac{\partial}{\partial x_{i1}} + k\frac{\partial}{\partial x_{i3}}.$$

The most striking aspect of these syzygies (to which we often refer to as "exceptional" syzygies) is the fact that they cannot be expressed directly in terms of the operators $\partial/\partial\bar{q}_1$, $\partial/\partial\bar{q}_2$, $\partial/\partial\bar{q}_3$. Adopting a terminology which we will formally introduce only in Chapter 4, we say that the syzygies do not belong to the "radial algebra" generated by $\partial/\partial\bar{q}_1$, $\partial/\partial\bar{q}_2$, $\partial/\partial\bar{q}_3$. The existence of these exceptional syzygies led to a modified version of the Fischer decomposition, see [192]. We should also point out that there are deep geometrical and dimensional reasons for the existence of theses syzygies. These reasons are fully discussed and explained in Chapter 4.

These early CoCoA computations, however, only allowed us to conjecture the number of first syzygies for the multivariable Cauchy–Fueter system. Before a fully developed theory could be obtained, we needed a few extra computer experiments to guide our intuition. When we consider that $f$ must satisfy $\partial f/\partial\bar{q}_t = 0$ for $t = 1, \ldots, n$, we actually obtain the following matrix of differential operators:

$$P_n(D) = \begin{bmatrix} U_1(D) \\ U_2(D) \\ \vdots \\ U_n(D) \end{bmatrix}$$

where each $U_t(D)$ is the $4 \times 4$ matrix given by

$$U_t(D) = \begin{bmatrix} \partial_{x_{t0}} & -\partial_{x_{t1}} & -\partial_{x_{t2}} & -\partial_{x_{t3}} \\ \partial_{x_{t1}} & \partial_{x_{t0}} & -\partial_{x_{t3}} & \partial_{x_{t2}} \\ \partial_{x_{t2}} & \partial_{x_{t3}} & \partial_{x_{t0}} & -\partial_{x_{t1}} \\ \partial_{x_{t3}} & -\partial_{x_{t2}} & \partial_{x_{t1}} & \partial_{x_{t0}} \end{bmatrix}$$

and $P_n(D)$ is therefore a $4n \times 4$ matrix which maps $[\mathcal{E}(\mathbb{R}^{4n})]^4$ to $[\mathcal{E}(\mathbb{R}^{4n})]^{4n}$. According to the algebraic theory which we described in Chapter 2, it is now sufficient to take the Fourier transform of $P_n(D)$ and consider the map induced by

$$P_n = \begin{bmatrix} U_1 \\ U_2 \\ \vdots \\ U_n \end{bmatrix}$$

where

$$U_t = \begin{bmatrix} \xi_{t0} & -\xi_{t1} & -\xi_{t2} & -\xi_{t3} \\ \xi_{t1} & \xi_{t0} & -\xi_{t3} & \xi_{t2} \\ \xi_{t2} & \xi_{t3} & \xi_{t0} & -\xi_{t1} \\ \xi_{t3} & -\xi_{t2} & \xi_{t1} & \xi_{t0} \end{bmatrix}$$

for $t = 1, \ldots, n$ and where the variables $\xi_{ij}$ are the dual variables of the variables $x_{ij}$ (more appropriately, one should make the substitution $-i\partial/\partial x_{rs} \to \xi_{rs}$; however, since $-i$ is a common factor, we can disregard it).

**Warning**. Note that from now on, for sake of simplicity and with an abuse of notation, we will denote with the same letter both the variables and their dual.

Using the CoCoA commands illustrated in Chapter 2, and setting

$$R = \mathbb{C}[x_{10}, x_{11}, x_{12}, x_{13}, \ldots, x_{n0}, x_{n1}, x_{n2}, x_{n3}],$$

the results obtained are (for $n = 3, 4$):

$$n = 3: \quad 0 \longrightarrow R^8(-6) \longrightarrow R^{36}(-5) \longrightarrow R^{60}(-4) \longrightarrow$$
$$\longrightarrow R^{40}(-3) \longrightarrow R^{12}(-1) \longrightarrow R^4 \longrightarrow M_3 \longrightarrow 0;$$
$$n = 4: \quad 0 \longrightarrow R^{12}(-8) \longrightarrow R^{80}(-7) \longrightarrow R^{224}(-6) \longrightarrow R^{336}(-5) \longrightarrow$$
$$\longrightarrow R^{280}(-4) \longrightarrow R^{112}(-3) \longrightarrow R^{16}(-1) \longrightarrow R^4 \longrightarrow M_4 \longrightarrow 0,$$

where $M_n$ is the module associated to the Cauchy–Fueter system in $n$ variables.

These resolutions naturally lead to the formulation of conjectures dealing with the length of the resolutions, the degree of the syzygies (which can be read from the homogeneity degrees in parentheses), and the Betti numbers of the resolution (i.e., the exponents which appear). All the major questions one could ask about these resolutions can be answered through the use of Gröbner bases. This has been, in some sense, our major contribution to the theory of regular functions and because the methods we used can be extended to other situations, we follow here [5] and [6] to show how to fully analyze the module associated to the sheaf of regular functions. As always, $R$ will denote the ring of polynomials $\mathbb{K}[x_{00}, \ldots, x_{n3}]$ over the field $\mathbb{K}$. Note that, in the applications, the field $\mathbb{K}$ will be the complex field $\mathbb{C}$. With the notation $\langle P_n^t \rangle$ we will denote the $R$-module generated by the columns of the matrix $P_n^t$.

**Proposition 3.4.1.** *The reduced Gröbner basis for the $R$-module $\langle P_n^t \rangle$ is given by the columns of $P_n^t$ together with the columns of the $\binom{n}{2}$ matrices*

$$B_{rs} = U_r U_s - U_s U_r, \qquad 1 \leq r < s \leq n.$$

*Moreover the module $Lt(P_n^t)$ generated by the leading terms of all the elements of $\langle P_n^t \rangle$ is*

$$Lt(P_n^t) = \langle x_{i0}\mathbf{e}_\ell, x_{r2}x_{s1}\mathbf{e}_\ell \rangle, \qquad i = 1, \ldots, n, \quad 1 \leq r < s \leq n, \quad \ell = 1, 2, 3, 4.$$

*Proof.* The statement can be verified for $n = 1, 2, 3, 4$ using CoCoA. Let us assume $n > 4$. The $S$-polynomials generated by any two columns of $P_n^t$ can be computed and reduced as in the case $n = 2$ and give rise to the columns of the matrices $B_{rs}$. In order to prove that the columns of $P_n^t$ and $B_{rs}$ form a reduced Gröbner basis for $\langle P_n^t \rangle$ we need to show that all the $S$-polynomials generated by them reduce to zero. An $S$-polynomial generated by a column of $P_n^t$ and a column of $B_{rs}$ is computed and reduced as in the case $n = 2$ or $n = 3$. An $S$-polynomial generated by two columns of $B_{rs}$ is computed and reduced as in the case $n = 3$ or $n = 4$. Finally, an $S$-polynomial generated by a column in $B_{rs}$ and a column in $B_{tu}$ is computed and reduced as in the case $n = 3, 4$ depending on whether one or none of the indices $r, s$ and $t, u$ is the same. The result about $Lt(P_n^t)$ follows from the fact that for $1 \leq r < s \leq n$ the matrix $B_{rs}$ is given by

$$\begin{bmatrix} 0 & x_{r2}x_{s3} - x_{r3}x_{s2} & x_{r1}x_{s3} - x_{r3}x_{s1} & x_{r2}x_{s1} - x_{r1}x_{s2} \\ x_{r3}x_{s2} - x_{r2}x_{s3} & 0 & x_{r2}x_{s1} - x_{r1}x_{s2} & x_{r3}x_{s1} - x_{r1}x_{s3} \\ x_{r1}x_{s3} - x_{r3}x_{s1} & x_{r2}x_{s1} - x_{r1}x_{s2} & 0 & x_{r3}x_{s2} - x_{r2}x_{s3} \\ x_{r2}x_{s1} - x_{r1}x_{s2} & x_{r3}x_{s1} - x_{r1}x_{s3} & x_{r2}x_{s3} - x_{r3}x_{s2} & 0 \end{bmatrix}$$

and by the definition of term order. □

This result can be used to compute the Hilbert–Poincaré series for $M_n$ as done in [5], which we follow closely for the next few pages. Let us set

$$M_n = \bigoplus_{\nu \geq 0} M_{\nu n}$$

where $M_{\nu n}$ is the $k$-vector space of elements of $M_n$ of degree $\nu$.

**Definition 3.4.1.** *Let us set*

$$H_n(\nu) = \dim_k M_{\nu n}.$$

*Then $H_n(\nu)$ is called the Hilbert function of $M_n$ while*

$$\mathcal{P}_n(t) = \sum_{\nu=0}^{\infty} H_n(\nu)t^\nu$$

*is called the Hilbert–Poincaré series for $M_n$.*

The Hilbert–Poincaré series can be explicitly computed.

**Theorem 3.4.1.** *The Hilbert–Poincaré series is given by*

$$\mathcal{P}_n(t) = 4\frac{1+(n-1)t}{(1-t)^{2n+1}}.$$

*Proof.* First, we recall that the Hilbert function and hence the Hilbert–Poincaré series are the same for $M_n = R^4/\langle P_n^t\rangle$ and for $R^4/\langle \mathrm{Lt}(P_n^t)\rangle$ by virtue of Macaulay's result [121]. Then from the symmetry of $\mathrm{Lt}(P_n^t)$ in each component, we see that

$$\mathcal{P}_n(t) = 4\mathcal{Q}_n(t)$$

where $\mathcal{Q}_n(t)$ is the Hilbert–Poincaré series for $R/I_n$, where $I_n$ is the ideal of $R$ defined by

$$I_n = \langle x_{i0}, x_{r2}x_{s1}\rangle_{i=1,\dots,n,1\le r<s\le n}.$$

Let us denote by $T_1, \dots, T_r$, $r$ monomials in $R$ and let $\mathcal{Q}_{T_1,\dots,T_r}(t)$ be the Hilbert–Poincaré series for $R/\langle T_1, \dots, T_r\rangle$. To compute $\mathcal{Q}_n(t)$ we need the following results (see [16]).

**Lemma 3.4.1.** *If $S$ is a subset of the variables, then*

$$\mathcal{Q}_S(t) = \frac{1}{(1-t)^{4n-|S|}}$$

*where $|S|$ denotes the cardinality of the set $S$.*

**Lemma 3.4.2.** *If $T_1, \dots, T_r, T$ are monomials with deg $T = d$, then*

$$\mathcal{Q}_{T_1,\dots,T_r,T}(t) = \mathcal{Q}_{T_1,\dots,T_r}(t) - t^q\mathcal{Q}_{\frac{T_1}{(T_1,T)},\dots,\frac{T_r}{(T_r,T)}}(t)$$

*where $(T_i, T)$ denotes the greatest common divisor of the two monomials $T_i$ and $T$.*

We now consider the $n+2$ variables that do not appear in the generating set for $I_n$, namely $x_{11}$, $x_{n2}$ and $x_{i3}$ for $i = 1, \dots, n$. Applying Lemma 3.4.2 above to these variables we get

$$\mathcal{Q}_A(t) = (1-t)^{n+2}\mathcal{Q}_n(t)$$

where

$$A = \{x_{i0}, x_{i3}, x_{11}, x_{n2}, x_{r2}x_{s1}\}_{i=1,\dots,n,\ \ 1\le r<s<\le n}.$$

It is then clear that the variables $x_{i0}$, $x_{i3}$ for $1 \le i \le n$ and $x_{11}$, $x_{n2}$ are no longer important to the computation. Thus we may replace $R$ by

$$S = \mathbb{K}[x_1, \dots, x_{n-1}, y_1, \dots, y_{n-1}],$$

letting $x_r = x_{r2}$ for $1 \le r \le n-1$ and $y_{s-1} = x_{s2}$ for $2 \le s \le n$ and $A$ can be replaced by

$$B = \{x_r y_s \mid 1 \le r \le s \le n-1\},$$

thus we have (with the obvious notation)

$$\mathcal{Q}_B(t) = \mathcal{Q}_A(t).$$

To prove the theorem we must show that

$$\mathcal{Q}_B(t) = \frac{1+(n-1)t}{(1-t)^{n-1}}.$$

We will prove the statement by induction on $n-1$. If $n-1 = 1$, then using first Lemma 3.4.1 and then Lemma 3.4.2 with $B = \emptyset$ we get

$$\mathcal{Q}_{x_1 y_1}(t) = \mathcal{Q}_{\emptyset}(t) - t^2 \mathcal{Q}_{\emptyset}(t) = \frac{1-t^2}{(1-t)^2} = \frac{1+t}{1-t},$$

as desired. For the induction, let $C = \{x_r y_s \mid 1 \le r \le s \le n-2\}$ so that

$$\begin{aligned} \mathcal{Q}_B(t) &= \mathcal{Q}_{C, x_1 y_{n-1}, \dots, x_{n-1} y_{n-1}}(t) \\ &= \mathcal{Q}_{C, x_1 y_{n-1}, \dots, x_{n-2} y_{n-1}}(t) - t^2 \mathcal{Q}_{C, x_1, \dots, x_{n-2}}(t) \\ &= \mathcal{Q}_{C, x_1 y_{n-1}, \dots, x_{n-2} y_{n-1}}(t) - t^2 \mathcal{Q}_{x_1, \dots, x_{n-2}}(t) \\ &= \mathcal{Q}_{C, x_1 y_{n-1}, \dots, x_{n-2} y_{n-1}}(t) - \frac{t^2}{(1-t)^{2(n-1)-(n-2)}} \\ &= \mathcal{Q}_{C, x_1 y_{n-1}, \dots, x_{n-2} y_{n-1}}(t) - \frac{t^2}{(1-t)^n} \end{aligned}$$

where we have used Lemma 3.4.2, the equality

$$\langle C, x_1, \dots, x_{n-2} \rangle = \langle x_1, \dots, x_{n-2} \rangle$$

and then Lemma 3.4.1. We repeat this reasoning on the first term,

$$\mathcal{Q}_{C, x_1 y_{n-1}, \dots, x_{n-2} y_{n-1}}(t)$$

with $T = x_{n-2} y_{n-1}$ in Lemma 3.4.2. This time there is a nontrivial greatest common denominator with a term in $C$. Using Lemma 3.4.1 we obtain

$$\begin{aligned} \mathcal{Q}_{C, x_1 y_{n-1}, \dots, x_{n-2} y_{n-1}}(t) &= \mathcal{Q}_{C, x_1 y_{n-1}, \dots, x_{n-3} y_{n-1}}(t) - t^2 \mathcal{Q}_{x_1, \dots, x_{n-3}, y_{n-2}}(t) \\ &= \mathcal{Q}_{C, x_1 y_{n-1}, \dots, x_{n-3} y_{n-1}}(t) - \frac{t^2}{(1-t)^n}. \end{aligned}$$

Continuing this procedure we get

$$\mathcal{Q}_B(t) = \mathcal{Q}_C(t) - (n-1)\frac{t^2}{(1-t)^n}.$$

We now can determine $\mathcal{Q}_C(t)$ from the induction assumption, however we note that $\mathcal{Q}_C(t)$ occurs in a ring with two more variables than the one in the induction assumption, and so we have

$$\mathcal{Q}_C(t) = \frac{1+(n-2)t}{(1-t)^{n-2+2}}.$$

Therefore

$$\mathcal{Q}_B(t) = \frac{1+(n-2)t}{(1-t)^n} - (n-1)\frac{t^2}{(1-t)^n} = \frac{1+(n-1)t}{(1-t)^{n-1}}.$$

which is the result we had to prove. □

Using this theorem it is easy to determine the Hilbert function of $M_n$. Moreover, since the degree of the numerator in $\mathcal{P}(t)$ is less than the degree of the denominator, we have that the Hilbert function corresponds to the Hilbert polynomial for all degrees $\nu$. So we have the following corollaries.

**Corollary 3.4.1.** *The Hilbert polynomial $H_n(\nu)$ is given by*

$$H_n(\nu) = 4\binom{\nu+2n}{2n} + 4(n-1)\binom{\nu+2n-1}{2n}.$$

We now recall that the Krull dimension of a ring is the supremum of the lengths of chains $P_0 \subset P_1 \subset P_2 \subset \ldots$ of prime ideals $P_j$ in it, see [70]. If we set $\dim M_n$ equal to the Krull dimension of $R/\mathrm{ann}(M_n)$, as a corollary of the previous theorem we also get:

**Corollary 3.4.2.** *The Krull dimension of $M_n$ is equal to $2n+1$.*

To compute the projective dimension of $M_n$, we need the following proposition:

**Proposition 3.4.2.** *The variables $x_{11}$, $x_{n2}$, $x_{i3}$, $i = 1, \ldots, n$ form an $M_n$-regular sequence of length $n+2$.*

*Proof.* This result is a consequence of the fact that the variables $x_{11}$, $x_{n2}$, $x_{i3}$, $i = 1, \ldots, n$ are exactly the only variables which do not appear in any of the leading terms of the elements of the reduced Gröbner basis of $\langle P_n^t \rangle$ given in Proposition 3.4.1. In general, if $D$ is a submodule of $R^4$ and a variable $x_{ij}$ does not appear in any of the leading terms in a Gröbner basis for $D$, then $x_{ij}$ is a nonzero divisor on $R^4/D$. This is because if $\mathbf{0} \neq \mathbf{g} \in R^4$ and $\mathbf{g}$ is reduced with respect to the Gröbner basis of $D$ and $x_{ij}\mathbf{g}$ is in $D$, then $x_{ij}\mathrm{lt}(\mathbf{g})$ must be divisible by the leading term of one of the elements of the given Gröbner basis. So $\mathrm{lt}(\mathbf{g})$ must also be divisible by the same leading term, contradicting the fact that $\mathbf{g}$ is reduced. □

To enlarge this regular sequence, we now consider the module

$$\begin{aligned} M_n^* &= M_n/\langle x_{11}, x_{n2}, x_{i3}, i = 1, \dots, n\rangle M_n \\ &\cong R^4/\langle P_n^t + \langle x_{11}, x_{n2}, x_{i3}, i = 1, \dots, n\rangle R^4\rangle \\ &\cong R^4/\langle U_i, B_{rs}, x_{11}\mathbf{e}_\ell, x_{n2}\mathbf{e}_\ell, x_{i3}\mathbf{e}_\ell\rangle_{i=1,\dots,n, 1\le r<s\le n, \ell=1,2,3,4} \end{aligned}$$

and we set

$$B = \langle U_i, B_{rs}, x_{11}\mathbf{e}_\ell, x_{n2}\mathbf{e}_\ell, x_{i3}\mathbf{e}_\ell\rangle_{i=1,\dots,n, 1\le r<s\le n, \ell=1,2,3,4}.$$

We note that the columns of $B_{rs} = [U_r, U_s]$ can be reduced, using $x_{r3}\mathbf{e}_\ell$ and $x_{s3}\mathbf{e}_\ell$, $\ell = 1, 2, 3, 4$, to the matrix

$$\begin{bmatrix} 0 & 0 & 0 & x_{r2}x_{s1} - x_{r1}x_{s2} \\ 0 & 0 & x_{r2}x_{s1} - x_{r1}x_{s2} & 0 \\ 0 & x_{r2}x_{s1} - x_{r1}x_{s2} & 0 & 0 \\ x_{r2}x_{s1} - x_{r1}x_{s2} & 0 & 0 & 0 \end{bmatrix}$$

so we have

$$B = \langle U_i, (x_{r2}x_{s1} - x_{r1}x_{s2})\mathbf{e}_\ell, x_{11}\mathbf{e}_\ell, x_{n2}\mathbf{e}_\ell,$$

$$x_{i3}\mathbf{e}_\ell\rangle_{i=1,\dots,n,\ 1\le r<s\le n,\ \ell=1,2,3,4}. \tag{3.45}$$

We note that the generators of $B$ given in (3.45) form a Gröbner basis for $B$. In fact, note that the only $S$-polynomials we need to consider are those computed using a column of $U_i$ and one of $(x_{r2}x_{s1} - x_{r1}x_{s2})\mathbf{e}_\ell$, $x_{11}\mathbf{e}_\ell$, $x_{n2}\mathbf{e}_\ell$, or $x_{i3}\mathbf{e}_\ell$. The leading terms of the columns of $U_i$ are $x_{i0}\mathbf{e}_\ell$ which are relatively prime to $x_{11}\mathbf{e}_\ell$, $x_{n2}\mathbf{e}_\ell$, $x_{i3}\mathbf{e}_\ell$, and it is easy to verify that the corresponding $S$-polynomials reduce to zero. The leading term of $(x_{r2}x_{s1} - x_{r1}x_{s2})\mathbf{e}_\ell$ is $x_{r2}x_{s1}\mathbf{e}_\ell$ and so it is relatively prime to $x_{i0}$. Again it is easy to verify that the corresponding $S$-polynomials reduce to zero.

**Proposition 3.4.3.** *The polynomials $x_{21}+x_{12}$, $x_{31}+x_{22}$, $\dots$, $x_{r,1}+x_{r-1,2}$, $\dots$, $x_{n1}+x_{n-1,2}$ form a maximal $M_n^*$-regular sequence in $\wp_n$, where $\wp_n$ denotes the ideal of the variables in $R$.*

*Proof.* In order to show that the polynomials $x_{21} + x_{12}$, $x_{31} + x_{22}$, $\dots$, $x_{n,1} + x_{n-1,2}$ form a $M_n^*$-regular sequence in $\wp_n$, we need to show that the polynomial $x_{\nu+1,1} + x_{\nu 2}$ is a nonzero divisor on $R^4/B_{\nu-1}$ (for $\nu = 1, 2, \dots, n-1$) where

$$B_{\nu-1} = \langle B, (x_{21} + x_{12})\mathbf{e}_\ell, (x_{31} + x_{22})\mathbf{e}_\ell, \dots, (x_{\nu,1} + x_{\nu-1,2})\mathbf{e}_\ell\rangle_{\ell=1,2,3,4}$$

and $B_0 = B$. Then to show that the sequence is maximal we will show that every element of $\wp_n$ is a zero divisor on $R^4/B_{n-1}$.

In order to do this we first construct a Gröbner basis $G_{\nu-1}$ for $B_{\nu-1}$, $1 \le \nu \le n$. This basis will consist of the following vectors:

(a) the columns of $U_i$ for $1 \leq i \leq n$;

(b) $x_{12}x_{s-1,2}\mathbf{e}_\ell$ for $2 \leq s \leq \nu$;

(c) $x_{12}x_{s1}\mathbf{e}_\ell$ for $\nu + 1 \leq s \leq n$;

(d) $x_{r2}x_{n1}\mathbf{e}_\ell$ for $1 \leq r < n$;

(e) $(x_{r2}x_{s-1,2} - x_{r-1,2}x_{s2})\mathbf{e}_\ell$ for $2 \leq r < s \leq \nu$;

(f) $(x_{r2}x_{s1} - x_{r-1,2}x_{s2})\mathbf{e}_\ell$ for $2 \leq r \leq \nu < s < n$;

(g) $(x_{r2}x_{s1} - x_{r1}x_{s2})\mathbf{e}_\ell$ for $\nu < r < s < n$;

(h) $x_{11}\mathbf{e}_\ell$;

(i) $x_{n2}\mathbf{e}_\ell$;

(j) $x_{i3}\mathbf{e}_\ell$ for $1 \leq i \leq n$;

(k) $(x_{r1} + x_{r-1,2})\mathbf{e}_\ell$ for $1 < r \leq \nu$,

where $\ell = 1, 2, 3, 4$. These vectors are obtained from the vectors in the generating set for $B$ given in equation (3.45) by substituting 0 for $x_{11}$ and $x_{n2}$ and $-x_{r-1,2}$ for $x_{r1}$. Thus the given vectors do form a generating set for the module $B_{\nu-1}$. That this set of vectors forms a Gröbner basis with respect to the given order can be verified by checking that all the corresponding $S$-polynomials in fact reduce to 0. Note that all the terms are written with their leading term first. Also note that in the extreme case for $\nu$, i.e., $\nu = 1$ and $\nu = n-1$, the ranges in many of the above contain no $r$ or $s$. We now verify that for $\nu = 1, 2, \ldots, n-1$, $x_{\nu+1,1} + x_{\nu 2}$ is a nonzero divisor on $R^4/B_{\nu-1}$. The verification will be made in the case where all the vectors in the above list appear.

The extreme cases of $\nu = 1$ and $\nu = n-1$ are the same but avoid some of the following complications. Assume that we have a vector $\mathbf{g}$ in $R^4 - B_{\nu-1}$ such that $(x_{\nu+1,1} + x_{\nu 2})\mathbf{g} \in B_{\nu-1}$. We may assume that $\mathbf{g}$ is reduced with respect to $G_{\nu-1}$. In particular this means that $\mathbf{g}$ can only contain the variables $x_{r1}$ for $\nu + 1 \leq r \leq n$ and $x_{s2}$ for $1 \leq s \leq n-1$. The fact that the variables $x_{11}$, $x_{n2}$, $x_{i3}$, $1 \leq i \leq n$ and $x_{r1}$, $1 < r \leq \nu$ do not appear follows from the vectors in (h), (i), (j) and (k) in the above list for the Gröbner basis for $B_{\nu-1}$. The variables $x_{i0}$ for $1 \leq i \leq n$ do not appear since in the matrices $U_i$ for $1 \leq i \leq n$ there is a leading term of the form $x_{i0}\mathbf{e}_\ell$ for $1 \leq i \leq n$ and $\ell = 1, 2, 3, 4$ and no $x_{i0}$ in other coordinate of that vector in $U_i$. We, of course, have that $(x_{\nu+1,1} + x_{\nu 2})\mathbf{g}$ reduces to zero by $G_{\nu-1}$. Only the vectors in (b), (c), (d), (e), (f), and (g) in the list of the Gröbner basis $G_{\nu-1}$ above can ever be used to reduce $(x_{\nu+1,1} + x_{\nu 2})\mathbf{g}$. Now $\mathbf{g}$ must have a nonzero coordinate, say $g\mathbf{e}_\ell$ for some $\ell = 1, 2, 3, 4$. Then, due to the nature of the vectors in the Gröbner basis $G_{\nu-1}$ that can be used to reduce $(x_{\nu+1,1} + x_{\nu 2})\mathbf{g}$ we see that $(x_{\nu+1,1} + x_{\nu 2})g$ must reduce to zero using polynomials in the list below:

(b) $x_{12}x_{s-1,2}$ for $2 \leq s \leq \nu$;

(c) $x_{12}x_{s1}$ for $\nu+1 \leq s \leq n$;

(d) $x_{r2}x_{n1}$ for $1 \leq r < n$;

(e) $x_{r2}x_{s-1,2} - x_{r-1,2}x_{s2}$ for $2 \leq r < s \leq \nu$;

(f) $x_{r2}x_{s1} + x_{r-1,2}x_{s2}$ for $2 \leq r \leq \nu < s < n$;

(g) $x_{r2}x_{s1} - x_{r1}x_{s2}$ for $\nu < r < s < n$.

We denote this new list of polynomials by $H_{\nu-1}$. Note that $g$ is reduced with respect to $H_{\nu-1}$ and only involves the variables $x_{r1}$ for $\nu+1 \leq r \leq n$ and $x_{s2}$ for $1 \leq s \leq n-1$. Thus one of the leading power products in $H_{\nu-1}$ must divide $\text{lp}((x_{\nu+1,1} + x_{\nu 2})g) = x_{\nu+1,1}\text{lp}(g)$ and cannot divide $\text{lp}(g)$. These polynomials come from polynomials in c) and f) in the list for $H_{\nu-1}$ above, and so we see that $x_{r2}$ must divide $\text{lp}(g)$ for one of $r = 1, \dots, \nu$. Since $g$ is reduced with respect to $H_{\nu-1}$ we see, using the polynomials in (c), (d) and (f) in the list for $H_{\nu-1}$, that no $x_{r1}$ can divide $\text{lp}(g)$. Thus

$$g = x_{12}^{a_1}x_{22}^{a_2}\cdots x_{n-1,2}^{a_{n-1}} + h,$$

where all of the terms in $h$ are smaller than

$$\text{lp}(g) = x_{12}^{a_1}\cdots x_{n-1,2}^{a_{n-1}}.$$

Moreover one of the $a_r$, for $1 \leq r \leq \nu$, is nonzero. Then

$$(x_{\nu+1,1} + x_{\nu 2})g$$

$$= x_{\nu+1,1}x_{12}^{a_1}\cdots x_{n-1,2}^{a_{n-1}} + x_{12}^{a_1}\cdots x_{\nu-1,2}^{a_{\nu-1}}x_{\nu,2}^{a_{\nu}+1}x_{\nu+1,2}^{a_{\nu+1}}\cdots x_{n-1,2}^{a_{n-1}} + (x_{\nu+1,1} + x_{\nu 2})h.$$

If $a_1 \geq 1$, then using monomials in (c) in the list for $H_{\nu-1}$ we have that $(x_{\nu+1,1} + x_{\nu 2})g$ reduces to

$$x_{12}^{a_1}\cdots x_{\nu-1,1}^{a_{\nu-1}}x_{\nu,2}^{a_{\nu}+1}x_{\nu+1,2}^{a_{\nu+1}}\cdots x_{n-1,2}^{a_{n-1}} + (x_{\nu+1,1} + x_{\nu 2})h.$$

If $a_1 = 0$, then one of $a_2, \dots, a_\nu$ is greater than zero, say $a_j \geq 1$ $(2 \leq j \leq \nu)$ and so using the polynomial $x_{\nu+1,1}x_{j2} + x_{j-1,2}x_{\nu+1,2}$ in the list for $H_{\nu-1}$, the term $(x_{\nu+1,1} + x_{\nu 2})g$ reduces to

$$-x_{12}^{a_1}\cdots x_{j-2,2}^{a_{j-2}}x_{j,2}^{a_{j-1}}x_{j-1,2}^{a_{j-1}+1}\cdots x_{\nu,2}^{a_\nu}x_{\nu+1,2}^{a_{\nu+1}+1}x_{\nu+2,2}^{a_{\nu+2}}\cdots x_{n-1,2}^{a_{n-1}}$$

$$+x_{12}^{a_1}\cdots x_{\nu-1,2}^{a_{\nu-1}}x_{\nu,2}^{a_\nu+1}x_{\nu+1,2}^{a_{\nu+1}}\cdots x_{n-1,2}^{a_{n-1}} + (x_{\nu+1,1} + x_{\nu,2})h.$$

We see that the second term in this last expression is larger than the first term in the degrevlex ordering, since $a_{\nu+1}+1 > a_{\nu+1}$ (the degrees of the two monomials are the same). Thus the leading power product in the reduced polynomial just obtained from $(x_{\nu+1,1} + x_{\nu 2})g$ is either

$$x_{12}^{a_1} \cdots x_{\nu-1,2}^{a_{\nu-1}} x_{\nu,2}^{a_\nu+1} x_{\nu+1,2}^{a_{\nu+1}} \cdots x_{n-1,2}^{a_{n-1}}$$

or

$$x_{\nu+1,1}\mathrm{lp}(h).$$

**Claim 3.4.1.** *Let $X$ be a term of $h$ such that*

$$x_{\nu+1,1}X > x_{12}^{a_1} \cdots x_{\nu-1,2}^{a_{\nu-1}} x_{\nu,2}^{a_\nu+1} x_{\nu+1,2}^{a_{\nu+1}} \cdots x_{n-1,2}^{a_{n-1}}.$$

*Assume that $x_{\nu+1,1}X$ can be reduced using $H_{\nu-1}$. Then, using $H_{\nu-1}$, we have that $x_{\nu+1,1}X$ can be reduced to a term $Y$ such that*

$$Y < x_{12}^{a_1} \cdots x_{\nu-1,2}^{a_{\nu-1}} x_{\nu,2}^{a_\nu+1} x_{\nu+1,2}^{a_{\nu+1}} \cdots x_{n-1,2}^{a_{n-1}}.$$

Assuming the claim, we show that $x_{\nu+1,1}+x_{\nu 2}$ is a nonzero divisor on $R^4/B_{\nu-1}$ as follows. We first observe that

$$x_{12}^{a_1} \cdots x_{\nu-1,2}^{a_{\nu-1}} x_{\nu,2}^{a_\nu+1} x_{\nu+1,2}^{a_{\nu+1}} \cdots x_{n-1,2}^{a_{n-1}}$$

cannot be reduced using $H_{\nu-1}$. In fact, since only the variables $x_{s2}$ appear we could only possibly use the polynomials in (d) or (e) in the list for $H_{\nu-1}$. Then since $g$ is reduced with respect to $H_{\nu-1}$, $x_{\nu 2}$ would have to appear in the polynomial used to do the reduction, but this variable does not appear in any of the polynomials in (b) and (e). Thus if

$$x_{12}^{a_1} \cdots x_{\nu-1,2}^{a_{\nu-1}} x_{\nu,2}^{a_\nu+1} x_{\nu+1,2}^{a_{\nu+1}} \cdots x_{n-1,2}^{a_{n-1}}$$

is the leading term, we have a contradiction. Otherwise $x_{\nu+1,1}\mathrm{lp}(h)$ is the leading term and so must be reducible using $H_{\nu-1}$.

Setting $X = \mathrm{lt}(h)$ in the Claim 3.4.1 and $h' = h - \mathrm{lt}(h)$, we reduce $(x_{\nu+1,1} + x_{\nu 2})g$ to

$$x_{12}^{a_1} \cdots x_{\nu-1,2}^{a_{\nu-1}} x_{\nu,2}^{a_\nu+1} x_{\nu+1,2}^{a_{\nu+1}} \cdots x_{n-1,2}^{a_{n-1}} + Y + x_{\nu 2}X + (x_{\nu+1,1} + x_{\nu 2})h'.$$

The leading term of this last expression is either

$$x_{12}^{a_1} \cdots x_{\nu-1,2}^{a_{\nu-1}} x_{\nu,2}^{a_\nu+1} x_{\nu+1,2}^{a_{\nu+1}} \cdots x_{n-1,2}^{a_{n-1}}$$

or $x_{\nu+1,1}\mathrm{lp}(h')$, since $\mathrm{lp}(g) > \mathrm{lp}(h)$. If the leading term is $x_{\nu+1,1}\mathrm{lp}(h')$, then $x_{\nu+1,1}\mathrm{lp}(h')$ must be reducible using $H_{\nu-1}$. Thus the argument may be repeated until we obtain an expression which must reduce to zero using $H_{\nu-1}$ but whose leading term is

$$x_{12}^{a_1} \cdots x_{\nu-1,2}^{a_{\nu-1}} x_{\nu,2}^{a_\nu+1} x_{\nu+1,2}^{a_{\nu+1}} \cdots x_{n-1,2}^{a_{n-1}}$$

and we have again a contradiction.

*Proof of Claim 3.4.1.* As above, we see that $x_{\nu+1,1}X$ can only be reduced using the polynomials in (c) and (f) in the list for $H_{\nu-1}$ above. So $x_{r2}$ must divide $X$ for some $r$, $1 \le r \le \nu$ and no variable $x_{r1}$ can divide $X$, in fact $X$ cannot be reduced using $H_{\nu-1}$. So we obtain that $X$ is of the form

$$X = x_{12}^{b_1} x_{22}^{b_2} \dots x_{n-1,2}^{b_{n-1}}.$$

If we can use the monomial $x_{12}x_{\nu+1,1}$ in (c) then $x_{\nu+1,1}X$ reduces to zero and the claim is true. Otherwise $x_{r2}$ divides $X$ for some $r$, $2 \le r \le \nu$. For the reduction of $x_{\nu+1,1}X$ we replace $x_{r2}x_{\nu+1,1}$ by $-x_{r-1,2}x_{\nu+1,2}$. Thus we need to show that

$$\frac{X}{x_{r2}} x_{r-1,2}x_{\nu+1,2} < x_{12}^{a_1} \dots x_{\nu-1,2}^{a_{\nu-1}} x_{\nu,2}^{a_\nu+1} x_{\nu+1,2}^{a_{\nu+1}} \dots x_{n-1,2}^{a_{n-1}} \tag{3.46}$$

under the hypotheses

$$X = x_{12}^{b_1} \dots x_{n-1,2}^{b_{n-1}} < x_{12}^{a_1} \dots x_{n-1,2}^{a_{n-1}}$$

and

$$x_{\nu+1,1}X > x_{12}^{a_1} \dots x_{\nu-1,2}^{a_{\nu-1}} x_{\nu,2}^{a_\nu+1} x_{\nu+1,2}^{a_{\nu+1}} \dots x_{n-1,2}^{a_{n-1}}.$$

These two hypotheses guarantee that all the terms have the same degree. From the first we choose $\ell$ such that

$$b_{\ell+1} = a_{\ell+1}, \dots, b_{n-1} = a_{n-1}, \quad b_\ell > a_\ell.$$

The second hypothesis assures that $\ell \le \nu$ but then the left side of (3.46) has exponent $b_{\nu+1}+1$ for $x_{\nu+1,2}$, while the right-hand side has exponent $a_{\nu+1} = b_{\nu+1}$ and both sides have equal exponents for all $x_{r2}$ for $r > \nu + 1$, thus (3.46) is true. This completes the proof of the claim.□

We finally need to show that every element of $\wp_n$ is a zero divisor of $R^4/B_{n-1}$. We first note that the Gröbner basis for $B_{n-1}$ is given by the following vectors:

(a) the columns of $U_i$ for $1 \le i \le n$;

(b) $x_{12}x_{s-1,2}\mathbf{e}_\ell$ for $2 \le r \le s < n$;

(d) $x_{r2}x_{n1}\mathbf{e}_\ell$ for $1 \le r < n$;

(e) $(x_{r2}x_{s-1,2} - x_{r-1,2}x_{s2})\mathbf{e}_\ell$ for $2 \le r < s \le n$;

(h) $x_{11}\mathbf{e}_\ell$;

(i) $x_{n2}\mathbf{e}_\ell$;

(j) $x_{i3}\mathbf{e}_\ell$ for $1 \le i \le n$;

(k) $(x_{r1} + x_{r-1,2})\mathbf{e}_\ell$ for $1 < r \leq n$,

where $\ell = 1, 2, 3, 4$. This basis however is obviously nonreduced since, for example, the vectors in (k) with $r = n$ can be used to reduce the vectors in (d) to $x_{r2}x_{n-1,2}\mathbf{e}_\ell$ for $1 \leq r < n$. The vectors in (e) can be reduced as follows: if $s = n-1$, then $x_{r-1,2}x_{n-1,2}\mathbf{e}_\ell$ can be used to reduce $(x_{r2}x_{n-2,2} - x_{r-1,2}x_{n-1,2})\mathbf{e}_\ell$ to $x_{r2}x_{n-2,2}\mathbf{e}_\ell$. With this last vector we reduce the vector in (e) with $s = n-2$ to $x_{r2}x_{n-3,2}\mathbf{e}_\ell$. Iterating this procedure, we obtain the reduced Gröbner basis $G$ for $B_{n-1}$. It consists of the following vectors:

(a′) the columns of $U_i$ for $1 \leq i \leq n$;

(b′) $x_{s2}x_{r2}\mathbf{e}_\ell$ for $1 \leq r \leq s \leq n-1$;

(c′) $x_{11}\mathbf{e}_\ell$;

(d′) $x_{n2}\mathbf{e}_\ell$;

(e′) $x_{i3}\mathbf{e}_\ell$ for $1 \leq i \leq n$;

(f′) $(x_{r1} + x_{r-1,2})\mathbf{e}_\ell$ for $1 < r \leq n$,

where $\ell = 1, 2, 3, 4$. It is now immediate to show that if $f \in \wp_n$ is nonzero, then $f$ reduces to zero using $G$. Indeed, if $f\mathbf{e}_1 \in B_{n-1}$, then $f(\mathbf{e}_1 + B_{n-1}) = 0$ and so $\mathbf{e}_1 \notin B_{n-1}$ implies that $f$ is a zero divisor in $R^4/B_{n-1}$. Let us assume $f\mathbf{e}_1 \notin B_{n-1}$, then $f(f\mathbf{e}_1 + B_{n-1}) = f^2\mathbf{e}_1 + B_{n-1}$, and so it is enough to show that $f^2\mathbf{e}_1$ reduces to zero using $G$ that is equivalent to show that $f^2\mathbf{e}_1 \in B_{n-1}$ for any $f \in \wp_n$. Since $f \in \wp_n$, every term in $f^2$ is of degree 2 or higher; using the columns of the $U_i$'s, we can however reduce $f^2\mathbf{e}_1$ to a vector $\mathbf{f}_1$ with no variables $x_{i0}$, $1 \leq i \leq n$, and containing only polynomials of degree 2 or higher. Using the last four types of vectors in $G$ listed above we can reduce $\mathbf{f}_1$ to a vector $\mathbf{f}_2$ containing only the variables $x_{r2}$, $1 \leq r \leq n-1$ and containing only polynomials of degree 2 or higher. Finally, using the vectors in b′) the vector $\mathbf{f}_2$ reduces to zero. This concludes the proof. □

**Corollary 3.4.3.** *The depth of the maximal ideal $\wp_n$ in $M_n$ is given by*

$$\text{depth}(\wp_n, M_n) = 2n + 1.$$

*Proof.* This is obtained using the definition of the depth of an ideal. Indeed, if $I$ is an ideal of $R$ and $M$ is an $R$-module, the depth of $I$ on $M$, denoted by depth$(I, M)$ is the length of any maximal $M$-regular sequence in $I$. The result therefore follows from Propositions 3.4.2 and 3.4.3. □

**Theorem 3.4.2.** *The module $M_n$ has projective dimension*

$$pd(M_n) = 2n - 1.$$

*Proof.* This follows from the Auslander–Buchsbaum formula and Corollary 3.4.3:

$$pd(M_n) = \text{depth}(\wp_n, R) - \text{depth}(\wp_n, M_n)$$

$$= 4n - \text{depth}(\wp_n, M_n) = 2n - 1. \qquad \square$$

Now we go back to the problem of finding the minimal free resolution of the module $\langle P_n^t \rangle$. As we have seen above in the explicit computations made with the use of CoCoA, the maps in the resolution are all linear except the first one. We will start with the description of the first syzygy module of $\langle P_n^t \rangle$.

**Theorem 3.4.3.** *A generating set for the first syzygy module of $P_n^t$ consists of the quadratic relations arising from the following identities:*

1. *for each of the $\binom{n}{2}$ pairs of indices $r, s$ with $1 \leq r, s \leq n$,*

$$[U_r, U_s U_s^t] = 0 \qquad \textit{and} \qquad [U_r U_r^t, U_s] = 0;$$

2. *for each of the $\binom{n}{3}$ triples of indices $p, r, s$ with $1 \leq p, r, s \leq n$,*

$$[U_r, U_s U_p^t + U_p U_s^t] = 0 \qquad \textit{and} \qquad [U_s, U_r U_p^t + U_p U_r^t] = 0;$$

3. *for each of the $\binom{n}{3}$ triples of indices $p, r, s$ with $1 \leq p, r, s \leq n$,*

$$[U_r, U_s]J[U_p, I] + [U_s, U_p]J[U_r, I] + [U_p, U_r]J[U_s, I] = 0,$$

*and*

$$[U_r, U_s]I[U_p, J] + [U_s, U_p]I[U_r, J] + [U_p, U_r]I[U_s, J] = 0,$$

*where $I$ and $J$ denote the matrices representing the imaginary units $i$ and $j$, respectively, and $[-,-]$ denotes the commutator. The generators thus defined form a minimal generating set consisting of $8\binom{n}{2} + 16\binom{n}{3}$ elements.*

*Proof.* The computation of a Gröbner basis for the module $\langle P_n^t \rangle$ in Proposition 3.4.1 shows that it consists of vectors involving one or two of the matrices $U_\ell$. Once we have a Gröbner basis $\mathcal{G}$ for a module, we can follow Proposition 1.1.9 (Chapter 1) to compute the syzygy module of $\mathcal{G}$ by computing and reducing the $S$-polynomial of any two elements in $\mathcal{G}$. Since the reduction process cannot involve variables not already in the $S$-polynomial, we see that the generators of the syzygy module of $\mathcal{G}$ involve at most four of the matrices $U_\ell$, i.e., the generating set for the syzygy of $\mathcal{G}$ can be considered as the union of the generating sets of the syzygies of four of the $U_\ell$ at a time. The syzygy module of $\langle P_n^t \rangle$ can be obtained as in Theorem 1.1.8. The procedure we have to follow is to consider the $4n \times (4n + 4\binom{n}{2})$ transformation matrices $T$ between vectors in $P_n^t$ and $\mathcal{G}$

that turns out to be a $4n \times 4n$ identity matrix in the first $4n$ columns while the remaining $4\binom{n}{2}$ columns form the following matrix

$$\begin{bmatrix} U_2 & U_3 & \dots & U_n & 0 & \dots & 0 & 0 & \dots & 0 & \dots \\ -U_1 & 0 & \dots & 0 & U_3 & \dots & U_n & 0 & \dots & 0 & \dots \\ 0 & -U_1 & \dots & 0 & -U_2 & \dots & 0 & U_4 & \dots & U_n & \dots \\ \vdots & \vdots & \ddots & \vdots & \vdots & \ddots & \vdots & \vdots & \ddots & \vdots & \ddots \\ 0 & 0 & \dots & -U_1 & 0 & \dots & -U_2 & 0 & \dots & -U_3 & \dots \end{bmatrix}.$$

The syzygies of $P_n^t$ can be obtained by multiplying the syzygies of $\mathcal{G}$ by this matrix, so that they do not involve more than four of the matrices $U_\ell$. Once again a minimal generating set for the syzygies of $P_n^t$ can be taken as the union of the minimal generating set of the syzygies of four of the $U_\ell$ taken at a time. Using CoCoA one can compute the minimal generating set for the syzygies of $P_n^t$ for $n = 2, 3, 4$, obtaining the result in the statement of the theorem and this completes the proof. □

To prove that the remaining syzygies are all linear requires a lemma which allows one to compute the Castelnuovo–Mumford regularity of the module $\langle P_n^t \rangle$. The Castelnuovo–Mumford regularity of a homogeneous submodule $M$ of a free $R$-module is a very well-known concept in algebraic geometry (see [70]), but we recall it here for the convenience of the reader.

**Definition 3.4.2.** *Let $M$ be a homogeneous submodule of a free $R$-module and let*

$$0 \longrightarrow \oplus_j R(-d_{rj}) \longrightarrow \dots \longrightarrow \oplus_j R(-d_{1j}) \longrightarrow \oplus_j R(-d_{0j}) \longrightarrow M \longrightarrow 0$$

*be a minimal graded free resolution. We say that $M$ is $m$-regular if $d_{ij} - i \leq m$ for all $i, j$. The Castelnuovo–Mumford regularity of $M$ denoted by $\operatorname{reg}(M)$ is the least integer $m$ for which $M$ is $m$-regular.*

In what follows we will use the following result, see [14] for a proof.

**Proposition 3.4.4.** (Bayer and Stillman). *An ideal $I$ in $R$ is $m$-regular if and only if there are linear forms $h_1, \dots, h_r \in R$, $r > 0$ such that*

$$(\langle I, h_1, \dots, h_{i-1} \rangle \ : \ h_i)_m = \langle I, h_1, \dots, h_{i-1} \rangle_m$$

*for $i = 2, \dots, r$ and*

$$\langle I, h_1, \dots, h_r \rangle_m = R_m.$$

**Theorem 3.4.4.** *We have* $\operatorname{reg}(\langle P_n^t \rangle) = \operatorname{reg}(Lt(P_n^t)) = 2$.

*Proof.* It is well known (see [14]) that $\operatorname{reg}(\langle P_n^t \rangle) \leq \operatorname{reg}(Lt(P_n^t))$. Moreover we have that $\operatorname{reg}(\langle P_n^t \rangle) \geq 2$ by Theorem 3.4.3 so it remains to show that

$\operatorname{reg}(Lt(P_n^t)) = 2$. Recalling Proposition 3.4.1 we have that $\operatorname{reg}(Lt(P_n^t)) = \operatorname{reg}(I_n)$ where the ideal $I_n$ is defined by

$$I_n = \langle x_{i0}, x_{r2}x_{s1}\rangle, \qquad i = 1, \dots, n, \quad 1 \leq n < s \leq n.$$

We will compute the regularity by applying Proposition 3.4.4 to the regular sequence we computed in Proposition 3.4.3. Let

$$h_1 = x_{11}, \ \ h_2 = x_{n2}, \ \ h_3 = x_{13}, \ \ h_4 = x_{23}, \ \ \dots, h_{n+2} = x_{n3},$$

$$h_{n+3} = x_{21} + x_{12}, \ \ h_{n+4} = x_{31} + x_{22}, \ \ \dots, h_{2n+1} = x_{n1} + x_{n-1,2}.$$

The first row of the sequence above consists of variables that do not appear in the generators for $I_n$ and so they form a regular sequence. To continue the process we add the variables in $I_n$. In the ideal they have no effect on the regular sequence or on the Bayer and Stillman criterion, so we can simplify the notation setting $x_j = x_{j1}$ for $j = 2, \dots, n$ and $y_j = x_{j2}$ for $j = 2, \dots, n-1$. To prove our statement we need the following lemma whose proof can be found in [5].

**Lemma 3.4.3.** *Let $S = \mathbb{K}[x_2, \dots, x_n, y_1, \dots, y_{n-1}]$ and let $I$be the ideal in $S$ defined by $I = \langle x_i y_j \mid 1 \leq j < i \leq n\rangle$. Then*

1. $\langle I, x_2 + y_1, \dots, x_r + y_{r-1}\rangle : (x_{r+1} + y_r) = \langle I, x_2 + y_1, \dots, x_r + y_{r-1}\rangle$ *for* $r = 1, \dots n-1$;
2. $\langle I, x_2 + y_1, \dots, x_n + y_{n-1}\rangle : f \neq \langle I, x_2 + y_1, \dots, x_n + y_{n-1}\rangle$ *for* $f \in \wp_n$;
3. $\langle I, x_2 + y_1, \dots, x_n + y_{n-1}\rangle_2 = S_2$.

Note that statements 1 and 2 in the lemma imply that $x_2 + y_1, \dots, x_n + y_{n-1}$ forms a maximal $I$-regular sequence in $\wp_n$, while statements 1 and 3 imply that this sequence satisfies the Bayer and Stillman criterion showing that the regularity of $I$ is 2 and this completes the proof. □

**Theorem 3.4.5.** *All the syzygy modules in the minimal resolution of $P_n^t$ after the first step are generated by linear polynomials.*

*Proof.* The proof is a corollary of Theorem 3.4.4. In fact, we know that the first syzygies are quadratic and thus, since the regularity of $\langle P_n^t\rangle$ is 2, all the other syzygies must be linear. □

We are now able to write the minimal free resolution of $M_n$ that has length $2n-1$ and to verify that the degrees of the syzygies are always one except at the first step. It only remains to compute all the Betti numbers $\beta_\nu$ of the $R$-modules in the resolution. The resolution is of the following type

$$0 \longrightarrow R^{\beta_{2n-1}}(-2n) \xrightarrow{Q^t} R^{\beta_{2n-2}}(-2n+1) \xrightarrow{T^t} \dots$$

$$\longrightarrow R^{\beta_3}(-4) \longrightarrow R^{\beta_2}(-3) \longrightarrow R^{4n}(-1) \xrightarrow{P_n^t} R^4 \longrightarrow M_n \longrightarrow 0. \qquad (3.47)$$

**Proposition 3.4.5.** *The Betti numbers $\beta_\nu$, $\nu = 2, \dots 2n-1$, of the module $M_n$ are given by*

$$\beta_\nu = 4\binom{2n-1}{\nu}\frac{n(\nu-1)}{\nu+1}.$$

*Proof.* The Hilbert–Poincaré series can be read from the minimal free resolution of the module as

$$\mathcal{P}_n(t) = \frac{\beta_0 - \beta_1 t + \beta_2 t^3 - \beta_3 t^4 - \dots + \beta_{2n-2}t^{2n-1} - \beta_{2n-1}t^{2n}}{(1-t)^{4n}}$$

where in our case $\beta_0 = 4$ and $\beta_1 = 4n$ (see [200]). From Theorem 3.4.1 we already know that

$$\mathcal{P}_n(t) = 4\frac{1+(n-1)t}{(1-t)^{2n+1}},$$

so by equating the coefficients in the two expressions of $\mathcal{P}_n(t)$ we obtain the result. □

We finally describe the characteristic variety $V(M_n)$.

**Theorem 3.4.6.** *The characteristic variety $V(M_n)$ has dimension $2n+1$.*

*Proof.* As we explained in the Chapter 2, the characteristic variety $V(M_n)$ of a module of differential operators with linear constant coefficients is the affine variety $V$ associated to the ideal of the minors of maximal order of the matrix which defines $M_n$. In our case, the characteristic variety is therefore the subset of points in $\mathbb{C}^{4n}$ where the rank of the matrix $P_n^t$ is strictly less than 4. If we set $\zeta = (\zeta_1, \dots, \zeta_n)$, we will write $\zeta_i = (\xi_{i0}, \xi_{i1}, \xi_{i2}, \xi_{i3}) \in \mathbb{C}^4$, $i = 1, \dots, n$. We will show that the algebraic set $V(M_n)$ has dimension $2n+1$ in a neighborhood of an arbitrary point $\zeta^0 \neq 0$ in $V(M_n)$. We write $\zeta = (\zeta_1, \dots, \zeta_n) \in V(M_n)$ where $(\zeta_1, \dots, \zeta_n) \in \mathbb{C}^{4n}$. We can consider each vector $\zeta_i$ as the element $\zeta_i = \xi_{i0} + \xi_{i1}i + \xi_{i2}j + \xi_{i3}k$ of the complexified quaternionic algebra $\mathbb{H}_\mathbb{C} = \mathbb{H} \otimes \mathbb{C}$, and we will denote by $\zeta_i^*$ the element $\zeta_i^* = \xi_{i0} - \xi_{i1}i - \xi_{i2}j - \xi_{i3}k$. The columns of the matrix $P_n^t(\zeta)$ correspond to the quaternions

$$\zeta_1^*, \zeta_1^* i, \zeta_1^* j, \zeta_1^* k, \dots, \zeta_n^*, \zeta_n^* i, \zeta_n^* j, \zeta_n^* k.$$

The determinant of the $i$-th $4\times 4$ block in $P_n^t(\zeta)$ is equal to $(\zeta_i^*\zeta_i)^2$ and the equation $\zeta_i^*\zeta_i = 0$ defines a quadratic cone $V_i$ of dimension three in $\mathbb{C}^4$. Now for $\eta \in \mathbb{H}_\mathbb{C}$, we define four complex subspaces of $\mathbb{H}_\mathbb{C}$ as follows:

$$L_\eta = \{\eta q \mid q \in \mathbb{H}_\mathbb{C}\}, \qquad L_\eta^\perp = \{q \in \mathbb{H}_\mathbb{C} \mid : \ \eta q = 0\}$$

and

$$R_\eta = \{q\eta \mid q \in \mathbb{H}_\mathbb{C}\}, \qquad R_\eta^\perp = \{q \in \mathbb{H}_\mathbb{C} \mid q\eta = 0\}.$$

The spaces $L_\eta$ and $R_\eta$ are the image of left and right multiplication by $\eta$ respectively, while the other two spaces are the kernels of these maps. It follows that

$$\dim_{\mathbb{C}} L_\eta + \dim_{\mathbb{C}} L_\eta^\perp = \dim_{\mathbb{C}} R_\eta + \dim_{\mathbb{C}} R_\eta^\perp = 4.$$

If $\eta \in V_1$ and $\eta \neq 0$, then $\eta^*\eta = 0$ and $\dim_{\mathbb{C}} L_\eta + \dim_{\mathbb{C}} L_\eta^\perp = 2$; in fact the map of the left multiplication by $\eta$ corresponds to the first four columns of $P_n^t$ with $\eta^*$ substituted in it. The $3 \times 3$ minors of this matrix are multiples of $\eta^*\eta$ and the fact that $\eta \neq 0$ implies that not all the $2 \times 2$ minors are zero. Since $L_\eta \subseteq L_{\eta^*}^\perp$, as a consequence of the dimension we get $L_\eta = L_{\eta^*}^\perp$ and similarly $\dim_{\mathbb{C}} R_\eta = 2$ and $R_\eta = R_{\eta^*}^\perp$. Now we show that

$$\zeta \in V(M_n) \Longleftrightarrow \zeta_1 \in V_1 \text{ and } \zeta_j \in R_{\zeta_1}, \quad j = 2, \dots, n.$$

Let us assume that $\zeta_1 \in V_1$ and $\zeta_j \in R_{\zeta_1}$ for $j = 2, \dots, n$. We have that $\zeta_j = q_j\zeta_1$ for a suitable $q_j \in \mathbb{H}_{\mathbb{C}}$, therefore $z_j^* e \in L_{\zeta_1^*}$ where $e = 1, i, j, k$ so that the space generated by the columns is contained in the two-dimensional space $L_{\zeta_1^*}$. The rank of $P_n^t(\zeta)$ is two and so $\zeta \in V(M_n)$. Conversely, let us suppose that $\zeta \in V(M_n)$. Since $\dim_{\mathbb{C}} L_{\zeta_1^*} = 2$ we may assume that $z_1^*$ and $z_1^* i$ form a basis for $L_{\zeta_1^*}$. Since the rank of $P_n^t(\zeta)$ is not maximal we have that, for any fixed $\ell$, the elements $\zeta_1^*, \zeta_1^* i, \zeta_\ell^*, \zeta_\ell^* i$ are linearly dependent. This means that there exist complex numbers $a_1, b_1, c_1, d_1$ not all zero such that

$$\zeta_1^*(a_1 + b_1 i) = \zeta_\ell^*(c_1 + d_1 i).$$

Our hypothesis implies that one of $c_1$ or $d_1$ is not zero. If we suppose $d_1 = 0$, then $z_\ell = q^*\zeta_1 \in R_{\zeta_1}$ with $q = c_1^{-1}(a_1 + b_1 i)$ and the result follows. So we can suppose $d_1 \neq 0$. The result would follow unless we have complex numbers $a_2, b_2, c_2, d_2$ $a_3, b_3, c_3, d_3$ with $d_\ell \neq 0$ for $\ell = 2, 3$ such that

$$\zeta_1^*(a_2 + b_2 i) = \zeta_\ell^*(c_2 + d_2 j) \quad \text{and} \quad \zeta_1^*(a_3 + b_3 i) = \zeta_\ell^*(c_3 + d_3 k).$$

Multiplying these last three equations on the left by $\zeta_1$ and recalling that $\zeta_1\zeta_1^* = 0$, we get $\zeta_1\zeta_\ell^*(c_1 + d_1 i) = 0$, $\zeta_1\zeta_\ell^*(c_2 + d_2 j) = 0$ and $\zeta_1\zeta_\ell^*(c_3 + d_3 k) = 0$, that is

$$(c_1 + d_1 i),\ (c_2 + d_2 j),\ (c_3 + d_3 k) \in L_{\zeta_1\zeta_\ell^*}^\perp.$$

Now $(\zeta_1\zeta_\ell^*)(\zeta_1\zeta_\ell^*)^* = 0$ so that $(c_1 + d_1 i)$, $(c_2 + d_2 j)$, $(c_3 + d_3 k)$ are linearly independent on $\mathbb{C}$ implies $\zeta_1\zeta_\ell^* = 0$ since the converse would imply $\dim_{\mathbb{C}} L_{\zeta_1\zeta_\ell^*}^* = 2$. Thus we have

$$\zeta_\ell^* \in L_{\zeta_1}^\perp = L_{\zeta_1^*}.$$

We conclude that $\zeta_\ell^* = \zeta_1^* q$ for some $q \in \mathbb{H}_{\mathbb{C}}$, thus $\zeta_\ell \in R_{\zeta_1}$. It follows immediately that

$$\dim\,(V(M_n)) = \dim_{\mathbb{C}} V_1 + (n-1)\dim_{\mathbb{C}} R_{\zeta_1}$$

$$= 3 + 2(n-1) = 2n + 1. \qquad \square$$

Considering the dual of the resolution (3.47) we obtain the complex

$$0 \longrightarrow R^4 \xrightarrow{P_n} R^{4n} \longrightarrow R^{\beta_2} \longrightarrow \dots \longrightarrow R^{\beta_{2n-2}} \xrightarrow{Q} R^{\beta_{2n-1}} \longrightarrow 0$$

whose cohomology groups are, by definition $\mathrm{Ext}^i(M_n, R)$. Those are related to the characteristic variety by the following proposition proved in [142], Chapter 8.

**Proposition 3.4.6.** *Let $R$ be the ring of polynomials in $m$ variables and let $M$ be an $R$-module. The characteristic variety of $M$ has dimension strictly less than $m-k$, $0 \le k \le m$, if and only if*

$$\mathrm{Ext}^i(M, R) = 0, \quad \textit{for all} \quad i = 0, \dots, k.$$

**Theorem 3.4.7.** *If $M_n$ is as above, then we have*

$$\mathrm{Ext}^i(M_n, R) = 0, \quad \textit{for all} \quad i = 0, \dots, 2n-2$$

*and*

$$\mathrm{Ext}^{2n-1}(M_n, R) \neq 0.$$

*Proof.* Since the characteristic variety $V(M_n)$ has dimension $2n+1$, from Proposition 3.4.6 we immediately obtain that $\mathrm{Ext}^i(M_n, R) = 0$, for all $i = 0, \dots, 2n-2$. Now consider the map $Q : R^{\beta_{2n-2}} \to R^{\beta_{2n-1}}$. If it is surjective, then we obtain a matrix $C : R^{\beta_{2n-1}} \to R^{\beta_{2n-2}}$ such that $QC$ is the identity matrix. Then also $C^tQ^t$ is the identity matrix and the map $Q^t$ corresponding to $Q$ in the resolution (3.47) splits, so $R^{\beta_{2n-2}} = \mathrm{im}\, Q^t \oplus \ker C^t$. But $\ker C^t$ is free and $T^t$ (defined in (3.47)) restricted to $\ker C^t$ is one-to-one so we have obtained a free resolution for $M_n$ shorter than (3.47) which is absurd. □

We now give an alternative, algebraic proof of Hartogs' Theorem 3.3.5 for regular functions of several quaternionic variables.

**Theorem 3.4.8.** *Let $K$ be a compact convex subset of $\mathbb{H}^n$, $n > 1$, and let $f$ be a (left) regular function on $\mathbb{H}^n \setminus K$. Then $f$ extends uniquely to an entire (left) regular function.*

*Proof.* In view of Theorem 2.1.15, it is sufficient to prove that $\mathrm{coker}(P_n)$ is torsion free. It is immediately verified that $P_n$ is of maximal rank. Because of its special form, on the other hand, it is obvious that its $4 \times 4$ minors are relatively prime, so we can apply Lemma 2.1.1 to conclude the proof. □

**Remark 3.4.1.** If $K$ is a bounded convex compact set, then Theorem 3.4.8 follows from the vanishing of $\mathrm{Ext}^1(M_n, R)$ and from Theorem 2.1.14. The vanishing of the Ext-modules proved in Theorem 3.4.7 and the results in Chapter 8 in [142] allows us to prove other propositions on the removability of other types of singularities.

**Proposition 3.4.7.** *Let $\Omega$ be a convex open set in $\mathbb{H}^n \cong \mathbb{R}^{4n}$ and let $K$ be a compact subset of $\Omega$. Let $\Sigma_1, \dots, \Sigma_{2n-2}$ be closed half spaces in $\mathbb{R}^{4n}$ and set $\Sigma = \Sigma_1 \cup \ldots \cup \Sigma_{2n-2}$. Then every function $F \in \mathcal{R}(\Omega\backslash(\Sigma \cup K))$ extends to a regular function on all $\Omega\backslash\Sigma$ and the extension coincides with $F$ on $\Omega\backslash(\Sigma \cup K')$ for $K'$ a compact subset of $\Omega$.*

*Proof.* It immediately follows from Theorem 3.4.7 and Theorem 4, page 405 in [142]. □

**Proposition 3.4.8.** *Let $L$ be a linear subspace of $\mathbb{H} = \mathbb{R}^{4n}$ of dimension $2n+2$. Then for every compact $K$ contained in $L$, and every connected open set $\Omega$, relatively compact set in $K$, every regular function defined in a neighborhood of $K\backslash\Omega$ can be extended to a regular function defined in a neighborhood of $K$.*

*Proof.* The result follows from Theorem 3.4.7 and Theorem 4, (p. 405 in [142]) if we can prove that none of the varieties associated to the module $\mathrm{Ext}^{2n-1}(M_n, R)$ is hyperbolic with respect to $L$. By Corollary 2, (p. 377 in [142]) the characteristic variety of $\mathrm{Ext}^{2n-1}(M_n, R)$ is contained in $V(M_n)$ and since $M_n$ is elliptic also every variety in the characteristic variety of $\mathrm{Ext}^{2n-1}(M_n, R)$ must be elliptic. □

We now collect some cohomological properties of the sheaf of regular functions. Note that we have considered $P_n(D)$ as an operator acting from $[\mathcal{E}(\mathbb{R}^{4n})]^4$ to $[\mathcal{E}(\mathbb{R}^{4n})]^{4n}$ so we can denote by $\mathcal{R} = \mathcal{E}^{P_n}$ the sheaf of $\mathcal{C}^\infty$ solutions of $P_n(D)$. By definition it follows that:

**Proposition 3.4.9.** *Regular functions of $n \geq 1$ variables form a sheaf $\mathcal{R}$.*

**Remark 3.4.2.** Note that $P_n(D)$ is an elliptic operator since it factors the Laplace operator.

Recalling the resolution (2.12), we immediately have the following result.

**Theorem 3.4.9.** *The sheaf $\mathcal{R}$ has flabby dimension equal to $2n-1$.*

*Proof.* Dualizing the resolution (3.47) we get the complex

$$0 \longrightarrow R^4 \xrightarrow{P_n} R^{4n} \longrightarrow R^{r_2} \longrightarrow \ldots \longrightarrow R^{\beta_{2n-2}} \xrightarrow{Q} R^{\beta_{2n-1}} \longrightarrow 0.$$

From the ellipticity of $P_n(D)$ and from the flabby resolution (2.12), we obtain

$$0 \longrightarrow \mathcal{R} \longrightarrow \mathcal{B}^4 \xrightarrow{P_n(D)} \mathcal{B}^{4n} \longrightarrow \ldots \mathcal{B}^{\beta_{2n-2}} \xrightarrow{Q(D)} \mathcal{B}^{\beta_{2n-1}} \longrightarrow 0.$$

This shows that $\mathrm{fl.dim}\,(\mathcal{R}) \leq 2n-1$. On the other hand the flabby dimension cannot be strictly less than $2n-1$ because this would imply $H_K^{2n-1}(\mathbb{H}^n, \mathcal{R}) = 0$ for every compact convex set $K$ in $\mathbb{H}^n$; in other words by Theorem 2.1.7 this would imply that $\mathrm{Ext}^{2n-1}(M_n, R) = 0$, thus contradicting Theorem 3.4.7. Therefore $\mathrm{fl.dim}\,(\mathcal{R}) = 2n-1$. □

**Remark 3.4.3.** Theorem 3.4.9 generalizes to the sheaf of germs of regular functions the fact that fl.dim $(\mathcal{O}) = n$.

As a consequence of Theorem 3.4.9 we obtain a generalization of the Malgrange Theorem 1.3.12:

**Corollary 3.4.4.** *Let $U$ be an open set in $\mathbb{H}^n$; then*

$$H^j(U, \mathcal{R}) = 0, \quad j \geq 2n - 1.$$

*Proof.* It is an immediate consequence of Theorem 3.4.9. □

Let us consider the sheaf $\mathcal{S}^{Q^t}$ of solutions of the system $Q^t(D)f = 0$ and $\mathcal{S}$ is, for example, the sheaf of distributions (recall theorems 2.1.4 and 2.1.5). Then, an application of Theorem 2.1.11 gives

**Theorem 3.4.10.** *Let $K$ be a compact convex set in $\mathbb{H}^n$. Then*

$$H_K^{2n-1}(\mathbb{H}^n, \mathcal{S}^{Q^t}) \cong [\mathcal{R}_l(K)]' \tag{3.48}$$

*and*

$$H_K^{2n-1}(\mathbb{H}^n, \mathcal{R}_l) \cong [\mathcal{S}^{Q^t}(K)]'.$$

**Remark 3.4.4.** The isomorphism (3.48) is the quaternionic analogue of Proposition 1.3.11 and generalizes the previous discussion in the case of one variable. It is then natural to think that the previous theorem will lead to the definition of hyperfunctions in several quaternionic variables, and we refer the reader to Section 6.1 for a more complete discussion on this issue.

As we have shown, the algebraic approach to the theory of regular functions is extremely powerful and allows us to conduct a rather deep analysis of their functional properties. The advantages of this method over the one used by Pertici in [145] appear now clearly, and they are similar to Malgrange's work versus the more elementary approach of Hartogs.

Moreover, in view of Ehrenpreis–Palamodov Fundamental Principle and of the fact that we have constructed the appropriate characteristic variety, we can give a completely general description of regular functions of several complex variables. Let $q = (q_1, \dots, q_n)$ and $q_\ell = x_{\ell 0} + x_{\ell 1}i + x_{\ell 2}j + x_{\ell 3}k$, $\ell = 1, \dots, n$ and $p = (p_1, \dots, p_n)$, $p_\ell = y_{\ell 0} + y_{\ell 1}i + y_{\ell 2}j + y_{\ell 3}k$. We define the bilinear form $pq := \sum x_{\ell r} y_{\ell r}$. According to [143], we can state the following representation theorem.

**Theorem 3.4.11.** *Let $f : U \subseteq \mathbb{H}^n \to \mathbb{H}$ be a regular function in the $n$ quaternionic variables $q_i$. Then*

$$f(q) = \int_V \exp(pq) \sum_{j=1}^n \bar{p}_j \omega_j$$

*where* $\omega_j$, $j = 1, \ldots, n$ *are* $\mathbb{H}_{\mathbb{C}}$*-valued densities in* $V = V(M_n)$ *satisfying the condition*

$$\cup_j \operatorname{supp} \omega_j \subset V \setminus \{0\}$$

*and the conditions (2.24).*

## 3.5 The Moisil–Theodorescu system

In this section we will study a variation of the Cauchy–Fueter system, namely the Moisil–Theodorescu system, both in one and several variables. Historically, the latter case was introduced before the former one (see [136]) by G. Moisil and N. Theodorescu; this system, bearing their names, is of independent interest for its applications to physics (see the preface of this book and [179], [180], [181]). The one-dimensional analysis on the Moisil–Theodorescu system is now quite developed, see for instance [83], [111], [112]. At the same time, we know of no other work so far for the case of several variables. Even though the Moisil–Theodorescu operator $\mathcal{D}_u$ is related to the Cauchy–Fueter operator by the formula

$$\frac{\partial}{\partial \bar{q}} = \frac{\partial}{\partial x_0} + \mathcal{D}_u$$

(note that this is the same type of formula relating the Dirac operator to the Weyl operator, see next chapter) there is no reason to suppose that they share the same algebraic analysis. We will show, for instance, that they have characteristic variety of different dimension. Let us denote by $\mathcal{D}_u$ the operator

$$\mathcal{D}_u := i\frac{\partial}{\partial x_1} + j\frac{\partial}{\partial x_2} + k\frac{\partial}{\partial x_3}. \tag{3.49}$$

Its right-hand side counterpart is given by

$$\mathcal{D}_q := \frac{\partial}{\partial x_1}i + \frac{\partial}{\partial x_2}j + \frac{\partial}{\partial x_3}k,$$

while $\overline{\mathcal{D}_u}$ and $\overline{\mathcal{D}_q}$ are their respective conjugate operators. Then we have

$$\mathcal{D}_u \cdot \overline{\mathcal{D}_u} = \overline{\mathcal{D}_u} \cdot \mathcal{D}_u = -\mathcal{D}_u^2 = \Delta_{\mathbb{R}^3}, \tag{3.50}$$

and analogous identities hold for $\mathcal{D}_q$. The matrix representation of the four real components of the quaternionic equation $\mathcal{D}_u f = 0$ is

$$\begin{bmatrix} 0 & \partial_{x_1} & \partial_{x_2} & \partial_{x_3} \\ \partial_{x_1} & 0 & -\partial_{x_3} & \partial_{x_2} \\ \partial_{x_2} & \partial_{x_3} & 0 & -\partial_{x_1} \\ \partial_{x_3} & -\partial_{x_2} & \partial_{x_1} & 0 \end{bmatrix} \begin{bmatrix} f_0 \\ f_1 \\ f_2 \\ f_3 \end{bmatrix} = 0. \tag{3.51}$$

In analogy with what has been done for holomorphic functions of several complex variables and for regular functions of several quaternionic variables, one can consider $\mathbb{H}$-valued functions of a variable $u = (u_1, \dots, u_n)$ in $\mathbb{R}^{3n}$ with $u_r = ix_{r1} + jx_{r2} + kx_{r3}$. Let $U$ be an open set. The space of functions $f : U \subseteq \mathbb{R}^{3n} \to \mathbb{H}$, satisfying $\mathcal{D}_{u_r} f = 0$ (resp. $\mathcal{D}_{q_r} f = 0$), $r = 1, \dots, n$, will be denoted by $\mathcal{R}_l(U)$ (resp. $\mathcal{R}_l(U)$) and will be said to be the space of left (resp. right) MT-regular functions. We now consider $n = 1$, i.e., left MT-regular functions in one variable $u = (x_1, x_2, x_3) \in \mathbb{R}^3$. One can prove the validity of the main theorems which hold for regular functions of one quaternionic variable, for example, the analogue of the Cauchy–Fueter theorem. If we introduce the Moisil–Theodorescu kernel

$$G(u) = \frac{u - u_0}{|u - u_0|^3}$$

and the form $\sigma = \sum_{k=1}^{3} (-1)^{k-1} i_k d\hat{x}_k$, we have:

**Theorem 3.5.1.** *Let $f : U \subseteq \mathbb{R}^3 \longrightarrow \mathbb{H}_\mathbb{C}$ be a left MT-regular function. Let $V \subset U$ be an open set in $U$ with smooth boundary, relatively compact in $U$. If $u_0 \in V$, then*

$$f(u_0) = \frac{1}{4\pi} \int_{\partial V} \frac{u - u_0}{|u - u_0|^3} \sigma f(u).$$

*More generally, for every smooth function $f$, we have the Borel–Pompeiu formula*

$$f(u) = \frac{1}{4\pi} \int_{\partial V} \frac{u - u_0}{|u - u_0|^3} \sigma f(u) + \int_V \frac{u - u_0}{|u - u_0|^3} \mathcal{D}_3 f(u) du,$$

*where $V$ is an open set, $V \subset U$.*

We do not repeat here the theory of MT-regular functions, for which we refer the reader to [182], provided the similarities with the theory of regular functions. We will only point out a difference in the series expansion of an MT-regular function. Let us introduce the following polynomials:

$$p_\nu(u) = \frac{(-1)^n}{n!} \sum_{2 \le \lambda_1, \dots, \lambda_n \le 3} [i(i_{\lambda_1} x_1 + i x_{\lambda_1})] \dots [i(i_{\lambda_n} x_1 + i x_{\lambda_1})]$$

where $i$ is the imaginary unit of the complex numbers and the sum is over all $\binom{n}{r}$ different orderings of $r$ 2's and $(n - r)$ 3's. Following standard arguments, see for example [206], it is possible to show that those polynomials are MT-regular and allow us to write a Taylor expansion as follows:

**Proposition 3.5.1.** *Let $f$ be a MT-regular function in a neighborhood of* 0. *Then there is a ball $B$ in which $f(u)$ is represented by a uniformly convergent series*

$$f(u) = \sum_{n=0}^{\infty} \sum_{2 \le \lambda_1, \dots, \lambda_n \le 3} p_\nu(u) a_\nu, \qquad a_\nu \in \mathbb{H}.$$

We now consider the case of $n > 1$ variables. Introducing the matrix

$$P_n(\mathcal{D}) := \begin{bmatrix} \mathcal{D}_{u_1} \\ \vdots \\ \mathcal{D}_{u_n} \end{bmatrix}, \tag{3.52}$$

we can write the condition of MT-regularity as

$$P_n(\mathcal{D})f = 0.$$

Note that sometimes we will write the imaginary units $i$, $j$, $k$ as $i_1$, $i_2$, $i_3$ respectively, and we set $i_0 = 1$.

Let $P_n^t$ be the polynomial matrix associated to the Moisil–Theodorescu system; then $P_n^t$ is of the form

$$P_n^t = \begin{bmatrix} U_1 & \dots & U_n \end{bmatrix}$$

where the matrices $U_i$ are defined by

$$U_i = \begin{bmatrix} 0 & x_{i1} & x_{i2} & x_{i3} \\ x_{i1} & 0 & -x_{i3} & x_{i2} \\ x_{i2} & x_{i3} & 0 & -x_{i1} \\ x_{i3} & -x_{i2} & x_{i1} & 0 \end{bmatrix}$$

for $i = 1, \dots, n$. We denote by $R$ the ring $\mathbb{C}[x_{11}, x_{12}, x_{13} \dots x_{n1}, x_{n2}, x_{n3}]$ so the module $M = M_n$ associated to the Moisil–Theodorescu system is given by

$$M_n = \frac{R^4}{\langle P_n^t \rangle},$$

where $\langle P_n^t \rangle$ denotes the module generated by the columns of $P_n^t$.

All the computations in this section are based on our paper [158] and mimic those ones already done in the previous section for the Cauchy–Fueter system. We start by computing the projective dimension $\mathrm{pd}(M_n)$ of the module $M_n$.

If $n = 1$, it is immediate to see that the syzygy module of $P_1$ is zero and therefore $\mathrm{pd}(M_1) = 1$. We will now assume that $n > 1$. The key tool in our approach is once again the theory of Gröbner bases. As usual, we will use the degree reverse lexicographic (degrevlex) term ordering on $R$ with the order on the variables

$$x_{11} > x_{21} > \dots > x_{n1} > x_{12} > x_{22} > \dots x_{n2} > x_{13} > \dots > x_{n3} \tag{3.53}$$

and the TOP ordering on $R^4$ with $\mathbf{e}_1 > \mathbf{e}_2 > \mathbf{e}_3 > \mathbf{e}_4$.

**Proposition 3.5.2.** *The reduced Gröbner basis for the R-module $\langle P_n^t \rangle$ is given by the columns of $P_n$ together with the columns of the $\binom{n}{2}$ matrices*

$$B_{rs} = x_{r3}x_{s2}I_4$$

*where $1 \leq r < s \leq n$ and $I_4$ is the $4 \times 4$ identity matrix. Moreover the module $Lt(P_n^t)$ generated by the leading terms of all the elements of $\langle P_n^t \rangle$ is*

$$\mathrm{Lt}(P_n^t) = \langle x_{i1}\mathbf{e}_l, x_{r3}x_{s2}\mathbf{e}_n \rangle, \qquad i = 1, \dots, n, \quad 1 \leq r < s \leq n, \quad l = 1, 2, 3, 4.$$

*Proof.* The statement can be verified for $n = 1, 2, 3$ using CoCoA. Let us assume $n > 4$. The $S$-polynomials generated by the columns of $P_n^t$ can be computed and reduced as in the case $n = 2$ and give rise to the columns of the matrices $B_{rs}$. In order to prove that the columns of $P_n^t$ and $B_{rs}$ form a reduced Gröbner basis for $\langle P_n^t \rangle$ we need to show that all the $S$-polynomials generated by them reduce to zero. An $S$-polynomial generated by a column of $P_n^t$ and a column of $B_{rs}$ is computed and reduced as in the case $n = 3$. An $S$-polynomial generated by two columns of $B_{rs}$ is computed and reduced as in the case $n = 2$. Finally, an $S$-polynomial generated by a column in $B_{rs}$ and a column in $B_{tu}$ is computed and reduced as in the case $n = 3, 4$ depending on whether one or none of the indices $r, s$ and $t, u$ is the same. The result about $\mathrm{Lt}(P_n^t)$ follows from the definitions of regular sequence and term ordering. □

**Proposition 3.5.3.** *The two variables $x_{12}$, $x_{n3}$ form an $M_n$-regular sequence.*

*Proof.* This result is an immediate consequence of the fact that the variables $x_{12}$ and $x_{n3}$ are exactly the only variables which do not appear in any of the leading terms of the elements of the reduced Gröbner basis of $\langle P_n^t \rangle$ given in Proposition 3.5.2. The proposition is then consequence of general properties of Gröbner bases and regular sequences. □

Let us now consider the module

$$M_n^* = M_n / \langle x_{12}, x_{n3} \rangle M_n = R^4 / \langle U_i, B_{rs}, x_{12}\mathbf{e}_l, x_{n3}\mathbf{e}_l \rangle, \qquad 1 \leq r < s \leq n,$$

and let $\wp_n$ be the maximal ideal in $R$ generated by the $3n$ variables.

**Proposition 3.5.4.** *The $(n-1)$ polynomials*

$$x_{22} + x_{13}, \quad x_{32} + x_{23}, \quad \dots, \quad x_{r,2} + x_{r-1,3}, \quad \dots, \quad x_{n,2} + x_{n-1,3}$$

*form a maximal $M_n^*$-regular sequence in $\wp_n$, where $\wp_n$ denotes the ideal of the variables in $R$.*

*Proof.* The proof follows the proof of Proposition 3.4.3; thus we will only point out the major differences that occur. Let us set

$$B = \langle U_i, x_{r3}x_{s2}\mathbf{e}_l, x_{12}\mathbf{e}_l, x_{n3}\mathbf{e}_l \rangle, \ i = 1, \dots, n, \ 1 \leq r < s \leq n, \ l = 1, 2, 3, 4$$

and

$$B_{\nu-1} = \langle B, (x_{22} + x_{13})\mathbf{e}_l, \dots, (x_{\nu 2} + x_{\nu-1,3})\mathbf{e}_l \rangle, \ \nu = 1, 2, \dots, n$$

with $B_0 = B$. By definition of maximal regular sequence we need to show that the polynomial $x_{\nu+1,2} + x_{\nu,3}$ is a nonzero divisor of $R^4/B_{\nu-1}$ and that every element of $\wp_n$ is a zero divisor on $R^4/B_{n-1}$. To begin with, we construct a Gröbner basis $G_{\nu-1}$ of $B_{\nu-1}$, $1 \leq \nu \leq n$ (we will actually split the construction depending on whether $\nu < n$ or $\nu = n$). Since the generators of $B$ given above form a Gröbner basis for $B$, the basis $G_{\nu-1}$ is obtained by considering, in addition to the vectors in the representation of $B$, the vectors $x_{r2} + x_{r-1,3}\mathbf{e}_l$ with $2 \leq r \leq \nu$ and then reducing. For the readers convenience, let us explicitly describe what happens for the case $n = 6$ and $\nu = 4$ (the general case is similar). The second degree monomials which appear in the Gröbner basis for $B$ are given by the following diagram:

$$\begin{matrix} x_{13}x_{22} & x_{13}x_{32} & x_{13}x_{42} & x_{13}x_{52} & x_{13}x_{62} \\ & x_{23}x_{32} & x_{23}x_{42} & x_{23}x_{52} & x_{23}x_{62} \\ & & x_{33}x_{42} & x_{33}x_{52} & x_{33}x_{62} \\ & & & x_{43}x_{52} & x_{43}x_{62}. \end{matrix}$$

After introducing the new elements $x_{22} + x_{13}$, $x_{32} + x_{23}$ and $x_{42} + x_{33}$ the diagram becomes

$$\begin{matrix} x_{13}^2 & x_{13}x_{23} & x_{13}x_{33} & x_{13}x_{52} & x_{13}x_{62} \\ & x_{23}^2 & x_{23}x_{33} & x_{23}x_{52} & x_{23}x_{62} \\ & & x_{33}^2 & x_{33}x_{52} & x_{33}x_{62} \\ & & & x_{43}x_{52} & x_{43}x_{62}. \end{matrix}$$

It is immediate to show that the basis which one obtains consists of the following vectors:
(a) the columns of $U_i$ for $1 \leq i \leq n$;
(b) $x_{r3}x_{s3}\mathbf{e}_l$ for $1 \leq r \leq s < \nu$;
(c) $x_{r3}x_{s2}\mathbf{e}_l$ for $\nu + 1 \leq s \leq n$, $1 \leq r < s$;
(d) $x_{r3}x_{n2}\mathbf{e}_l$ for $1 \leq r < n$; (e) $x_{12}\mathbf{e}_l$;
(f) $x_{n3}\mathbf{e}_l$;
(g) $(x_{r2} + x_{r-1,3})\mathbf{e}_l$ for $2 \leq r \leq \nu$,
and $l = 1, 2, 3, 4$.

The proof that, for $\nu = 1, 2, \dots, n-1$, $x_{\nu+1,2} + x_{\nu,3}$ is a nonzero divisor on $R^4/B_{\nu-1}$ can be carried out exactly as in the proof of Proposition 3.4.3 with the obvious modifications and therefore we do not repeat the details here. We finally need to show that every element of $\wp_n$ is a zero divisor of $R^4/B_{n-1}$. We first note that the Gröbner basis for $B_{n-1}$ is given by the following vectors:
(a) the columns of $U_i$, for $1 \leq i \leq n$;
(b) $x_{r3}x_{s3}\mathbf{e}_l$ for $1 \leq r \leq s < n$;
(c) $x_{r3}x_{n2}\mathbf{e}_l$ for $1 \leq r < n$;
(d) $x_{12}\mathbf{e}_l$;
(e) $x_{n3}\mathbf{e}_l$;
(f) $(x_{r2} + x_{r-1,3})\mathbf{e}_l$ for $2 \leq r \leq n$, and $l = 1, 2, 3, 4$.

This basis however is obviously nonreduced since, for example, the vectors in (f) with $r = n$ can be used to reduce the vectors in (c). By applying the usual reduction procedure we obtain the reduced Gröbner basis $G$ for $B_{n-1}$ consisting of the following vectors:
(a) the columns of $U_i$, for $1 \leq i \leq n$;
(b) $x_{r3}x_{s3}\mathbf{e}_l$ for $1 \leq r \leq s < n$;
(c) $x_{12}\mathbf{e}_l$;
(d) $x_{n3}\mathbf{e}_l$;
(e) $(x_{r2} + x_{r-1,3})\mathbf{e}_l$ for $2 \leq r \leq n$, and $l = 1, 2, 3, 4$.

It is now immediate to show that if $f \in \wp_n$ is nonzero, then $f$ reduces to zero using $G$. Indeed, if $f\mathbf{e}_1 \in B_{n-1}$, then $f(\mathbf{e}_1 + B_{n-1}) = 0$ and therefore $f$ is a zero divisor in $R^4/B_{n-1}$. Then $f(f\mathbf{e}_1 + B_{n-1}) = f_2\mathbf{e}_1 + B_{n-1}$, and so it is enough to show that $f_2\mathbf{e}_1$ reduces to zero using $G$. Since $f \in \wp_n$, every term in $f_2$ is of degree 2 or higher; using the columns of the $U_i$'s, we can however reduce $f_2\mathbf{e}_1$ to a vector with no variables $x_{i1}$ and containing only polynomials of degree 2 or higher. The terms containing $x_{i2}$ can be reduced using the vectors in (c) or (e). Finally, using (b) and (d) we can reduce $f_2$ to zero. This concludes the proof. □

**Corollary 3.5.1.** *The depth of the maximal ideal $\wp_n$ in $M_n$ is given by*

$$\operatorname{depth}(\wp_n, M_n) = n + 1.$$

*Proof.* As in Corollary 3.4.3. □

**Theorem 3.5.2.** *The module $M_n$ has projective dimension*

$$pd(M_n) = 2n - 1.$$

*Proof.* This follows immediately from the Auslander–Buchsbaum formula

$$pd(M_n) = \operatorname{depth}(\wp_n, R) - \operatorname{depth}(\wp_n, M_n) = 3n - \operatorname{depth}(\wp_n, M_n) = 2n - 1.$$

□

**Remark 3.5.1.** The module $M_n$ in the case of the Moisil–Theodorescu system admits a free resolution having the same length of the resolution in the Cauchy–Fueter case.

One may wonder if also the Betti numbers appearing in the resolution are the same as those ones obtained for the Cauchy–Fueter resolution. The answer is positive and one may follow the lines of the proof given in the Cauchy–Fueter case to get the result in this case. Instead, we give a direct proof, at least in the case of the first syzygies.

**Theorem 3.5.3.** *The matrix of the first syzygies of the Moisil–Theodorescu system in $n$ variables has a number of rows equal to*

$$r_n = 2\binom{n}{2} + 4\binom{n}{3}.$$

*Proof.* In this proof we denote by $A$ the $n$-dimensional Cauchy–Fueter matrix and let $B$ be the matrix representing the first syzygies. Obviously, $A$ is a $4n \times 4$ matrix while $B$ is a $r_n \times 4n$ matrix and in the ring $R$, it is im $A$=ker $B$. Denote by $A_0$ and $B_0$ the matrices obtained from $A$ and $B$ by replacing $x_{i0}$ with 0. $A_0$ is obviously the Moisil–Theodorescu matrix and we claim that $B_0$ is its syzygies matrix. To prove this, notice that $BA = 0$ immediately implies that $B_0 A_0 = 0$ (we are just restricting polynomial equations to $\mathbb{C}^{3n} \subseteq \mathbb{C}^{4n}$) and therefore im $A_0 \subseteq$ ker $B_0$. On the other hand, we will now prove that if $B_0$ is the syzygies matrix for $A_0$, then it can be obtained from $B$ by setting $x_{i0} = 0$. Let $B_0$ be such that $B_0 A_0 = 0$; we now look for a polynomial matrix $Q$ such that if we define $B = B_0 + Q$ then $BA = 0$. This is obviously possible, but our statement will be proved if we show that $Q$ is such that all of its entries are divisible by $x_{i0}$ for some $i$. Write $A$ as $A_0 + P$ so that $P = [x_{10}I|\dots|x_{n0}I]^t$ then $BA = 0$ is equivalent to $B_0 A_0 + B_0 P + QA = 0$ and since $B_0 A_0 = 0$ we are looking for the solutions $Q$ to the equation $B_0 P = -QA$. If now $X$ is an 1-inverse (see [22]) for $A$, we have that the solutions $Q$ are of the form $B_0 PX$. By the definition of $P$ it is immediate to see that $Q$ has the required form. Note that by the explicit formulas which are known for the construction of $X$, see [22], the matrix $X$ may have rational entries. However by multiplying both $Q$ and $B_0$ by the least common multiple of these denominators, we still obtain the syzygy relations for $A$. □

Now we wish to study the characteristic variety $V(M_n)$ of the module associated to the Moisil–Theodorescu system that is the subset of points in $\mathbb{C}^{3n}$ where the rank of the matrix $P_n^t$ is strictly less than 4. If we set $\zeta = (\zeta_1, \dots, \zeta_n)$, we will write $\zeta_i = (x_{i_0}, x_{i_1}, x_{i2}, x_{i3}) \in \mathbb{C}^4$, $i = 1, \dots, n$ and therefore since $\mathbb{C}^{3n}$ can be also represented as the linear variety

$$\mathbb{C}^{3n} = \{\zeta = (\zeta_1, \dots, \zeta_n) \in \mathbb{C}^{4n} \ : x_{i0} = 0, \ \forall i = 1, \dots, n\}$$

the variety $V(M_n)$ can be thought of as a subvariety of $\mathbb{C}^{3n} \subset \mathbb{C}^{4n}$.

**Theorem 3.5.4.** *The characteristic variety $V(M_n)$ has dimension $n + 1$.*

*Proof.* We will show that the algebraic set $V(M_n)$ has dimension $n$+1 in a neighborhood of an arbitrary point $\zeta^0 \neq 0$ in $V(M_n)$. We write $\zeta = (\zeta_1, \dots, \zeta_n) \in V(M_n)$ where $(\zeta_1, \dots, \zeta_n) \in \mathbb{C}^{4n}$. We can consider each vector $\zeta_i$ as the element $\zeta_i = x_{i0} + x_{i1}i + x_{i2}j + x_{i3}k$ of the complexified quaternionic algebra $\mathbb{H}_\mathbb{C} = \mathbb{H} \otimes \mathbb{C}$, and we will denote by $\zeta_i^*$ its conjugate $\zeta_i^* = x_{i0} - x_{i1}i - x_{i2}j - x_{i3}k$. We denote respectively by $v_i$ and $u_i$ the scalar and the vector part of $\zeta_i$. It is immediate to see that the variety $V(M_n)$ now corresponds to the matrix

$$\tilde{P}_n^t = [\ U_1| \ \ \dots| \ \ U_n| \ \ \zeta_{10}I_4| \ \ \dots| \ \ \zeta_{n0}I_4 \ ]$$

whose columns correspond to the quaternions $u_1^*, u_1^*i, u_1^*j, u_1^*k, \dots, u_n^*, u_n^*i, u_n^*j, u_n^*k, v_1, v_1i, v_1j, v_1k, \dots, v_n, v_ni, v_nj, v_nk$. The determinant of the first four columns of $\tilde{P}_n^t$ is easily seen to be $(u_1^* u_1)^2$. Let $V_1$ be the two-dimensional variety in $\mathbb{C}^4$ defined by the system

$$\begin{cases} (\zeta_1 \zeta_1^*)^2 = 0 \\ \zeta_{10} = 0. \end{cases}$$

Now for $\eta \in \mathbb{H}_\mathbb{C}$, we define the four complex subspaces of $\mathbb{H}_\mathbb{C}$ as follows:

$$L_\eta = \{\eta q \ : \ q \in \mathbb{H}_\mathbb{C}\}, \qquad L_\eta^\perp = \{q \in \mathbb{H}_\mathbb{C} \ : \ \eta q = 0\},$$

and

$$R_\eta = \{q\eta \ : \ q \in \mathbb{H}_\mathbb{C}\}, \qquad R_\eta^\perp = \{q \in \mathbb{H}_\mathbb{C} \ : \ q\eta = 0\}.$$

It is immediate to see that if $\eta \neq 0$ and $\eta \in V_1$, then these are all bidimensional spaces since $L_\eta \subset L_{\eta^*}^\perp$. On the other hand, similar to what it is shown also in [6],

$$\zeta \in V(M_n) \iff \zeta_1 \in V_1 \text{ and } u_j \in R_{\zeta_1} \cap \{\zeta_{j0} = 0\}, \ j = 2, \ldots, n.$$

It follows immediately that

$$\dim\,(V(M_n)) = \dim V_1 + (n-1)\dim\,(R_{\zeta_1} \cap \{\zeta_{j0} = 0\}) = 2 + (n-1) = n+1.$$

□

From Theorem 3.5.4, one gets the following:

**Theorem 3.5.5.**

$$\mathrm{Ext}^j(M_n, R) = 0, \qquad j = 0, \ldots, 2n-2$$

*and*

$$\mathrm{Ext}^{2n-1}(M_n, R) \neq 0.$$

*Proof.* The statement on the vanishing is an immediate consequence of our Theorem 3.5.4 and Corollary 1, (p. 377 [142]). □

Since the Moisil–Theodorescu system is elliptic, see [102], we have results analogous to Theorems 2.12 and 3.4.9.

**Theorem 3.5.6.** *The sheaf $\mathcal{R}_l$ of MT-regular functions in $n$ variables has flabby dimension $2n-1$.*

*Proof.* This is an immediate and standard consequence of the exactness properties of the Moisil–Theodorescu system, dual of the free resolution of $M_n$ and of the ellipticity of the Moisil–Theodorescu operator. □

An immediate consequence of previous theorem is the following MT-regular version of the well-known Malgrange theorem:

**Corollary 3.5.2.** *If $U$ is any open set in $\mathbb{R}^{3n}$, then $H^p(U, \mathcal{R}_l) = 0$ for all $p \geq 2n-1$.*

As an application of duality arguments, see Theorem 2.1.11, we obtain the following corollary, in which $Q^t$ will denote the last map in the minimal free resolution of $P_n^t$ and $\mathcal{S}^{Q^t}$ denotes the sheaf of solutions in $\mathcal{S}$ to the system associated to $Q^t$.

**Corollary 3.5.3.** *Let $K$ be a compact convex set in $\mathbb{R}^{3n}$. Then*

$$H_K^{2n-1}(\mathbb{R}^{3n}, \mathcal{S}^{Q^t}) \cong [H^0(K, \mathcal{R}_l)]',$$

*as well as*

$$H_K^{2n-1}(\mathbb{R}^{3n}, \mathcal{R}_l) \cong [H^0(K, \mathcal{S}^{Q^t})]',$$

*where the dual indicates the space of complex linear functionals.*

As before, we can obtain Hartogs' phenomenon for MT-regular functions as well as two other results.

**Corollary 3.5.4.** *Let $K$ be a compact convex set in $\mathbb{R}^{3n}$, $n \geq 2$. Then every MT-regular function $F$ on $\mathbb{R}^{3n}\backslash K$ extends uniquely to an MT-regular function everywhere.*

**Corollary 3.5.5.** *Let $\Omega$ be a convex connected open set in $\mathbb{R}^{3n}$ and let $K$ be a compact subset of $\Omega$. Let $\Sigma_1, \ldots, \Sigma_{2n-2}$ be closed half spaces in $\mathbb{R}^{3n}$ and set $\Sigma = \Sigma_1 \cup \ldots \cup \Sigma_{2n-2}$. Then every function $F \in \mathcal{R}(\Omega\backslash(\Sigma \cup K))$ extends to a MT-regular function on all $\Omega\backslash\Sigma$ and the extension coincides with $F$ on $\Omega\backslash(\Sigma \cup K')$ for $K'$ a compact subset of $\Omega$.*

**Corollary 3.5.6.** *Let $L$ be a subspace of $\mathbb{R}^{3n}$ of dimension $n+2$. Then for every compact $K$ contained in $L$, and every connected open set $\Omega$, relatively compact in $K$, every MT-regular function defined in a neighborhood of $K\backslash\Omega$ can be extended to a MT-regular function defined in a neighborhood of $K$.*

The results listed up to here give us information, both geometric and analytic, on the system. For example, the well-known Mittag-Leffler theorem on the vanishing of $H^1(U, \mathcal{R})$, where $U$ is an open set in $\mathbb{R}^3$ is nothing but a particular case of Corollary 3.5.2.

A consequence of the Mittag–Leffler theorem is the isomorphism, for $V$ open and $K \subset V$ compact,

$$H_K^1(V, \mathcal{R}_l) \cong \frac{\mathcal{R}_l(V\backslash K)}{\mathcal{R}_l(V)}.$$

In the case of one variable, the duality theorem given in Corollary 3.5.3 can be described more explicitly in the following form:

**Theorem 3.5.7.** *Let $K$ be a compact in $\mathbb{R}^3$ and let $V$ be an open set containing $K$. Then*

$$[\mathcal{R}_r(K)]' \cong \frac{\mathcal{R}_l(V\backslash K)}{\mathcal{R}_l(V)}, \tag{3.54}$$

$$[\mathcal{R}_l(K)]' \cong \frac{\mathcal{R}_r(V\backslash K)}{\mathcal{R}_r(V)}.$$

In particular, when $K$ lies in a 2-dimensional subspace of $\mathbb{R}^3$, the space $[H^0(K,\mathcal{R})]'$ is isomorphic to a suitably defined space of MT-hyperfunctions with support in $K$. In fact, let us consider any plane in $\mathbb{R}^3$, for example $\{x_1 = 0\}$. If $\Omega$ is an open set in this plane and $V$ is an open set in $\mathbb{R}^3$ in which $\Omega$ is relatively compact, then an MT-hyperfunction is defined as an element in the quotient

$$\mathcal{F}(\Omega) = \frac{\mathcal{R}_r(V\backslash\Omega)}{\mathcal{R}_r(V)}.$$

Obviously, by the Mittag–Leffler theorem, the definition does not depend on the choice of $V$. The correspondence

$$\Omega \subseteq \mathbb{R}^3 \longrightarrow \mathcal{F}(\Omega)$$

gives rise to a flabby sheaf called the sheaf of MT-hyperfunctions. In this framework, Corollary 3.5.3 becomes:

**Corollary 3.5.7.** *If $K$ is a compact in $\mathbb{R}^2 = \mathbb{R}^3 \cap \{x_1 = 0\}$, then*

$$[\mathcal{R}_r(K)]' \cong \mathcal{F}[K],$$

*where $\mathcal{F}[K]$ denotes the space of $K$-supported MT-hyperfunctions.*

# 4

# Special First Order Systems in Clifford Analysis

## 4.1 Introduction to Clifford algebras

### 4.1.1 Standard Clifford algebras

In this section we recall the fundamental notions about Clifford algebras. The reader interested in a more thorough treatment can consult [40], [55], [78], [118].

Let us consider a real vector space $V$, of finite dimension $m$, with a nondegenerate symmetric bilinear form $\mathcal{B}: V \times V \to \mathbb{R}$. It is always possible to determine a basis $e_1, e_2, \ldots, e_m$ such that $\mathcal{B}(e_i, e_j) = 0$ if $i \neq j$, $\mathcal{B}(e_i, e_i) = +1$ for $i = 1, \ldots, p$, and $\mathcal{B}(e_i, e_i) = -1$ for $i = p+1, \ldots, m$. Such a basis is called an orthonormal basis. The number $p$ depends only on the space $V$ and the pair $(p, q)$, $q = m - p$, is called the signature of $V$. A model for a real vector space with signature $(p, q)$ is the vector space $\mathbb{R}^{p,q}$ of $m$-tuples of real numbers $(x_1, x_2, \ldots, x_m)$ with the scalar product $\mathcal{B}(x, y) = \sum_{i=1}^{p} x_i y_i - \sum_{i=p+1}^{m} x_i y_i$.

**Definition 4.1.1.** *The Clifford algebra $C(V)$ over a real finite dimensional vector space $V$ with a nondegenerate bilinear form $\mathcal{B}$ is the free algebra with unit 1 containing $V$ and $\mathbb{R}$ as a linear subspaces and such that*

1. *it is generated as a real algebra by 1 and $V$,*

2. *for all $x \in V$ it is*

$$x^2 = \mathcal{B}(x, x). \tag{4.1}$$

If we consider an orthonormal basis of the vector space $V$ with signature $(p, q)$ and the Clifford algebra $\mathcal{C}(V)$, the relation (4.1) becomes equivalent to

$$e_i e_j + e_j e_i = 2\mathcal{B}(e_i, e_j),$$

and leads to another equivalent formulation of the definition: the real Clifford algebra $\mathcal{C}(V)$ is the associative algebra over $\mathbb{R}$ generated by $m$-basis elements $e_1, \ldots, e_m$, the so-called units, together with the defining relations

$$e_i^2 = +1 \text{ for } i = 1, \ldots, p, \quad e_i^2 = -1 \text{ for } i = p+1, \ldots, m$$

$$e_i e_j + e_j e_i = 0, \quad i \neq j.$$

Note that when $V$ is $\mathbb{R}^{p,q}$, we will denote its universal Clifford algebra by $\mathbb{R}_{p,q}$. In particular, if no confusion arises, we will write $\mathbb{R}_m$ instead of $\mathbb{R}_{0,m}$.

**Remark 4.1.1.** The relations above imply that, when considering products of two elements in the basis, it suffices to look at $e_i e_j$ with $i < j$. The algebra $\mathcal{C}(V)$ is generated by all the possible products $e_A := e_{a_1} \ldots e_{a_k}$ of the units $e_{a_i}$, where $A = \{a_1, \ldots, a_k\}$ is a subset of $\{1, 2, \ldots, m\}$, $1 \leq a_1 \leq \ldots \leq a_k \leq m$, and with the assumption $e_\emptyset = 1$. It is then clear that $\dim \mathcal{C}(V) \leq 2^m$ and when the equality holds, $\mathcal{C}(V)$ is called a universal Clifford algebra because it satisfies a universal property, as we will explain in the sequel. The reader interested in nonuniversal Clifford algebras may consult, for example, [21]. For our purposes, it suffices to know that nonuniversal Clifford algebras can only occur when $m$ is odd and $p - q - 1 \equiv 0 \bmod 4$.

We point out that there is more than one way to define Clifford algebras. Even though we will usually work using the definition given above, we will give a short description of other possible approaches. First, let us note that Clifford algebras can be defined without introducing a basis, following Clifford's original ideas [41], [42].

**Definition 4.1.2.** *Let $V$ be a real finite dimensional vector space with a nondegenerate symmetric bilinear form $\mathcal{B}$. Let $Q(x)$ be the nondegenerate quadratic form defined by $Q(x) = \mathcal{B}(x, x)$. We will call Clifford algebra the real associative algebra generated by $V$ and not generated by any proper subspace of $V$, and such that for any $x \in V$ it is*

$$x^2 = Q(x).$$

**Definition 4.1.3.** *Let $A$ be any associative real algebra with unity $1_A$ and let $C$ be the Clifford algebra generated by $V$ modulo the relations $x^2 = Q(x)$. A linear map $V \to A$ sending $x \to \phi_x$ and such that*

$$\phi_x^2 = Q(x)\, 1_A, \qquad \textit{for all } x \in V$$

*is called a Clifford map.*

**Example 4.1.1.** If $\mathcal{C}$ is a Clifford algebra, we obviously have a Clifford map $V \to \mathcal{C}$, defined by sending $x \to \gamma(x)$, where $\gamma$ denotes the canonical embedding of $V$ into $\mathcal{C}$.

**Remark 4.1.2.** The condition in Definition 4.1.2 that the Clifford algebra is not generated by any proper subspace of $V$ is crucial to obtain an algebra with dimension $2^m$, $m = \dim V$. This is equivalent to require the following universal property.

The Clifford algebra $\mathcal{C}$ is such that for any Clifford map $\phi: \ V \to A$ there is a unique algebra homomorphism $\psi : \mathcal{C} \to A$ making the following diagram commutative:

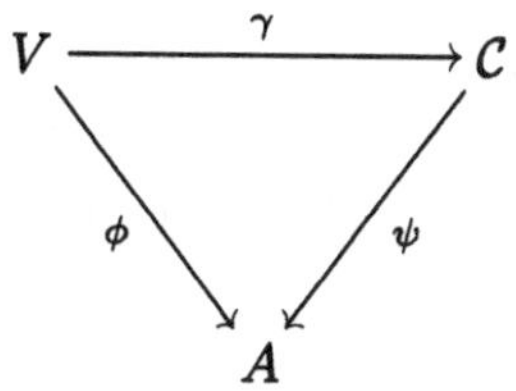


i.e., $\phi(x) = \psi(\gamma(x))$. The diagram implies that all Clifford maps can be obtained from the canonical embedding $\gamma: \ V \to \mathcal{C}$. Note that according to the definition of universality given in the appendix to Chapter 1, the Clifford algebra is the initial universal object in the category of the so called quadratic algebras that are the subalgebras of $A$ generated by the images of $\mathbb{R}$ and $V$ in $A$.

**Remark 4.1.3.** In 1954, Chevalley (see [39], [40]) constructed a universal Clifford algebra for $V$ with the quadratic form $Q$, as the quotient $T(V)/I$ of the tensor algebra

$$T(V) = \oplus_{j=0}^{\infty} V^{\otimes j}$$

with respect to the two sided ideal $I$ generated by

$$x \otimes x - Q(x), \quad x \in V.$$

It is clear that Chevalley's definition is equivalent to the previous ones.

**Remark 4.1.4.** So far, we have considered a vector space over the field of real numbers, but it is possible to repeat all the discussion replacing $\mathbb{R}$ by $\mathbb{C}$. Given any nondegenerate symmetric bilinear form $\mathcal{B}$, it is possible to find an orthonormal basis $\{e_1, \dots, e_m\}$ of $V$ such that $\mathcal{B}(e_i, e_j) = -\delta_{ij}$. Chevalley's construction shows the existence of a universal Clifford algebra for $V$ which, when $V = \mathbb{C}^m$, will be denoted by $\mathbb{C}_m$. For any $p$, $q$ with $m = p + q$, there is a real algebra isomorphism

$$\mathbb{C}_m \cong \mathbb{R}_{p,q} \otimes_{\mathbb{R}} \mathbb{C}.$$

In what follows, it will be useful to distinguish the elements in an orthonormal basis according to the sign of their squares; we will denote by $\epsilon_i$ and $e_i$ the units with positive and negative squares, respectively. With this notation, the defining relations for the Clifford algebra $\mathbb{R}_{p,q}$ are

$$\epsilon_j\epsilon_k + \epsilon_k\epsilon_j = 2\delta_{jk}, \quad e_je_k + e_ke_j = -2\delta_{jk}$$

$$e_j\epsilon_k + \epsilon_ke_j = 0,$$

where $\delta_{jk}$ denotes Kronecker's delta.

**Remark 4.1.5.** In the case of the Clifford algebra $\mathbb{R}_{m,m}$ it is possible to introduce an alternative basis. Let us consider the basis $\epsilon_j$, $e_j$ for $j = 1, \dots, m$. It is immediate to see that the elements

$$f_j = \frac{1}{2}(e_j - \epsilon_j), \qquad f_j' = \frac{1}{2}(e_j + \epsilon_j)$$

generate the whole algebra $\mathbb{R}_{m,m}$. Moreover, they satisfy the relations

$$f_jf_k + f_kf_j = 0, \qquad f_j'f_k' + f_k'f_j' = 0, \qquad f_j'f_k + f_kf_j' = -\delta_{jk}.$$

**Definition 4.1.4.** *The elements $f_j$, $f_j'$, $j = 1, \dots, m$ form the Witt basis for $\mathbb{R}_{m,m}$.*

**Remark 4.1.6.** Sometimes, it is useful to consider $V = \mathbb{R}^m$ with a degenerate quadratic form $Q(x)$. In this case, it is always possible to find a basis of the form

$$\epsilon_1, \dots, \epsilon_p, e_1, \dots, e_q, g_1, \dots, g_r$$

satisfying the anticommutativity of different basis elements together with

$$\epsilon_j^2 = 1, \qquad e_j^2 = -1, \qquad g_j^2 = 0.$$

**Remark 4.1.7.** Let us consider the Clifford algebra $\mathbb{R}_{p,q}$: if we consider new elements $e_{j+q}$, $j = 1, \dots, p$ and the identification

$$\epsilon_j \to ie_{j+q},$$

we get an embedding of $\mathbb{R}_{p,q}$ as a real subalgebra of $\mathbb{C}_m$, $m = p + q$. A Clifford algebra associated to a degenerate quadratic form with a basis of the form $\epsilon_1, \dots, \epsilon_p, e_1, \dots, e_q, g_1, \dots, g_r$, may be embedded into $\mathbb{C}_{p+q+2r}$ as the real subalgebra generated by the elements

$$e_1, \dots, e_q, ie_{q+1}, \dots, ie_{q+p},$$

$$\frac{1}{2}(e_{q+p+1} + ie_{q+p+r+1}), \dots, \frac{1}{2}(e_{q+p+r} + ie_{q+p+2r}).$$

**Example 4.1.2.** We now describe some examples of real Clifford algebras of low dimension.

- For $m = 1$ (and $p = 0$) we have that $\mathbb{R}_{0,1} = \mathbb{R}_1$ is the algebra generated by $e_1$ over $\mathbb{R}$ with the relation $e_1^2 = -1$. Hence we have the natural isomorphism $\mathbb{R}_1 \cong \mathbb{C}$.

- For $m = 2$ (and $p = 0$) we have that $\mathbb{R}_2$ is the Clifford algebra generated by $e_1$ and $e_2$, so that, by making the identification

$$e_1 \to i, \quad e_2 \to j,$$

(and the consequent $e_1 e_2 \to k$), one has the isomorphism $\mathbb{R}_2 \cong \mathbb{H}$.

- The Clifford algebra $\mathbb{R}_{1,1}$ is generated by the elements $e_1$ and $\epsilon_1$ such that $e_1^2 = -1$ and $\epsilon_1^2 = +1$, while the Clifford algebra $\mathbb{R}_{2,0}$ is generated by elements $\varepsilon_1$ and $\varepsilon_2$ both having square $+1$. Those two Clifford algebras are isomorphic. In fact, let us consider the matrices

$$\eta_0 = \begin{bmatrix} 1 & 0 \\ 0 & 1 \end{bmatrix} \qquad \eta_1 = \begin{bmatrix} 0 & 1 \\ 1 & 0 \end{bmatrix}$$

$$\eta_2 = \begin{bmatrix} 0 & -1 \\ 1 & 0 \end{bmatrix} \qquad \eta_3 = \begin{bmatrix} 1 & 0 \\ 0 & -1 \end{bmatrix}.$$

They form a basis for the vector space $M(2, \mathbb{R})$ of $2 \times 2$ real matrices. The map

$$\varphi : \mathbb{R}_{1,1} \longrightarrow M(2, \mathbb{R})$$

defined by $\varphi(\epsilon_1) = \eta_1$, $\varphi(e_1) = \eta_2$ can be extended to an isomorphism for which $\varphi(1) = \eta_0$, and $\varphi(\epsilon_1 e_1) = \eta_3$. The map

$$\psi : \mathbb{R}_{2,0} \longrightarrow M(2, \mathbb{R})$$

defined by $\psi(\varepsilon_1) = \eta_1$, $\psi(\varepsilon_2) = \eta_3$ can be extended to an isomorphism for which $\psi(1) = \eta_0$, $\psi(\varepsilon_1 \varepsilon_2) = \eta_2$. Thus the Clifford algebras $\mathbb{R}_{1,1}$ and $\mathbb{R}_{2,0}$ are isomorphic.

Let us consider $\mathbb{R}_m$ as a real vector space with basis $e_1, \ldots, e_m$ and let us look at the linear subspaces generated by the $\binom{m}{k}$ elements of the form $e_A = e_{a_1} \ldots e_{a_k}$. For $k = 0$, we have the subspace $\mathbb{R}$ of scalars; for $k = 1$ the subspace $\mathbb{R}_{m;1}$ of 1-vectors (also called vectors) with basis $e_1, \ldots, e_m$; then the subspace $\mathbb{R}_{m;2}$ of bivectors, with basis $e_{12}, e_{13}, \ldots, e_{ij} = e_i e_j$, $i < j$, and so on. In general, for any subset $A = \{a_1, \ldots, a_k\}$ of $M = \{1, \ldots, m\}$ of cardinality $|A| = k$, we consider the elements $e_A = e_{a_1} \ldots e_{a_k}$ which form a basis for the $\binom{m}{k}$-dimensional vector space $\mathbb{R}_{m;k}$ of the $k$-vectors. Note that in the literature the symbol $\mathbb{R}_m^k$ is often used instead of $\mathbb{R}_{m;k}$ (see [55]).

Every element $a \in \mathbb{R}_m$ may be written in a unique way as

$$a = [a]_0 + [a]_1 + \ldots + [a]_k + \ldots + [a]_m$$

where $[\cdot]_k : \mathbb{R}_m \to \mathbb{R}_{m;k}$ denotes the projection of $\mathbb{R}_m$ onto the space of $k$-vectors. An analogous decomposition can be done in $\mathbb{R}_{p,q}$:

$$\mathbb{R}_{p,q} = \mathbb{R}_{p,q;0} \oplus \mathbb{R}_{p,q;1} \oplus \ldots \oplus \mathbb{R}_{p,q;m},$$

where $\mathbb{R}_{p,q;k}$ denotes the space of $k$-vectors in $\mathbb{R}_{p,q}$.

**Definition 4.1.5.** *The product $e_M = e_1 \ldots e_m$ is called pseudoscalar.*

For $m = 2n + 1$ the pseudoscalar satisfies

$$e_j e_M = e_M e_j,$$

so $e_M$ commutes with any element of the algebra $\mathbb{R}_m$, while when $m = 2n$ it fulfills

$$e_j e_M = -e_M e_j.$$

The following result justifies the name pseudoscalar.

**Proposition 4.1.1.** *The center of a Clifford algebra $\mathbb{R}_m$ is $\mathbb{R}$ for $m$ even, while it is $\mathbb{R} \oplus e_M \mathbb{R} = \{x + e_M y \mid x, y \in \mathbb{R}\}$ for $m$ odd.*

It is interesting to note that one also has the commutation relations

$$e_{ij} e_M = e_M e_{ij},$$

which indicates that $e_M$ belongs to the center of the subalgebra generated by the bivectors $e_{ij}$.

**Remark 4.1.8.** Every element $a \in \mathbb{R}_m$ may uniquely be written in the form

$$a = a_+ + a_-$$

where $[a]_+ = [a]_0 + [a]_2 + \ldots$, and $[a]_- = [a]_1 + [a]_3 + \ldots$. We hence have a direct sum decomposition

$$\mathbb{R}_m = \mathbb{R}_{m,+} \oplus \mathbb{R}_{m,-}$$

where $\mathbb{R}_{m,+}$ is the even subalgebra that coincides with the real subalgebra generated by the bivectors $e_{ij}$, while $\mathbb{R}_{m,-}$ contains all the elements $a$ that may be written in the form $a = -e_1(e_1 a)$ , $e_1 a \in \mathbb{R}_{m,+}$. We point out that $\mathbb{R}_{m,-}$ is not an algebra.

**Theorem 4.1.1.** *There is an algebra isomorphism*

$$\mathbb{R}_{m,+} \cong \mathbb{R}_{m-1}.$$

*Proof.* By setting

$$e_1 e_m \to E_1, \quad e_2 e_m \to E_2, \quad \ldots, \quad e_{m-1} e_m \to E_{m-1}$$

and taking into account that the elements $e_j e_m$, $j = 1, \ldots, m-1$ generate the $2^{m-1}$-dimensional algebra $\mathbb{R}_{m,+}$, it follows that $\mathbb{R}_{m,+}$ is generated by the $m-1$ elements $E_1, \ldots, E_{m-1}$ satisfying the defining relations

$$E_i E_j + E_j E_i = -2\delta_{ij},$$

which proves the isomorphism $\mathbb{R}_{m,+} \cong \mathbb{R}_{m-1}$. □

Let us give a few examples.

**Example 4.1.3.**

- For $m = 1$ we simply have that $\mathbb{R}_{1,+} = \mathbb{R}$.
- For $m = 2$ we have that $\mathbb{R}_{2,+}$ is generated by $e_{12}$, so it is isomorphic to $\mathbb{R}_1$ after identifying $e_{12}$ with $E_1$.
- For $m = 3$ we have that $\mathbb{R}_{3,+}$ is generated by $e_{23}$, $e_{31}$, $e_{12}$ which by the identifications
$$e_{23} \to i, \quad e_{31} \to j, \quad e_{12} \to k,$$
is isomorphic to the algebra of quaternions $\mathbb{H}$. Note that if we consider the elements $O_\pm = \frac{1}{2}(1 \pm e_{123})$, every element $a \in \mathbb{R}_3$ may be uniquely written in the form
$$a = O_+ a_1 + O_- a_2, \qquad a_1, a_2 \in \mathbb{R}_{3,+} \cong \mathbb{H}.$$
The elements $O_\pm$ are in the center of $\mathbb{R}_3$ and satisfy the relations
$$O_+^2 = O_+, \quad O_-^2 = O_-, \quad O_\pm O_\mp = 0, \quad O_+ + O_- = 1.$$
In particular, this shows that $\mathbb{R}_3$ is isomorphic to the algebra
$$\mathbb{H} \oplus \mathbb{H} = \{(q_1, q_2) : q_j \in \mathbb{H}\}$$
with the product $(q_1, q_2) * (q_1', q_2') = (q_1 q_1', q_2 q_2')$.

**Remark 4.1.9.** What happens for $m = 3$ actually takes place in all dimensions of the form $m = 4n - 1$, because in those cases the pseudoscalar $e_M = e_1 \ldots e_m$ belongs to the center of the algebra and satisfies $e_M^2 = 1$. It is then possible to construct the zero divisors $O_\pm = \frac{1}{2}(1 \pm e_M)$ and write every Clifford algebra element into the form $a = O_+ a_+ + O_- a_-$, for a unique choice of $a_\pm \in \mathbb{R}_{4n-1,+} \cong \mathbb{R}_{4n-2}$, so that $\mathbb{R}_{4n-1} \cong \mathbb{R}_{4n-2} \oplus \mathbb{R}_{4n-2}$, with the product defined componentwise. When $m = 4n + 1$ this technique does not apply since $e_M$ belongs to the center but $e_M^2 = -1$. If one needs some reductions in the dimension of the algebra, it is better to consider complex Clifford algebras and to note that there is an algebra isomorphism $\mathbb{R}_{4n+1} \cong \mathbb{C} \otimes \mathbb{R}_{4n+1,+} \cong \mathbb{C}_{4n}$ where $\mathbb{C}$ is identified with the algebra generated by 1, $e_M$.

**Remark 4.1.10.** For the complex Clifford algebra $\mathbb{C}_m$ with $m = 2n+1$ we can always set $\omega = e_M$ or $\omega = ie_M$, so that $\omega$ is central and satisfies $\omega^2 = 1$. Hence, putting $O_\pm = \frac{1}{2}(1 \pm \omega)$ we have the isomorphism

$$\mathbb{C}_{2n+1} \cong O_+\mathbb{C}_{2n+1,+} \oplus O_-\mathbb{C}_{2n+1,+}.$$

In other words, $\mathbb{C}_{2n+1}$ is isomorphic to the direct sum $\mathbb{C}_{2n} \oplus \mathbb{C}_{2n}$, with the product defined componentwise.

**Theorem 4.1.2.** *We have the isomorphisms*

$$\mathbb{C}_{2m} \cong M(2^m, \mathbb{C}), \qquad \mathbb{C}_{2m+1} \cong M(2^m, \mathbb{C}) \oplus M(2^m, \mathbb{C}).$$

*Proof.* Let us set $n = 2m$ or $n = 2m+1$ according to the case we are considering. Let us introduce in $\mathbb{C}^n$ and $\mathbb{C}^2$ the orthonormal bases $\{e_1, \dots, e_n\}$ and $\{E_1, E_2\}$, respectively. The elements

$$e_1' = 1 \otimes E_1, \quad e_2' = 1 \otimes E_2, \quad e_{\ell+2}' = ie_\ell \otimes E_1E_2, \quad \ell = 1, \dots, n$$

belong to $\mathbb{C}_n \otimes_{\mathbb{C}} \mathbb{C}_2$ and satisfy the relations

$$(e_j')^2 = -1, \quad r = 1, \dots, n+2,$$

$$e_r'e_s' + e_s'e_r' = 0, \quad r \neq s.$$

Computing the product $\prod_{j=1}^{n+2} e_j'$, we find that, by construction, it never coincides with $\pm 1 \otimes 1$ so the elements $e_j'$ generate the universal complex Clifford algebra $\mathbb{C}_{n+2}$. The construction above shows that $\mathbb{C}_{n+2} \subseteq \mathbb{C}_n \otimes_{\mathbb{C}} \mathbb{C}_2$ and because of the dimension we have

$$\mathbb{C}_{n+2} \cong \mathbb{C}_n \otimes_{\mathbb{C}} \mathbb{C}_2. \tag{4.2}$$

In the particular case $n = 1$, we have $\mathbb{C}_1 \cong \mathbb{C} \cdot 1 \oplus \mathbb{C} \cdot e_1$ so the statement holds. In the case $n = 2$, we can introduce the map $\varphi : \mathbb{C}_2 \to M(2, \mathbb{C})$ defined by

$$\varphi(e_1) = \begin{bmatrix} i & 0 \\ 0 & -i \end{bmatrix} \qquad \varphi(e_2) = \begin{bmatrix} 0 & 1 \\ -1 & 0 \end{bmatrix}.$$

The map $\varphi$ can be extended to a homomorphism between $\mathbb{C}_2$ and $M(2, \mathbb{C})$ that turns out to be an isomorphism. Recalling the well-known result on tensor products of algebras of matrices: $M(n, \mathbb{C}) \otimes_{\mathbb{C}} M(m, \mathbb{C}) \cong M(nm, \mathbb{C})$, starting from the cases $\mathbb{C}_1$ and $\mathbb{C}_2$ and by applying recursively (4.2), we get the statement. □

**Remark 4.1.11.** The algebra $\mathbb{C}_m$ is simple when $m$ is even, while it is semisimple when $m$ is odd.

The following automorphisms are of fundamental importance for complex Clifford algebras: the main involution, the conjugation and the reversion.

**Definition 4.1.6.** *The (main) involution is defined by*

$$\widetilde{(ab)} = \tilde{a}\tilde{b}, \qquad \tilde{e}_j = -e_j.$$

It follows that for any $a \in \mathbb{R}_m$, $a = \sum a_A e_A$, $a_A \in \mathbb{R}$,

$$\tilde{a} = \sum a_A \tilde{e}_A = [a]_0 - [a]_1 + [a]_2 - [a]_3 + \ldots = [a]_+ - [a]_-$$

so that $\mathbb{R}_{m,+}$ and $\mathbb{R}_{m,-}$ are the eigenspaces of the involution $\sim$.

**Definition 4.1.7.** *The conjugation is defined by*

$$\overline{ab} = \bar{b}\bar{a}, \qquad \bar{e}_j = -e_j.$$

For any $a \in \mathbb{R}_m$, $a = \sum a_A e_A$, $a_A \in \mathbb{R}$,

$$\bar{a} = \sum a_A \bar{e}_a = [a]_0 - [a]_1 - [a]_2 + [a]_3 + [a]_4 - \ldots$$

i.e., for any $a \in \mathbb{R}_{m;k}$ we have the 4-periodicity

$$\bar{a} = a \quad \text{for} \quad k \equiv 0, 3 \bmod 4,$$

$$\bar{a} = -a \quad \text{for} \quad k \equiv 1, 2 \bmod 4.$$

Conjugation can be used to define a positive definite inner product on $\mathbb{R}_m$ by setting

$$< a, b >= [\bar{a}b]_0 = [b\bar{a}]_0 = [\bar{b}a]_0.$$

In other words, the Clifford-conjugation plays a role similar to that of the complex conjugation.

**Definition 4.1.8.** *The reversion is the composition of the conjugation and the main involution i.e.*

$$(ab)^* = b^* a^* \quad \textit{and} \quad e_j^* = e_j.$$

This implies that for $e_A = e_{a_1} \ldots e_{a_k}$ it is

$$e_A^* = e_{a_k} e_{a_{k-1}} \cdots e_{a_1},$$

i.e., $*$ reverses the order of the Clifford product. For any $a \in \mathbb{R}_m$, $a = \sum a_A e_A$, $a_A \in \mathbb{R}$, we have that

$$a^* = \sum a_a e_a^* = [a]_0 + [a]_1 - [a]_2 - [a]_3 + \ldots,$$

i.e. for any $a \in \mathbb{R}_{m,k}$ we have the 4-periodicity

$$a^* = a \quad \text{for} \quad k \equiv 0, 1 \bmod 4,$$

$$a^* = -a \quad \text{for} \quad k \equiv 2, 3 \bmod 4.$$

The previous automorphisms can be defined also for $\mathbb{C}_m$ but in that case one can define another antiinvolution called Hermitian conjugation by $a^+ = \sum \bar{a}_A \bar{e}_A$ where $\bar{a}_A$ denotes the complex conjugate of $a_A \in \mathbb{C}$. The corresponding Hermitian inner product is given by

$$(a, b) = [a^+b]_0 = [ba^+]_0 = \overline{[ab^+]}_0.$$

Note that if we consider the basis for the Clifford algebra $\mathbb{R}_{p,q}$ given by

$$\epsilon_1, \dots, \epsilon_p, e_1, \dots, e_q,$$

then in view of the identification $\epsilon_j = ie_{j+q}$ one has that $\epsilon_j^+ = \epsilon_j$, while $\bar{\epsilon}_j = -\epsilon_j$ and $e_j^+ = \bar{e}_j = -e_j$, $j = 1, \dots, q$.

In a Clifford algebra it is possible to define the dot and the wedge product.

**Definition 4.1.9.** *The dot product of two vectors $x$ and $y$ is defined by*

$$x \cdot y = \frac{1}{2}(xy + yx)$$

*and the wedge product is defined by*

$$x \wedge y = \frac{1}{2}(xy - yx).$$

**Remark 4.1.12.** Note that for vectors the dot product coincides with the scalar product. The wedge product represents the directed and oriented surface measure of the parallelogram individuated by $x$ and $y$.

More generally we have the following definition.

**Definition 4.1.10.** *Let $u$ be a $k$-vector and $v$ a $h$-vector. The dot product is defined by $u \cdot v = [uv]_{|k-h|}$ if $k$ and $h$ are both zero or both nonzero while $u \cdot v = 0$ if only one between $k$, $h$ is zero. The wedge product is given by $u \wedge v = [uv]_{k+h}$. The general cases are defined by linearity.*

## 4.1.2 Endomorphisms and spinor spaces

In addition to the projection operators $a \to [a]_k$, and the automorphisms which we have just introduced, there are various other important endomorphisms in Clifford analysis, such as the left and right multiplication operators

$$e_j \cdot : \; a \to e_j a, \qquad \cdot e_j : a \to a e_j,$$

which play a role if one considers Dirac-type operators acting from two sides on Clifford algebra valued functions. It is in fact possible to provide a full characterization of the space of endomorphisms of $\mathbb{R}_m$.

**Theorem 4.1.3.** *We have the following isomorphism:*

$$\mathrm{End}(\mathbb{R}_m) \cong \mathbb{R}_{m,m}.$$

*Proof.* We consider the endomorphisms

$$a \to e_j a, \qquad a \to \tilde{a} e_j,$$

where as before $\tilde{a}$ denotes the main involution. With an abuse of notation, we will denote by $e_j$ the endomorphism sending $a$ to $e_j a$ and by $\epsilon_j$ the endomorphism sending $a$ to $\tilde{a} e_j$.

It is then easy to check that the endomorphisms $e_j$, $\epsilon_j$ satisfy the defining relations for the Clifford algebra $\mathbb{R}_{m,m}$, namely

$$e_j e_k + e_k e_j = -2\delta_{jk},$$

$$\epsilon_j \epsilon_k + \epsilon_k \epsilon_j = 2\delta_{jk},$$

$$e_j \epsilon_k + \epsilon_k e_j = 0.$$

The above endomorphisms generate a subalgebra of $\mathrm{End}(\mathbb{R}_m)$. Since an algebra generated by an even number of units is always universal (see Remark 4.1.1), we have that the subalgebra has dimension $2^{2m}$ so, on the one hand it coincides with $\mathrm{End}(\mathbb{R}_m)$, and on the other hand it is isomorphic to $\mathbb{R}_{m,m}$. □

**Remark 4.1.13.** An immediate consequence of this result is the fact that any endomorphism on $\mathbb{R}_m$ may be written in terms of the basis elements $e_j$, $\epsilon_j$. Moreover, the above isomorphism gives a description of the Clifford algebra $\mathbb{R}_{m,m}$ as the algebra $M(2^m, \mathbb{R})$ of real $2^m \times 2^m$ matrices.

In what follows, we will introduce the notion of spinor space and spinor representation. Let us start with an example.

**Example 4.1.4.** In the nonrelativistic theory of one-half spin particles, such as electrons, one considers the column vector containing the two wave functions $\psi_1$, $\psi_2$, satisfying the Schrödinger equation, representing particles with spin $1/2$ and $-1/2$, see for example [81]. Those vectors are called Pauli spinors and, by definition, they are of the type

$$\begin{bmatrix} \psi_1 \\ \psi_2 \end{bmatrix}$$

where $\psi_i$, $i = 1, 2$ are complex valued functions. One can obtain a complex vector space isomorphic to the vector space of spinors by considering the space of square matrices of the type

$$\psi = \begin{bmatrix} \psi_1 & 0 \\ \psi_2 & 0 \end{bmatrix}.$$

The matrices of this type can be also obtained by multiplying, on the right, any $2 \times 2$ matrix $A$, whose entries are complex valued functions, by the matrix

$$J = \begin{bmatrix} 1 & 0 \\ 0 & 0 \end{bmatrix},$$

so the matrix spinors $\psi$ are the elements of the left ideal $M(2, \mathbb{C})J$ of $M(2, \mathbb{C})$. Note also that $M(2, \mathbb{C})$, considered as a real algebra, is isomorphic to the Clifford algebra $\mathbb{R}_{3,0}$ by setting $\sigma_1 \to \epsilon_1$, $\sigma_2 \to \epsilon_2$, $\sigma_3 \to \epsilon_3$ where the $\sigma_i$ are the Pauli matrices defined in (5.16). Let us consider any element in $\mathbb{R}_{3,0}$, i.e., a matrix in $M(2, \mathbb{C})$, and let us multiply it by $\psi \in M(2, \mathbb{C})J$. The result is still an element in $M(2, \mathbb{C})J$ which is a spinor. This shows that the set

$$S := \left\{ \psi = \begin{bmatrix} \psi_1 & 0 \\ \psi_2 & 0 \end{bmatrix}, \ \psi_i \in \mathbb{C} \right\}$$

is a left ideal of $\mathbb{R}_{3,0}$. Moreover, $S$ contains no proper left ideal of $\mathbb{R}_{3,0}$, i.e., $S$ (whose elements correspond to column vectors) is a minimal left ideal in $\mathbb{R}_{3,0}$. The set $S$ can also be seen as a real vector space of dimension four with the basis

$$\begin{aligned} I_3 &= (1 + \epsilon_3)/2, \\ \epsilon_2 I_3 &= (\epsilon_2 + \epsilon_{23})/2, \\ \epsilon_{31} I_3 &= (\epsilon_{31} - \epsilon_1)/2, \\ \epsilon_{12} I_3 &= (\epsilon_{12} + \epsilon_{123})/2, \end{aligned}$$

where the element $I_3$ is the idempotent that corresponds to $J$.

**Remark 4.1.14.** The Clifford algebra $\mathbb{R}_{3,0} \cong M(2, \mathbb{C})$ contains a division algebra $\mathcal{J}$, isomorphic to $\mathbb{C}$, formed by all the matrices of the type

$$\begin{bmatrix} z & 0 \\ 0 & 0 \end{bmatrix} = zJ, \qquad z \in \mathbb{C}.$$

The minimal left ideal $S$ has a natural right $\mathcal{J}$ linear product defined by

$$S \times \mathcal{J} \longrightarrow S, \qquad (\psi, zJ) \longrightarrow \psi z J.$$

We call spinor space the minimal left ideal $S$ of $\mathbb{R}_{3,0}$ with this right $\mathcal{J}$ linear structure.

Let us now generalize the concepts of the previous example to a more general setting. Consider the Clifford algebra $\mathbb{R}_{m,m}$ with the basis $\{e_1, \dots, e_m, \epsilon_1, \dots, \epsilon_m\}$. Since we have proved that $\mathbb{R}_{m,m}$ is isomorphic to $M(2^m, \mathbb{R})$, we can consider the left ideal corresponding to $\mathbb{R}_{m,m} I$ where

$$I = I_1 \cdot \ldots \cdot I_m,$$

and the elements

$$I_j = \frac{1}{2}(1 - e_j \epsilon_j)$$

are idempotent elements which mutually commute. Note that also $I$ turns out to be idempotent. We have the following result.

**Proposition 4.1.2.** *For every element $a \in \mathbb{R}_{m,m}$ there is a uniquely determined $\hat{a} \in \mathbb{R}_m$ such that*

$$aI = \hat{a}I.$$

*Proof.* First, note that $e_j I = \epsilon_j I$, so that if we set $a = \sum_{j=1}^m a_j e_j + \sum_{j=1}^m b_j \epsilon_j$ we have

$$aI = \sum_{j=1}^m a_j e_j I + \sum_{j=1}^m b_j \epsilon_j I = \sum_{j=1}^m a_j e_j I + \sum_{j=1}^m b_j e_j I$$
$$= \sum_{j=1}^m (a_j + b_j) e_j I := \hat{a}I.$$

The uniqueness follows from the fact that $\hat{a}I = 0$ if and only if $\hat{a} = 0$. □

It is immediate to verify that:

**Corollary 4.1.1.** *The left ideal $\mathbb{R}_{m,m}I$ is equal to $\mathbb{R}_m I$.*

It is also possible to prove the following result (see [118]).

**Proposition 4.1.3.** *The left ideal $\mathbb{R}_m I$ is minimal.*

We are now in the position to give the definition of spinor space.

**Definition 4.1.11.** *The minimal left ideal $\mathbb{R}_m I$ of $\mathbb{R}_{m,m}$ is called the spinor space of $\mathbb{R}_{m,m}$.*

Finally, as in the proof of Theorem 4.1.3, let us denote by $e_j[a] := e_j a$, $\epsilon_j[a] := \tilde{a} e_j$ the actions of the endomorphisms $e_j$, $\epsilon_j \in \mathrm{End}(\mathbb{R}_m)$, on a general element $a \in \mathbb{R}_m$. Then we clearly have that (for any $a \in \mathbb{R}_m$):

$$e_j[a]I = e_j a I,$$

$$\epsilon_j[a]I = \tilde{a} e_j I = \tilde{a}\epsilon_j I = \epsilon_j a I,$$

i.e., the action of these elements as endomorphisms on $a \in \mathbb{R}_m$ multiplied with $I$ is simply the left multiplication of $aI$ with these elements. It follows that for every $b \in \mathbb{R}_{m,m} = \mathrm{End}(\mathbb{R}_m)$, and $a \in \mathbb{R}_m$, we have

$$b[a]I = baI.$$

**Remark 4.1.15.** Using Theorem 4.1.2 and the ideas already applied in Proposition 4.1.2, it is possible to reduce the dimension of the complex Clifford algebra $\mathbb{C}_{2m}$. Consider the basis $e_j, \epsilon_j$ for $j = 1, \dots, m$; we can identify it with the basis $e_j, -ie_{j+m}$, $j = 1, \dots, m$, so that we can introduce the Witt basis as

$$f_j = \frac{1}{2}(e_j - ie_{j+m}), \qquad f_j' = \frac{1}{2}(e_j + ie_{j+m}).$$

Let us introduce the idempotent

$$I = I_1 \dots I_m,$$

where

$$I_j = -f_j f_j' = \frac{1}{2}(1 - ie_j e_{j+m}), \qquad j = 1, \ldots, m.$$

One has

$$ie_{j+m}I = e_j I, \qquad e_j I = f_j' I$$

and, more generally, for any $a \in \mathbb{C}_{2m}$ there exists a unique $\hat{a} \in \mathbb{C}_m$ such that

$$e_j a I = e_j \hat{a} I;$$

moreover

$$e_{j+m} a I = e_{j+m} \hat{a} I = \tilde{\hat{a}} e_{j+m} I = -i\tilde{\hat{a}} e_j I.$$

In particular for $a \in \mathbb{C}_{2m,+}$ we simply obtain that

$$e_j a I = e_j \hat{a} I, \qquad e_{j+m} a I = -i\hat{a} e_j I.$$

### 4.1.3 Classifications of real Clifford algebras

We now come to the classification of real Clifford algebras. We will show that the Clifford algebra $\mathbb{R}_{p,q}$ is a simple algebra, i.e., it has no proper two-sided ideals and thus it is isomorphic to an algebra of matrices over $\mathbb{R}$ or $\mathbb{C}$ or $\mathbb{H}$ when $p - q \neq 1 + 4k$, $k \in \mathbb{Z}$. On the other hand, when $p - q = 1 + 4k$, $k \in \mathbb{Z}$, there exist two idempotents belonging to the center of the algebra that allows one to project it onto two copies of an algebra of matrices over $\mathbb{R}$ or $\mathbb{H}$. So any Clifford algebra turns out to be an algebra of matrices of suitable size with entries in $\mathbb{R}$, $\mathbb{C}$, $\mathbb{H}$ or in the rings $\mathbb{R} \oplus \mathbb{R}$, $\mathbb{H} \oplus \mathbb{H}$ with the product defined componentwise.

**Proposition 4.1.4.** (Dimension reduction principle). *For any $p$ and $q$ the following isomorphism*

$$\mathbb{R}_{p+1,q+1} \cong M(2, \mathbb{R}_{p,q})$$

*holds.*

*Proof.* Consider the basis $\epsilon_j$, $e_k$ of $\mathbb{R}_{p,q}$ and the $2 \times 2$ matrices

$$\mathcal{E}_{p+1} = \begin{bmatrix} 0 & 1 \\ 1 & 0 \end{bmatrix}, \qquad E_{q+1} = \begin{bmatrix} 0 & -1 \\ 1 & 0 \end{bmatrix},$$

$$\mathcal{E}_j = \begin{bmatrix} \epsilon_j & 0 \\ 0 & -\epsilon_j \end{bmatrix}, \qquad E_k = \begin{bmatrix} e_k & 0 \\ 0 & -e_k \end{bmatrix},$$

with $j = 1, \ldots, p$ and $k = 1, \ldots, q$. These matrices anticommute and satisfy $\mathcal{E}_j^2 = I$, $E_k^2 = -I$, and therefore they generate a Clifford algebra isomorphic to $\mathbb{R}_{p+1,q+1}$. □

**Proposition 4.1.5.** (Symmetry property). *For any $p \geq 1$ and $q \geq 0$ the following isomorphism*

$$\mathbb{R}_{p,q} \cong \mathbb{R}_{q+1,p-1}$$

*holds.*

*Proof.* Consider the basis $\epsilon_j$, $e_k$, with $j = 1, \dots, p$ and $k = 1, \dots, q$, for $\mathbb{R}_{p,q}$. We now set $\epsilon = \epsilon_p$ and

$$\mathcal{E}_1 = e_1\epsilon, \dots, \mathcal{E}_q = e_q\epsilon, \quad \mathcal{E}_{q+1} = \epsilon,$$

$$E_1 = \epsilon_1\epsilon, \dots, E_{p-1} = \epsilon_{p-1}\epsilon.$$

This is obviously a generating set for $\mathbb{R}_{q+1,p-1}$. □

The following result is due to E. Cartan (see [36]).

**Proposition 4.1.6.** (Periodicity property I). *The following isomorphism*

$$\mathbb{R}_{p,q} \cong \mathbb{R}_{p-4,q+4}$$

*holds for* $p \geq 4$.

*Proof.* As usual we consider a basis for $\mathbb{R}_{p,q}$ and we set $\omega = \epsilon_1\epsilon_2\epsilon_3\epsilon_4$. Let us now introduce the units

$$\mathcal{E}_j = \epsilon_j\omega, \qquad j = 1, \dots, 4.$$

The set $\{\mathcal{E}_1, \dots, \mathcal{E}_4, \dots, \epsilon_p, e_1, \dots, e_q\}$ is such that $\mathcal{E}_j^2 = -1$ and all the units anticommute. The isomorphism follows. □

In the proof of the next theorem it is necessary to know that $\mathbb{R}_{0,8} \cong M(16, \mathbb{R})$. The reader interested in its direct proof may consult [36].

**Proposition 4.1.7.** (Periodicity property II). *The following isomorphisms*

$$\mathbb{R}_{p,q+8} \cong \mathbb{R}_{p,q} \otimes M(16, \mathbb{R}) \cong M(16, \mathbb{R}_{p,q})$$

*hold.*

*Proof.* Consider the basis $\epsilon_j$, $e_k$, with $j = 1, \dots, p$ and $k = 1, \dots, q+8$, for $\mathbb{R}_{p,q+8}$. Let us define $\omega = e_{q+1}e_{q+2}\cdots e_{q+8}$ and introduce the elements $\mathcal{E}_j = \epsilon_j\omega$ for $j = 1, \dots, p$ and $E_k = e_k\omega$, $k = 1, \dots, q$. The subset $\{\mathcal{E}_j, E_k\}$ of $\mathbb{R}_{p,q+8}$ generates a subalgebra isomorphic to $\mathbb{R}_{p,q}$ while $\{e_{q+k}\}_{k=1,\dots,8}$ generates an algebra isomorphic to $\mathbb{R}_{0,8} \cong M(16, \mathbb{R})$. The algebras $\mathbb{R}_{p,q}$ and $M(16, \mathbb{R})$ commute componentwise and generate $\mathbb{R}_{p,q+8}$. □

**Theorem 4.1.4.** *The following algebras are sufficient to classify all the real Clifford algebras:*

- $\mathbb{R}_{0,0} = \mathbb{R}$;
- $\mathbb{R}_{0,1} = \mathbb{C}$;
- $\mathbb{R}_{1,0} = \mathbb{R} \oplus \mathbb{R}$ *with the product defined componentwise;*
- $\mathbb{R}_{0,2} = \mathbb{H}$;

- $\mathbb{R}_{0,3} = \mathbb{H} \oplus \mathbb{H}$ *with the product defined componentwise.*

*Proof.* All the Clifford algebras $\mathbb{R}_{p,q}$, $p+q \leq 8$ can be classified and turn out to be either an algebra in the list above or an algebra of matrices with entries in the list above (see [55], page 81 for the table of $\mathbb{R}_{p,q}$ with $p+q \leq 8$). The result follows from the previous propositions. □

We finish this subsection by introducing a group, $\mathrm{Spin}(p,q)$, which will be particularly important when discussing the properties of monogenic functions in the next section.

Let us consider the Clifford algebra $\mathbb{R}_{p,q}$. We can define the following transformation

$$\psi(s): \mathbb{R}_{p,q} \to \mathbb{R}_{p,q}$$

defined by

$$\psi(s)[\underline{x}] = s\underline{x}\tilde{s}^{-1}$$

for every $s \in \mathbb{R}_{p,q}$ for which $\tilde{s}^{-1}$ exists.

**Definition 4.1.12.** *The Pin group is defined as*

$$\mathrm{Pin}(p,q) = \{\ s \in \mathbb{R}_{p,q} \mid [\bar{s}s]_0 = \pm 1 \ \text{and}\ \forall\ \underline{x} \in \mathbb{R}_{p,q;1},\ \ \psi(s)[\underline{x}] \in \mathbb{R}_{p,q;1}\}.$$

In other words, the Pin group is the set of unit elements $s \in \mathbb{R}_{p,q}$ such that $[\bar{s}s]_0 = \pm 1$ for which the action of $\psi(s)$ leaves the vector space of 1-vector invariant. The Pin group is related to the orthogonal group $O(p,q)$ by the following result.

**Theorem 4.1.5.** *There is a surjective map* $\chi: \ \mathrm{Pin}(p,q) \longrightarrow O(p,q)$ *with kernel equal to* $\{-1,1\}$.

**Remark 4.1.16.** The statement of the previous theorem can be improved by saying that $\mathrm{Pin}(p,q)$ is a double covering of $O(p,q)$.

**Definition 4.1.13.** *The group Spin$(p,q)$ is the subgroup on Pin$(p,q)$ consisting of the products of an even numbers of unit vectors.*

**Proposition 4.1.8.** *Spin$(p,q)$ is a double covering of $SO(p,q)$.*

We can now come back to the notion of spinor space. We first give the notion of representation space.

**Definition 4.1.14.** *Let $V$ be a real linear space; then, $V$ is said a representation space of* $\mathrm{Spin}(p,q)$ *if there exists a map*

$$\rho:\ \mathrm{Spin}(p,q) \longrightarrow \mathrm{End}(V), \quad \forall s,t \in \mathrm{Spin}(p,q)$$

*such that $\rho(st) = \rho(s)\rho(t)$ and $\rho(1) = 1_V$.*

**Definition 4.1.15.** *A spinor space $V$ is a representation of the group* $\mathrm{Spin}(p,q)$ *which is a module over $\mathbb{R}_{p,q}$ and such that the action $\rho$ is the restriction of the left multiplication of the module.*

We consider a primitive idempotent $I \in \mathbb{R}_{p,q}$, i.e., an idempotent element that is not the sum of two mutually annihilating nonzero idempotents. Then the division ring $\Lambda = I\mathbb{R}_{p,q}I$ is isomorphic to

$$\begin{array}{lll} \mathbb{R} & \text{for} & p-q = 0,1,2 \mod 8, \\ \mathbb{C} & \text{for} & p-q = 3 \mod 4, \\ \mathbb{H} & \text{for} & p-q = 4,5,6 \mod 8. \end{array}$$

We define a minimal left ideal $S$ by setting $S = \mathbb{R}_{p,q}I$ and a map $S \times \Lambda \to S$ by $(\psi, \lambda) \to \psi\lambda$. According to Definition 4.1.11, the left ideal $S$ endowed with this $\Lambda$-linear structure is a spinor space. The spinor space $S$ gives an irreducible representation

$$\begin{aligned} \mathbb{R}_{p,q} &\longrightarrow \mathrm{End}_\Lambda(S) \\ x &\longrightarrow \gamma(x) \end{aligned}$$

such that $\gamma(x)\psi = x\psi$. This representation is faithful if $p - q \neq 1 + 4k$.

# 4.2 Introduction to Clifford analysis

## 4.2.1 Dirac operators

Clifford analysis largely consists of the study of the Dirac operator and some of its variations (see [29], [55]). Classically, the Dirac operator is defined on spinor valued functions. However, it is convenient to consider it as acting on functions with values in a Clifford algebra $C$: the main reason is that the kernel of the Dirac operator is a $C$-module; moreover some formulas do not hold when considering only spinor valued nullsolutions. The two approaches are related by decomposing $C$ as the sum of minimal left ideals and by writing a $C$-valued function $f$ as the sum of spinor valued functions.

**Definition 4.2.1.** *The operator $\partial_{\underline{x}} = \sum_{j=1}^m e_j \partial_{x_j}$ acting on functions $f(\underline{x}) = f(x_1, \dots, x_m)$ defined on the m-dimensional Euclidean space and taking values in the real (or complex) Clifford algebra $\mathbb{R}_n$ (resp. $\mathbb{C}_n$), $m \leq n$, is called Dirac operator.*

**Definition 4.2.2.** *A real differentiable function $f : \mathbb{R}^m \to \mathbb{R}_n$, $m \leq n$, is called (left) monogenic in some domain $U$ if it satisfies $\partial_{\underline{x}} f(\underline{x}) = 0$ on $U$. An analogous definition can be given for a real differentiable function $f : \mathbb{R}^m \to \mathbb{C}_n$.*

**Remark 4.2.1.** A variation of the Dirac operator is the Weyl operator

$$\partial_{x_0} + \partial_{\underline{x}},$$

whose nullsolutions $f : \mathbb{R}^{m+1} \to \mathbb{R}_n$, $m \leq n$ are still called (left) monogenic.

**Remark 4.2.2.** In this book we will consider the case $n = m$.

**Remark 4.2.3.** Of particular importance in physics is the space-time Dirac operator which we will write in the form

$$-i\partial_t + \sum_{j=1}^{m} e_j \partial_{x_j}.$$

It is obtained by replacing $x_0$ by the product $-it$ in the Weyl operator.

**Remark 4.2.4.** In the case of a single Dirac operator $\sum_{j=1}^{m} e_j \partial_{x_j}$ acting on functions $f(\underline{x})$, $\underline{x} = \sum_{j=1}^{m} e_j x_j \in \mathbb{R}^m$, we consider the action of the group $\mathrm{Spin}(m)$ given by

$$s \in \mathrm{Spin}(m) \to L(s) \ : \ f(\underline{x}) \to s f(\bar{s}\underline{x}s)$$

and one can easily verify the relation (see also [55], [78]):

$$\partial_{\underline{x}} L(s) = L(s) \partial_{\underline{x}}.$$

The previous remark motivates the following definition.

**Definition 4.2.3.** *A partial differential operator with constant coefficients is called* Spin$(m)$*-invariant if it commutes with* $L(s)$.

**Remark 4.2.5.** Note that the general inhomogeneous Dirac equation

$$\partial_{\underline{x}} f = g$$

is equivalent to the system of two equations

$$\begin{cases} \partial_{\underline{x}}[f]_+ = [g]_- \\ \partial_{\underline{x}}[f]_- = [g]_+, \end{cases}$$

where the symbol $[\cdot]_\pm$ denotes the projectors onto $\mathbb{R}_{m,\pm}$.

**Example 4.2.1.** Following the previous remarks, we can show that the 4-dimensional Dirac equation can be written, through suitable manipulations, as a Cauchy–Fueter equation. For this purpose, let us start with the 4-dimensional Dirac equation

$$(e_1\partial_{x_1} + e_2\partial_{x_2} + e_3\partial_{x_3} + e_4\partial_{x_4})f = g.$$

Multiply both sides by $-e_4$ (and call $g$ the right hand side). Setting $E_j = -e_4 e_j$, $j = 1, 2, 3$, we obtain the equation

$$(\partial_{x_4} + E_1\partial_{x_1} + E_3\partial_{x_2} + E_3\partial_{x_3})f = g. \tag{4.3}$$

Assume now, with no loss of generality because $\mathbb{R}_4 = \mathbb{R}_{4,+} \oplus \mathbb{R}_{4,-}$, that $f$ has values in $\mathbb{R}_{4,+} = \mathbb{R}_3$.

The pseudoscalar $E_1E_2E_3$ belongs to the center of $\mathbb{R}_3$ and has square 1, so that we may consider the mutually annihilating idempotents

$$O_\pm = \frac{1}{2}(1 \pm E_1E_2E_3).$$

If we multiply (4.3) by $O_+$ we get

$$O_+(\partial_{x_0} - E_2E_3\partial_{x_1} - E_3E_1\partial_{x_2} - E_1E_2\partial_{x_3})f = O_+g$$

and we may now consider both $f$, $g$ to take values in $\mathbb{R}_{3,+}$. By putting

$$i = -E_2E_3, \quad j = -E_3E_1, \quad k = -E_1E_2$$

and relabelling $x_4$ as $x_0$, we obtain the inhomogeneous Fueter equation

$$(\partial_{x_0} + i\partial_{x_1} + j\partial_{x_2} + k\partial_{x_3})f = g.$$

This is not the only equation which may be reobtained from the Dirac equation in Clifford analysis. For example, suppose that the dimension $m = 2n$ is even and consider a Clifford basis $e_1, \dots, e_n, e_{n+1}, \dots, e_{2n}$. Let

$$\partial_{\underline{X}} = \sum_{j=1}^{2n} e_j \partial_{x_j}$$

be the Dirac operator acting on functions $f : \mathbb{R}^m \to \mathbb{C}_m$. Through the identification

$$e_{j+n} \to -i\epsilon_j,$$

the Dirac operator may be written into the form

$$\partial_{\underline{X}} = \sum_{j=1}^{m} e_j\partial_{x_j} + \sum_{j=1}^{m} e_{j+n}\partial_{x_{j+n}} = \sum_{j=1}^{m} e_j\partial_{x_j} - i\epsilon_j\partial_{y_j},$$

where we put $x_{j+n} = y_j$ for $j = 1, \dots, n$. Let us consider the primitive idempotent $I = I_1 \dots I_n$, $I_j = \frac{1}{2}(1 - \epsilon_j e_j)$: multiplying on the right the function $f$ by this idempotent, we know that there is a unique $\mathbb{C}_n$-valued function $\hat{f}$ such that $fI = \hat{f}I$ (see Proposition 4.1.2). Then the Dirac equation $\partial_{\underline{X}} f = g$ may be rewritten as

$$\begin{aligned}\partial_{\underline{X}}\hat{f}I &= \sum_{j=1}^{m} e_j\partial_{x_j}\hat{f}I - i\sum_{j=1}^{m}\epsilon_j\partial_{y_j}\hat{f}I \\ &= \sum_{j=1}^{m} e_j\partial_{x_j}\hat{f}I - i\sum_{j=1}^{m}\partial_{y_j}\tilde{\hat{f}}e_jI = \hat{g}I,\end{aligned}$$

where $\tilde{f}$ denotes the main involution of $f$. In view of Remark 4.2.5, it suffices to consider the previous equation in the case in which $\hat{f}$ takes values in the even subalgebra $\mathbb{C}_{n,+}$ so that the nonhomogeneous Dirac equation may be essentially transformed into an equation of the form

$$\partial_{\underline{x}}[\hat{f}]_+ - i[\hat{f}]_+\partial_{\underline{y}} = [\hat{g}]_-,$$

where $[\hat{g}]_-$ takes values in $\mathbb{C}_{n,-}$. If $\hat{f}$ takes values in $\mathbb{R}_{n,-}$, the same Dirac equation leads to the equation

$$\partial_{\underline{x}}[\hat{f}]_- + i[\hat{f}]_-\partial_{\underline{y}} = [\hat{g}]_+,$$

where $[\hat{g}]_+$ takes values in $\mathbb{C}_{n,+}$.

**Remark 4.2.6.** In certain problems, one form of the Dirac equation is preferred over another. The form already introduced is preferable if we are interested in reducing the dimension of the matrices representing the system corresponding to a Dirac equation. On the other hand, if we want to investigate the Spin$(m)$-invariance it is better to work with the original Dirac operator $\partial_{\underline{X}}$.

**Remark 4.2.7.** Another important dimensional reduction can be obtained using the following method. If $m$ is even, one may divide the dimension by 2 by noting that the pseudoscalar $E_2 \cdot \ldots \cdot E_m$ is a central element and that the identity decomposes as a sum of two mutually annihilating central zero divisors

$$O_\pm = \frac{1}{2}(1 \pm i^s E_2 \ldots E_m),$$

where we set $s = 0$ or $1$, so that one only has to consider the $\mathbb{C}_{m-2}$ valued equation

$$\mathcal{D}_{\underline{x}} f O_+ = g O_+,$$

where

$$\mathcal{D}_{\underline{x}} = \partial_{x_1} + \sum_{j=2}^{m} E_j \partial_{x_j}.$$

**Example 4.2.2.** Let us consider $f : \mathbb{R}^6 \to \mathbb{C}_6$ and the Dirac operator

$$\partial_{\underline{X}} = \sum_{i=1}^{6} e_i \partial_{x_i}.$$

Multiplying by the primitive idempotent $I = I_1 I_2 I_3$, $I_j = \frac{1}{2}(1 - \epsilon_j e_j)$, we can reduce the 6-dimensional Dirac equation to an equation in $\mathbb{C}_3$ of the form

$$D_{\underline{x}} f - i f D_{\underline{y}} = g$$

where $f$ and $g$ are $\mathbb{C}_3$-valued. Now we can split the equations into its even and odd parts obtaining equations of the form

$$D_{\underline{x}}[f]_+ - i[f]_+ D_{\underline{y}} = [g]_-,$$

where

$$D_{\underline{x}} = \sum_{i=1}^{3} e_i \partial_{x_i}, \quad D_{\underline{y}} = \sum_{i=1}^{3} e_i \partial_{y_i}.$$

Since $n$ is odd, we may multiply this equation by the zero divisor $(1+e_{123})$ and in view of the equality

$$e_1(1+e_{123}) = -e_{23}(1+e_{123})$$

one has that, writing $i_1 = -e_{23}$, $i_2 = -e_{31}$, $i_3 = -e_{12}$, we arrive at the complex quaternion valued system

$$\tilde{D}_{\underline{x}} f - i f \tilde{D}_{\underline{y}} = g,$$

where $\tilde{D}_{\underline{x}}$ and $\tilde{D}_{\underline{y}}$ are Moisil–Theodorescu operators. Hence a system of the form

$$\begin{cases} \tilde{D}_{\underline{x}_1} f - i f \tilde{D}_{\underline{y}_1} = g_1 \\ \tilde{D}_{\underline{x}_2} f - i f \tilde{D}_{\underline{y}_2} = g_2 \\ \tilde{D}_{\underline{x}_3} f - i f \tilde{D}_{\underline{y}_3} = g_3, \end{cases} \tag{4.4}$$

corresponds to a nonhomogeneous system of 3 Dirac operators in $\mathbb{C}_6$.

In the same way, one could also consider Dirac operators on spaces $\mathbb{R}_{p,q}$ of mixed signature

$$\partial_{\underline{X}} = \sum_{j=1}^{p} e_j \partial_{x_j} + \sum_{j=1}^{q} \epsilon_j \partial_{y_j}$$

and, in particular, the Dirac operator on the phase space $\mathbb{R}_{m,m}$, given by

$$\partial_{\underline{X}} = \sum_{j=1}^{m} (e_j \partial_{x_j} + \epsilon_j \partial_{y_j}).$$

The theory of monogenic functions is extremely rich in results and many of them are similar to those recalled in the section about regular functions of one quaternionic variable, so we do not repeat them here and we refer the reader to the book [29], which is a thorough introduction to the topic.

### 4.2.2 Radial algebra

As we have noted in Chapter 3, the resolutions associated to the Cauchy–Fueter system cannot, in general, be described only in terms of the Cauchy–Fueter operators. This fact is, in a sense, a rather unpleasant characteristic of the quaternionic setting. In the general Clifford case, the theory of Dirac complexes sometimes leads to a much simpler structure, which can be described in terms of what we call "radial algebra." This notion will be used in Section 4.3.

**Definition 4.2.4.** *Let $S$ be a set of objects which we call "abstract vector variables." The radial algebra $R(S)$ is defined to be the associative algebra generated by $S$ over a field with the defining relations*

$$[\{x,y\},z] = xyz + yxz - zxy - zyx = 0, \quad \textit{for} \quad x,y,z \in S. \tag{4.5}$$

Let $T(S)$ be the tensor algebra (free associative algebra) generated by the elements of $S$, and let $I(S)$ be the two-sided ideal generated by the polynomials

$$[\{x,y\},z].$$

Then we have

$$R(S) = T(S)/I(S).$$

We denote by $r(S)$ the subalgebra of $R(S)$ generated by the *scalar elements* $\{x,y\} = xy + yx$, $x$, $y \in S$. We call $r(S)$ the *scalar subalgebra* of $R(S)$. The reason for this terminology lies in the fact that when $x$ and $y$ are not just abstract variables, but in fact represent Clifford variables, then $\{x,y\}$ is real. By $R_k(S)$ we denote the subspace of $k$-vectors in $R(S)$ which is generated over $r(S)$ by wedge products of length $k$ of the form

$$x_1 \wedge \cdots \wedge x_k := \frac{1}{k!}\sum_{\pi} sgn(\pi)x_{\pi(1)}\cdots x_{\pi(k)},$$

summed over all permutations $\pi$ of the indices. There is a direct sum decomposition

$$R(S) = R_0(S) \oplus R_1(S) \oplus \ldots \oplus R_k(S) \oplus \ldots,$$

where $R_0(S) = r(S)$ are the scalar elements. The vector derivatives $\partial_x$, $\partial_y$, $\partial_z, \ldots$, associated to the elements $x, y, z, \ldots \in S$ are defined as the endomorphisms on $R(S)$ which satisfy the following axioms: for $x \in S$, $f \in r(S)$, $F \in R(S)$, and $G \in R(S \setminus \{x\})$ one has

$$\begin{array}{lll}
(D_1) & \partial_x[fF] = \partial_x[f]F + f\partial_x[F], & [fF]\partial_x = F\partial_x[f] + f[F]\partial_x, \\
(D_2) & \partial_x[FG] = \partial_x[F]G, & [GF]\partial_x = G[F]\partial_x, \\
(D_3) & \partial_x[F\partial_y] = [\partial_x F]\partial_y, & \\
(D_4) & \partial_x x^2 = 2x, & \partial_x\{x,y\} = 2y.
\end{array}$$

It may be shown, see [188], that the objects $\partial_x[x]$ are independent of the choice $x \in S$. Moreover, the equality $\partial_x[x] = [x]\partial_x$ holds and $M = \partial_x[x]$ is a commuting object which belongs to the field over which the algebra $R(S)$ is generated. In the case of the field $\mathbb{R}$, $M$ is a real number.

Next, let $x \in S$ be a chosen vector variable and consider the other variables of $S$ as parameters; then an element $F \in R(S)$ may be written as a formal function $F(x)$ with respect to the variable $x$. $F(x)$ is called left-monogenic if it solves the equation $\partial_x F(x) = 0$. Similarly let us select a finite subset $\{x_1, \ldots, x_n\}$ of $S$, called the vector variables, and let us call the remaining vectors in $S$ "parameter vectors." Then $F \in R(S)$ may be written into the form $F(x_1, \ldots, x_n)$ and it is called monogenic if it satisfies the system

$$\partial_{x_j} F(x_1, \ldots, x_n) = 0, \qquad j = 1, \ldots, n.$$

One may consider an inhomogeneous system of the form

$$\begin{cases} \partial_{x_1} f = g_1 \\ \dots \\ \partial_{x_n} f = g_n \end{cases}$$

in the radial algebra setting, and look for the compatibility equations to be satisfied by $g_1, \dots, g_n$ for this system to be solvable. Necessary conditions for this are certainly the relations arising from the fact that the algebra generated by $\{\partial_{x_1}, \dots, \partial_{x_n}\}$ is a radial algebra with relations

$$\partial_{x_j}\{\partial_{x_k}, \partial_{x_l}\} = \{\partial_{x_k}, \partial_{x_l}\}\partial_{x_j}, \quad j, k, l = 1, \dots, n,$$

namely

$$\partial_{x_j}(\partial_{x_k} g_l + \partial_{x_l} g_k) = \{\partial_{x_k}, \partial_{x_l}\} g_j, \quad j, k, l = 1, \dots, n.$$

It is also clear that in this radial algebra setting there can only be compatibility equations that are expressible solely in terms of the abstract vector derivatives $\partial_{x_i}$. Using a suitable Clifford algebra representation

$$x \to \underline{x} = \sum e_j x_j$$

of sufficiently high dimension $m$, and by identifying

$$\partial_{x_j} \to -\partial_{\underline{x}_j} = -\sum e_k \partial_{x_{jk}},$$

one can look at the problem in the Clifford analysis setting. This can in turn be written in terms of real coordinates on which computer calculations with CoCoA can be used. For further details on how this concept can provide useful insights in the study of Dirac complexes, we refer the reader to [163], [188] and [191].

### 4.2.3 Fischer decomposition

We begin this section by recalling some general facts about the classical Fischer decomposition. As usual, $z = (z_1, \dots, z_n)$ is an element in $\mathbb{C}^n$, $z^\alpha = z_1^{\alpha_1} \cdots z_n^{\alpha_n}$ and

$$P(z) = \sum_\alpha a_\alpha z^\alpha$$

is a polynomial with complex coefficients. We set

$$P^*(z) = \sum_\alpha \bar{a}_\alpha z^\alpha = \overline{P(\bar{z})}$$

and we introduce the operators

$$P(D) = \sum_\alpha a_\alpha D^\alpha$$

and

$$P^*(D) = \sum_\alpha \bar{a}_\alpha D^\alpha.$$

It is possible to define an inner product for polynomials in $n$ variables, such that all the monomials are mutually orthogonal by setting

$$\langle z^\alpha, z^\beta \rangle = \alpha! \delta_{\alpha\beta}.$$

For two polynomials $P(z) = \sum_\alpha a_\alpha z^\alpha$ and $Q(z) = \sum_\alpha b_\alpha z^\alpha$ we have

$$\begin{aligned} \langle P, Q \rangle &= \sum_\alpha \alpha! a_\alpha \bar{b}_\alpha = \sum_\alpha \frac{1}{\alpha!} P^{(\alpha)}(0) \overline{Q^{(\alpha)}}(0) \\ &= [P(D) Q^*(z)]_{z=0} = [Q^*(D) P(z)]_{z=0}. \end{aligned}$$

In the particular case in which $P$ and $Q$ are homogeneous polynomials of the same degree we have the equality

$$\langle P, Q \rangle = P(D) Q^*(z) = Q^*(D) P(z).$$

Considering three arbitrary polynomials $P$, $f$, $\varphi$ we have

$$\langle P(z) f(z), \varphi(z) \rangle = \langle f(z), P^*(D) \varphi(z) \rangle,$$

showing that $P^*(D)$ is the natural candidate for the adjoint of the multiplication operator $P(z)$. We now give the following definition.

**Definition 4.2.5.** *Let $P(z)$ be a homogeneous polynomial in $n$ variables. We define the Fischer decomposition of a polynomial $f$ the splitting*

$$f = h + g$$

*with $h$, $g$ polynomials such that $P^*(D)h = 0$ and $g(z)$ divisible by $P(z)$.*

This definition is well posed since, using the inner product defined above, Fischer proved the following result, see [73].

**Theorem 4.2.1.** *Let $P(z)$ be a homogeneous polynomial in $n$ variables. Then every polynomial $f$ can be uniquely decomposed as*

$$f(z) = h(z) + g(z)$$

*where $h$, $g$ are polynomials such that*

$$P^*(D)h = 0, \quad \text{and} \quad P(z) \text{ divides } g(z).$$

*Moreover, if $f$ is a homogeneous polynomial of degree $k$, so are $h$ and $g$ unless they are zero.*

Note however that this setting can be further generalized: following Shapiro [178], we can consider a pair $(P, Q)$ of polynomials and a space of functions $\mathcal{X}$ and we will say that $(P, Q)$ is a Fischer pair with respect to $\mathcal{X}$, if every function $f \in \mathcal{X}$ can be uniquely decomposed as $f = h+g$, with $h$, $g \in \mathcal{X}$ where $Q(D)h = 0$ and $P$ divides $g$. Generalizing the discussion above, we call this splitting the Fischer decomposition. An example of space $\mathcal{X}$ that can be considered, besides the vector spaces of polynomials in $n$ variables, is the Fischer–Fock space $\mathbb{F}$. The space $\mathbb{F}$ contains entire functions of $n$ complex variables that are square integrable with respect to the Gauss measure $\pi^{-n}\exp(-|z|^2)\, d\mu(z)$, where $d\mu$ is the Lebesgue measure. It is endowed with the inner product

$$\langle f, g\rangle = \int_{\mathbb{C}^n} f(z)\overline{g(z)}d\sigma(z) := \frac{1}{\pi^n}\int_{\mathbb{C}^n} f(z)\overline{g(z)}e^{-|z|^2}d\mu(z).$$

The space $\mathbb{F}$ is the closure, with respect to the norm induced by the inner product, of the space of polynomials; moreover all entire functions of exponential type belong to $\mathbb{F}$.

Let us give an idea as to how the Fischer pair allows the decomposition in the case of the Fischer–Fock space $\mathbb{F}$. We suppose that the polynomial $Q(z)$ acts on the space $\mathbb{F}$ as

$$f(z) \mapsto Q(z)f(z),$$

while $P(D)$ acts on $\mathbb{F}$ as

$$f(z) \mapsto P(D)f(z).$$

We will write $\mathbb{F}$ as a subscript of an operator to denote the space of functions on which it acts. Newman and Shapiro proved the following result.

**Theorem 4.2.2.** *If $Q = P^*$ then $Q(z)_{\mathbb{F}} = P^*(z)_{\mathbb{F}}$ and $P(D)_{\mathbb{F}}$ are adjoint of each other. Moreover, the following orthogonal decomposition*

$$\mathbb{F} = \ker(P(D)_{\mathbb{F}}) \oplus \operatorname{im}(Q(z)_{\mathbb{F}})$$

*holds.*

Another space $\mathcal{X}$ in which the Fischer decomposition holds is, for example, the space of all entire functions (see [133]). However, in general, it is a difficult question to determine for which pairs $(P, Q)$ and in which function spaces the previous decomposition holds, and we refer the reader to [9] for the state of the art about this topic and for more recent results.

Our purpose is now to generalize what we introduced above into the Clifford setting. In what follows all the function will be $\mathbb{C}_m$-valued.

**Definition 4.2.6.** *A homogeneous monogenic polynomial $P_k : \mathbb{R}^m \to \mathbb{C}_m$ of degree $k$ is called a spherical monogenic of degree $k$.*

The set of all spherical monogenic of degree $k$ is denoted by $M(k)$, while the set of all homogeneous polynomials $\mathbb{C}_m$-valued of degree $k$ is denoted by $\mathcal{P}(k)$. In it, we can give the notion of Fischer inner product, see [55].

**Definition 4.2.7.** *Let $P(\underline{x}) = \sum_\alpha \underline{x}^\alpha a_\alpha$ and $Q(\underline{x}) = \sum_\alpha \underline{x}^\alpha b_\alpha$ belong to $\mathcal{P}(k)$. The Fischer inner product $\langle P, Q\rangle$ is given by*

$$\langle P, Q\rangle = \left[\sum_{|\alpha|=k} \alpha! \bar{a}_\alpha b_\alpha\right]_0 = \left[P^+(\partial_{\underline{x}})Q(\underline{x})|_{\underline{x}=0}\right]_0,$$

*i.e., one takes the action of the operator $P^+$ on $Q$ evaluated at the origin and then one takes the scalar part $[.]_0$ of this (i.e., the zero-vector part).*

Moreover, we have (see [48], [55]) the following result.

**Theorem 4.2.3.** (Fischer decomposition). *Let $k \in \mathbb{N}$. Then $\mathcal{P}(k)$ can be decomposed as*

$$\mathcal{P}(k) = \sum_{s=0}^{k} \underline{x}^s M(k-s).$$

*Proof.* For $P \in \mathcal{P}(k-1)$ and $Q \in \mathcal{P}(k)$ we have

$$\langle \underline{x}P, Q\rangle = \langle P, (-\partial_{\underline{x}})Q\rangle = -\langle P, \partial_{\underline{x}}Q\rangle$$

where the the inner product on the right-hand side is taken in $\mathcal{P}(k-1)$. So the multiplication operator $\underline{x}: \mathcal{P}(k-1) \to \mathcal{P}(k)$ has $-\partial_{\underline{x}}$ as its adjoint. It follows that

$$\mathcal{P}(k) = M(k) \oplus \underline{x}\mathcal{P}(k-1)$$

where the sum is orthogonal. In fact if for some $Q \in \mathcal{P}(k)$ and for all $P \in \mathcal{P}(k-1)$ the equality $\langle \underline{x}P, Q\rangle = 0$ holds, then we also have $\langle P, \partial_{\underline{x}}Q\rangle = 0$, so $\partial_{\underline{x}}Q = 0$ which means that the orthogonal complement of $\underline{x}\mathcal{P}(k-1)$ is a subspace of $M(k)$. Let us suppose that $S = \underline{x}P \in M(k) \cap \underline{x}\mathcal{P}(k-1)$; then we have

$$\langle S, S\rangle = \langle \underline{x}P, S\rangle = \langle P, \partial_{\underline{x}}S\rangle = 0,$$

hence $S = 0$. So any $R \in \mathcal{P}(k)$ can be uniquely decomposed as

$$R(\underline{x}) = M_0(R)(\underline{x}) + \underline{x}P_1(R)(\underline{x})$$

where $M_0(R) \in M(k)$ and $P_1(R) \in \mathcal{P}(k-1)$. Iterating the procedure we obtain the statement. □

This discussion can be generalized to general Clifford polynomials $P(\underline{x}_1, \ldots, \underline{x}_l)$ of several vector variables, but the proofs are much more complicated. The Fischer inner product $\langle P, Q\rangle$ of two Clifford polynomials is given by

$$\langle P, Q\rangle = [P^+Q|_{\underline{x}_1=\ldots=\underline{x}_l=0}]_0,$$

i.e., one considers the action of operator $P^+$ on $Q$ evaluated at the origin and then one takes the scalar part $[.]_0$. The inner product thus obtained is Hermitian.

Note that polynomials of different degree of homogeneity are always orthogonal so that from now on we restrict our attention to homogeneous polynomials of degree $k_j$ in the variable $\underline{x}_j$, i.e.

$$R(\underline{x}_1, \dots, a\underline{x}_j, \dots, \underline{x}_l) = a^{k_j} R(\underline{x}_1, \dots, \underline{x}_l), \quad j = 1, \dots, l.$$

One now has the basic orthogonal decomposition which follows directly from the definition of the Fischer inner product, see [187], along the same lines of the proof of the previous theorem:

**Lemma 4.2.1.** *Every Clifford polynomial admits an orthogonal decomposition of the form*

$$R(\underline{x}_1, \dots, \underline{x}_l) = M(R)(\underline{x}_1, \dots, \underline{x}_l) + M^*(R)(\underline{x}_1, \dots, \underline{x}_l),$$

*where*

*(i)* $\partial_{\underline{x}_j} M(R)(\underline{x}_1, \dots, \underline{x}_l) = 0$, $j = 1, \dots, l$,

*(ii)* $M^*(R)(\underline{x}_1, \dots, \underline{x}_l) = \sum \underline{x}_j R_j(\underline{x}_1, \dots, \underline{x}_l)$ *for some Clifford polynomials* $R_j$.

Next one may iterate this decomposition by applying the above decomposition to the polynomials $R_j$ and prove the following, see [187].

**Theorem 4.2.4.** (Rough Fischer decomposition). *Every Clifford polynomial admits an orthogonal decomposition of the form*

$$R(\underline{x}_1, \dots, \underline{x}_l) = M_0(R)(\underline{x}_1, \dots, \underline{x}_l) + \dots + M_j(R)(\underline{x}_1, \dots, \underline{x}_l) + \dots$$

*where*

$$M_0(R)(\underline{x}_1, \dots, \underline{x}_l) = M(R)(\underline{x}_1, \dots, \underline{x}_l)$$

*is left monogenic in each vector variable* $\underline{x}_j$ *and in general*

$$M_j(R)(\underline{x}_1, \dots, \underline{x}_l)$$

*may be written as linear combination of products of the form*

$$\underline{x}_{s_1} \cdots \underline{x}_{s_j} P(\underline{x}_1, \dots, \underline{x}_l), \qquad \partial_{\underline{x}_i} P(\underline{x}_1, \dots, \underline{x}_l) = 0, \qquad i = 1, \dots, l.$$

Now if we consider

$$M_1(R) = \underline{x}_1 P_1 + \dots + \underline{x}_l P_l, \qquad \partial_{\underline{x}_i} P_j(\underline{x}_1, \dots, \underline{x}_l) = 0, \qquad i = 1, \dots, l,$$

then we can show that for $m > 2$ the "monogenic coefficients" $P_1, \dots, P_l$ are uniquely determined, see [187]. One may wonder whether a similar result holds for the higher terms and the terms in a sum of the form

$$\sum \underline{x}_{s_1} \cdots \underline{x}_{s_j} P_{s_1 \dots s_j}.$$

Unfortunately, this decomposition is usually not unique as the following example shows:

**Example 4.2.3.** We consider the case of the 3-dimensional Cauchy–Fueter system. We have the relations (see Chapter 3):

$$(D'_{q_r}\partial_{\bar{q}_s} - D'_{q_s}\partial_{\bar{q}_r})\partial_{\bar{q}_h} + (D'_{q_s}\partial_{\bar{q}_h} - D'_{q_h}\partial_{\bar{q}_s})\partial_{\bar{q}_r}$$
$$+(D'_{q_h}\partial_{\bar{q}_r} - D'_{q_r}\partial_{\bar{q}_h})\partial_{\bar{q}_s} = 0$$

and

$$(D''_{q_r}\partial_{\bar{q}_s} - D''_{q_s}\partial_{\bar{q}_r})\partial_{\bar{q}_h} + (D''_{q_s}\partial_{\bar{q}_h} - D''_{q_h}\partial_{\bar{q}_s})\partial_{\bar{q}_r}$$
$$+(D''_{q_h}\partial_{\bar{q}_r} - D''_{q_r}\partial_{\bar{q}_h})\partial_{\bar{q}_s} = 0,$$

which are the Fischer conjugate of the polynomial identities

$$\underline{x}_h(\underline{x}_s\underline{x}'_r - \underline{x}_r\underline{x}'_s) + \underline{x}_r(\underline{x}_h\underline{x}'_s - \underline{x}_s\underline{x}'_h) + \underline{x}_s(\underline{x}_r\underline{x}'_h - \underline{x}_h\underline{x}'_r) = 0,$$

$$\underline{x}_h(\underline{x}_s\underline{x}''_r - \underline{x}_r\underline{x}''_s) + \underline{x}_r(\underline{x}_h\underline{x}''_s - \underline{x}_s\underline{x}''_h) + \underline{x}_s(\underline{x}_r\underline{x}''_h - \underline{x}_h\underline{x}''_r) = 0,$$

where

$$\underline{x}'_r = kx_{r3} - jx_{r2}, \qquad \underline{x}''_r = kx_{r3} - ix_{r1}, \qquad r = 1, 2, 3,$$

are clearly left monogenic polynomials. These relations are clear counterexamples to this claim.

One can hope to have uniqueness in the decomposition only using a suitable basis for the space of all polynomials of given total degree, generated by the vector variables $\underline{x}_1, \dots, \underline{x}_l$. Such a basis is given by the set of polynomials

$$\underline{x}_1^{s_1} \dots \underline{x}_l^{s_l} \prod_{i<j} < \underline{x}_i, \underline{x}_j >^{s_{ij}}, \qquad \sum(2s_i + 2s_{kj}) = s.$$

It seems natural to make the following conjecture.

**Conjecture 4.2.1.** (Complete Fischer decomposition). *Every Clifford polynomial in $l$ variables in a Clifford algebra over $m$ units, with $m \geq 2l - 1$ admits a unique decomposition of the form*

$$R(\underline{x}_1, \dots, \underline{x}_l) = \sum_{s_i, s_{ij}} \underline{x}_1^{s_1} \dots \underline{x}_l^{s_l} \prod_{i<j} \langle \underline{x}_i, \underline{x}_j \rangle^{s_{ij}} M_{s_i, s_{ij}}(\underline{x}_1, \dots, \underline{x}_l)$$

*where the sum runs over all ordered tuples*

$$s_i; s_{ij} \ : \ i = 1, \dots, l, \ j = i + 1, \dots, l,$$

*and the functions $M_{s_i, s_{ij}}(\underline{x}_1, \dots, \underline{x}_l)$ are monogenic.*

**Remark 4.2.8.** A possible proof of this conjecture is discussed in [192], but the main difficulty for the general statement is the fact that the procedure described works when the number of variables and the degree of the polynomials have been fixed. A general method has still to be found.

# 4.3 The Dirac complex for two, three and four operators

In Chapter 3 we have studied the Cauchy–Fueter complex, giving its description in the case of $n$ Cauchy–Fueter operators. In this section, we wish to carry out a similar analysis in the case of the Dirac operators acting on functions $f : \mathbb{R}^m \to \mathbb{C}_m$. The main feature is the complete description of all the maps in the complex, thus giving the possibility to apply the duality theorems in Chapter 2, Section 2.1.

## 4.3.1 Case of two operators

Let us start our study with the simplest case: we consider the system

$$\begin{cases} \partial_{\underline{x}_1} f &= g_1 \\ \partial_{\underline{x}_2} f &= g_2 \end{cases} \tag{4.6}$$

where $f, g_r : \mathbb{R}^m \longrightarrow \mathbb{R}_m$ (or $\mathbb{C}_m$), $r = 1, 2$, belong to the sheaves $\mathcal{E}$, $\mathcal{A}$, $\mathcal{D}'$ or $\mathcal{B}$.

**Theorem 4.3.1.** *When $m = 2$, the Dirac complex is*

$$0 \longrightarrow R^4(-2) \longrightarrow R^8(-1) \longrightarrow R^8 \longrightarrow M_2 \to 0.$$

*Proof.* When $m = 2$ we have

$$\partial_{\underline{x}_r} = e_1 \frac{\partial}{\partial_{x_{r1}}} + e_2 \frac{\partial}{\partial_{x_{r2}}},$$

for $r = 1, 2$ and it is immediate to verify the equality

$$\left(e_1 \frac{\partial}{\partial_{x_{11}}} - e_2 \frac{\partial}{\partial_{x_{12}}}\right)\left(e_1 \frac{\partial}{\partial_{x_{21}}} + e_2 \frac{\partial}{\partial_{x_{22}}}\right)$$

$$= \left(e_1 \frac{\partial}{\partial_{x_{21}}} - e_2 \frac{\partial}{\partial_{x_{22}}}\right)\left(e_1 \frac{\partial}{\partial_{x_{11}}} + e_2 \frac{\partial}{\partial_{x_{12}}}\right),$$

so there is only one compatibility condition on the data of the system:

$$\left(e_1 \frac{\partial}{\partial_{x_{11}}} - e_2 \frac{\partial}{\partial_{x_{12}}}\right) g_2 - \left(e_1 \frac{\partial}{\partial_{x_{21}}} - e_2 \frac{\partial}{\partial_{x_{22}}}\right) g_1 = 0,$$

and the complex ends after one step. □

When $m > 2$, there are no more commutation relation among the operators, not even allowing some variations of the signs in front of the units $e_i$. One can observe that the Laplacian operator $\Delta_r = -\partial_{\underline{x}_r}\partial_{\underline{x}_r}$, being real, commutes with any Dirac operator $\partial_{\underline{x}_s}$ so one has the following relations:

$$-\Delta_1\partial_{\underline{x}_2} + (\partial_{\underline{x}_2}\partial_{\underline{x}_1})\partial_{\underline{x}_1} = 0,$$

$$-\Delta_2\partial_{\underline{x}_1} + (\partial_{\underline{x}_1}\partial_{\underline{x}_2})\partial_{\underline{x}_2} = 0,$$

which imply the following compatibility conditions:

$$-\Delta_1 g_2 + (\partial_{\underline{x}_2}\partial_{\underline{x}_1})g_1 = 0,$$

$$-\Delta_2 g_1 + (\partial_{\underline{x}_1}\partial_{\underline{x}_2})g_2 = 0.$$

Considering the nonhomogeneous system

$$\begin{cases} -\Delta_1 g_2 + (\partial_{\underline{x}_2}\partial_{\underline{x}_1})g_1 = h_1 \\ -\Delta_2 g_1 + (\partial_{\underline{x}_1}\partial_{\underline{x}_2})g_2 = h_2 \end{cases}$$

one easily finds the relation which closes the complex:

$$\partial_{\underline{x}_2}h_1 + \partial_{\underline{x}_1}h_2 = 0.$$

The conditions listed are obviously necessary. The fact that they are also sufficient relies on the validity of the Fischer decomposition for polynomials in two variables (compare with Proposition 4.3.6). In this case the resolution turns out to be

$$0 \longrightarrow R^4(-4) \longrightarrow R^8(-3) \longrightarrow R^8(-1) \longrightarrow R^8 \longrightarrow M_2 \to 0.$$

### 4.3.2 Case of three operators

We study the case of three Dirac operators, considering a system of the type

$$\begin{cases} \partial_{\underline{x}_1} f &= g_1 \\ \partial_{\underline{x}_2} f &= g_2 \\ \partial_{\underline{x}_3} f &= g_3 \end{cases} \tag{4.7}$$

where $f, g_r : \mathbb{R}^m \longrightarrow \mathbb{C}_m$, $r = 1, 2, 3$ belong to the sheaves $\mathcal{E}$, $\mathcal{A}$, $\mathcal{D}'$ or $\mathcal{B}$, and

$$\partial_{\underline{x}_r} = \sum_{i=1}^{m} e_i \frac{\partial}{\partial x_{ri}}, \qquad r = 1, 2, 3.$$

The associated matrix $P(D)$ is

$$\begin{bmatrix} U_1(D) \\ U_2(D) \\ U_3(D) \end{bmatrix}$$

where $U_r(D)$, $r = 1, 2, 3$, is the matrix which describes $\partial_{\underline{x}_r}$ in real coordinates. The system can be rewritten as

$$P(D)f = \begin{bmatrix} g_1 \\ g_2 \\ g_3 \end{bmatrix}$$

with the obvious meaning of the symbols. Note that if $m \leq 4$, the complex is known: when $m = 2$, the complex can be constructed along the same lines of the Koszul complex, even though the setting is noncommutative, by virtue of the 3 relations (compare with the case of two Dirac operators)

$$\left(e_1\frac{\partial}{\partial_{x_{r1}}} - e_2\frac{\partial}{\partial_{x_{r2}}}\right)\left(e_1\frac{\partial}{\partial_{x_{s1}}} + e_2\frac{\partial}{\partial_{x_{s2}}}\right)$$
$$= \left(e_1\frac{\partial}{\partial_{x_{s1}}} - e_2\frac{\partial}{\partial_{x_{s2}}}\right)\left(e_1\frac{\partial}{\partial_{x_{r1}}} + e_2\frac{\partial}{\partial_{x_{r2}}}\right), \qquad r \neq s, \quad r, s \in \{1, 2, 3\}.$$

The complex ends in one step with the relation associated to the operator $P(D)^t$. The number of the independent relations and their degree is exactly the one predicted with the use of CoCoA. For $m = 3, 4$, it is a direct consequence of the discussion in Subsection 4.2.1 which shows that the Dirac operator can be reduced to the Cauchy–Fueter one.

To understand the general case, we begin with an example.

**Example 4.3.1.** Let us consider $f : \mathbb{R}^6 \to \mathbb{C}_6$ and the Dirac operator

$$\partial_{\underline{x}} = \sum_{i=1}^{6} e_i \partial_{x_i}.$$

As we have shown in example 4.2.2, the system

$$\begin{cases} D_{\underline{X}_1} f - i f D_{\underline{Y}_1} = g_1 \\ D_{\underline{X}_2} f - i f D_{\underline{Y}_2} = g_2 \\ D_{\underline{X}_3} f - i f D_{\underline{Y}_3} = g_3, \end{cases} \tag{4.8}$$

corresponds to a nonhomogeneous system of 3 Dirac operators in $\mathbb{C}_6$. The symbol of (4.8) is the matrix $P = [U_1, U_2, U_3]^t$ with

$$U_r^t = \begin{bmatrix} X_r & Y_r \\ -Y_r & X_r \end{bmatrix}, \tag{4.9}$$

$$X_r = \begin{bmatrix} 0 & -X_{r1} & -X_{r2} & -X_{r3} \\ X_{r1} & 0 & -X_{r3} & X_{r2} \\ X_{r2} & X_{r3} & 0 & -X_{r1} \\ X_{r3} & -X_{r2} & X_{r1} & 0 \end{bmatrix},$$

$$Y_r = \begin{bmatrix} 0 & -Y_{r1} & -Y_{r2} & -Y_{r3} \\ Y_{r1} & 0 & Y_{r3} & -Y_{r2} \\ Y_{r2} & -Y_{r3} & 0 & Y_{r1} \\ Y_{r3} & Y_{r2} & -Y_{r1} & 0 \end{bmatrix}.$$

The minimal resolution which one obtains through the use of CoCoA is

$$0 \longrightarrow R^8(-9) \longrightarrow R^{24}(-8) \longrightarrow R^{64}(-6) \longrightarrow R^{48}(-4) \oplus R^{48}(-5) \longrightarrow$$

$$\longrightarrow R^{64}(-3) \longrightarrow R^{24}(-1) \longrightarrow R^8 \longrightarrow M_3 \longrightarrow 0, \tag{4.10}$$

and gives:

- 8 quadratic syzygies at the first step;
- 6 quadratic and 6 linear syzygies at the second step;
- 8 quadratic syzygies at the third step;
- 3 quadratic syzygies at the fourth step;
- 1 linear syzygy at the last step.

In the following pages we want to solve the following problems.

**Problem 4.3.1.** *Give the explicit description of all the maps appearing in the resolution of three Dirac operators acting on functions $f: \mathbb{R}^m \to \mathbb{R}_m$ or $\mathbb{C}_m$, for $m \geq 5$.*

**Problem 4.3.2.** *Give the explicit description of all the maps appearing in the abstract complex associated to the system*

$$\begin{cases} \partial_{x_1} f = g_1 \\ \partial_{x_2} f = g_2 \\ \partial_{x_3} f = g_3 \end{cases}$$

*in the radial algebra setting, where $f, g_r \in R(S)$, $r = 1, 2, 3$.*

**Remark 4.3.1.** We remind the reader that the symbol $\partial_x$ denotes abstract operators whose formal properties are in the definition of the radial algebra, while the symbol $\partial_{\underline{x}}$ denotes the Dirac operators.

We now solve Problem 4.3.1 in the case $m = 6$. Analogous computations show that the same description holds for $m = 5$, but since no spinor reduction is available in this case, we prefer to carry out the calculations for $m = 6$ so that the size of the matrices involved in the various proofs is smaller.

**Proposition 4.3.1.** *The 8 compatibility conditions for the system*

$$\begin{cases} \partial_{\underline{x}_1} f = g_1 \\ \partial_{\underline{x}_2} f = g_2 \\ \partial_{\underline{x}_3} f = g_3 \end{cases}$$

*where $f, g_r : \mathbb{R}^6 \longrightarrow \mathbb{C}_6$, $\partial_{\underline{x}_r} = \sum_{i=1}^6 e_i \partial_{x_{ri}}$, $r = 1, 2, 3$, are given by the 6 relations:*

$$\partial_{\underline{x}_i} \partial_{\underline{x}_j} g_j - \partial^2_{\underline{x}_j} g_i = 0$$

*for each of the ordered pairs of indices* $1 \le i, j \le 3$, *together with the* 3 *relations*

$$\{\partial_{\underline{x}_i}, \partial_{\underline{x}_j}\}g_k = \partial_{\underline{x}_k}\partial_{\underline{x}_i} g_j + \partial_{\underline{x}_k}\partial_{\underline{x}_j} g_i$$

$1 \le i, j, k \le 3$, *of which only two are independent.*

*Proof.* The first relations come from the commutators $[\partial_{\underline{x}_i}, \partial^2_{\underline{x}_j}] = 0$ while the relations in the second group come from $[\{\partial_{\underline{x}_i}, \partial_{\underline{x}_j}\}, \partial_{\underline{x}_k}] = 0$. It can be verified directly that the syzygies in the statement are independent. Equivalently, one can write the block–matrix corresponding to the real counterpart of those relations and the module generated by the rows of this matrix. Then, using CoCoA applied to the matrix with entries $U_r$, one can verify that the number of generators of this module is 64 and corresponds to the number of these rows. This shows the independence. □

**Remark 4.3.2.** Note that the compatibility conditions obtained in the above proposition are in fact a restatement of the relations

$$[\partial_{\underline{x}_i}, \{\partial_{\underline{x}_j}, \partial_{\underline{x}_k}\}] = 0, \qquad 1 \le i, j, k \le 3$$

which are similar to the defining relations $[x_i, \{x_j, x_k\}] = 0$ of the radial algebra in 3 variables. The only operators which appear in the syzygies are the ones in the algebra generated by $\partial_{\underline{x}_1}, \partial_{\underline{x}_2}, \partial_{\underline{x}_3}$, i.e., "radial syzygies."

The 8 relations found in the above proposition lead to the new nonhomogeneous system

$$\begin{cases} \partial_{\underline{x}_2}\partial_{\underline{x}_1} g_1 - \partial^2_{\underline{x}_1} g_2 = h_{12} \\ \partial_{\underline{x}_3}\partial_{\underline{x}_1} g_1 - \partial^2_{\underline{x}_1} g_3 = h_{13} \\ \partial_{\underline{x}_1}\partial_{\underline{x}_2} g_2 - \partial^2_{\underline{x}_2} g_1 = h_{21} \\ \partial_{\underline{x}_3}\partial_{\underline{x}_2} g_2 - \partial^2_{\underline{x}_2} g_3 = h_{23} \\ \partial_{\underline{x}_1}\partial_{\underline{x}_3} g_3 - \partial^2_{\underline{x}_3} g_1 = h_{31} \\ \partial_{\underline{x}_2}\partial_{\underline{x}_3} g_3 - \partial^2_{\underline{x}_3} g_2 = h_{32} \\ \{\partial_{\underline{x}_2}, \partial_{\underline{x}_3}\} g_1 - \partial_{\underline{x}_1}\partial_{\underline{x}_2} g_3 - \partial_{\underline{x}_1}\partial_{\underline{x}_3} g_2 = a_1 \\ \{\partial_{\underline{x}_1}, \partial_{\underline{x}_3}\} g_2 - \partial_{\underline{x}_2}\partial_{\underline{x}_1} g_3 - \partial_{\underline{x}_2}\partial_{\underline{x}_3} g_1 = a_2 \\ \{\partial_{\underline{x}_1}, \partial_{\underline{x}_2}\} g_3 - \partial_{\underline{x}_3}\partial_{\underline{x}_1} g_2 - \partial_{\underline{x}_3}\partial_{\underline{x}_2} g_1 = a_3 \end{cases} \tag{4.11}$$

with the constraint $a_1 + a_2 + a_3 = 0$ that reduces the number of independent relations to 8. The compatibility conditions of (4.11) give the syzygies at the second step.

**Proposition 4.3.2.** *The* 6 *linear compatibility conditions of system (4.11) are given by*

$$\begin{cases} \partial_{\underline{x}_2} h_{12} + \partial_{\underline{x}_1} h_{21} = 0 \\ \partial_{\underline{x}_3} h_{13} + \partial_{\underline{x}_1} h_{31} = 0 \\ \partial_{\underline{x}_3} h_{23} + \partial_{\underline{x}_2} h_{32} = 0 \\ \partial_{\underline{x}_3} h_{12} + \partial_{\underline{x}_2} h_{13} = \partial_{\underline{x}_1} a_1 \\ \partial_{\underline{x}_1} h_{23} + \partial_{\underline{x}_3} h_{21} = \partial_{\underline{x}_2} a_2 \\ \partial_{\underline{x}_2} h_{31} + \partial_{\underline{x}_1} h_{32} = \partial_{\underline{x}_3} a_3. \end{cases}$$

*The* 6 *quadratic compatibility conditions are given by the cyclic permutations of* $(1,2,3)$ *in*

$$\{\partial_{\underline{x}_1}, \partial_{\underline{x}_2}\}h_{23} + \partial^2_{\underline{x}_2} a_3 = \partial_{\underline{x}_3}\partial_{\underline{x}_2} h_{21}$$

*of which only* 3 *conditions are independent, together with the permutations of the condition*

$$\partial^2_{\underline{x}_1} h_{23} - \partial^2_{\underline{x}_2} h_{13} = \partial_{\underline{x}_3}\partial_{\underline{x}_1} h_{21}$$

*which gives* 3 *more quadratic relations.*

*Proof.* Let us note that the first relation can be obtained as

$$\partial_{\underline{x}_2} h_{12} = \partial^2_{\underline{x}_2}\partial_{\underline{x}_1} g_1 - \partial_{\underline{x}_2}\partial^2_{\underline{x}_1} g_2 = \partial_{\underline{x}_1}(\partial^2_{\underline{x}_2} g_1 - \partial_{\underline{x}_1}\partial_{\underline{x}_2} g_2) = -\partial_{\underline{x}_1} h_{21},$$

and in a similar way one can obtain the other two relations. Next we have

$$\begin{aligned}\partial_{\underline{x}_1} a_1 &= \partial_{\underline{x}_2}\partial_{\underline{x}_3}\partial_{\underline{x}_1} g_1 + \partial_{\underline{x}_3}\partial_{\underline{x}_2}\partial_{\underline{x}_1} g_1 - \partial_{\underline{x}_2}\partial^2_{\underline{x}_1} g_3 - \partial_{\underline{x}_3}\partial^2_{\underline{x}_1} g_2 \\ &= \partial_{\underline{x}_2} h_{13} + \partial_{\underline{x}_3} h_{12}\end{aligned}$$

which, together with the two similar relations gives the linear syzygies. Even though there is no way to evaluate $\partial_{\underline{x}_1} a_1$ or $\partial_{\underline{x}_1} h_{12}$, we have that

$$\partial_{\underline{x}_3}\partial_{\underline{x}_1} h_{12} = \partial_{\underline{x}_3}\partial_{\underline{x}_1}\partial_{\underline{x}_2}\partial_{\underline{x}_1} g_1 - \partial_{\underline{x}_3}\partial^3_{\underline{x}_1} g_2, \tag{4.12}$$

$$\partial^2_{\underline{x}_1} a_3 = \{\partial_{\underline{x}_1}, \partial_{\underline{x}_2}\}\partial^2_{\underline{x}_1} g_3 - \partial_{\underline{x}_3}\partial^3_{\underline{x}_1} g_2 - \partial_{\underline{x}_3}\partial^2_{\underline{x}_1}\partial_{\underline{x}_2} g_1 \tag{4.13}$$

and subtracting (4.13) from (4.12) we obtain

$$\partial_{\underline{x}_3}\partial_{\underline{x}_1} h_{12} - \partial^2_{\underline{x}_1} a_3 = \{\partial_{\underline{x}_1}, \partial_{\underline{x}_2}\}\partial_{\underline{x}_3}\partial_{\underline{x}_1} g_1 - \{\partial_{\underline{x}_1}, \partial_{\underline{x}_2}\}\partial^2_{\underline{x}_1} g_3 = \{\partial_{\underline{x}_1}, \partial_{\underline{x}_2}\} h_{13}.$$

This together with its permutations leads to 6 quadratic relations of which only 3 are independent. Finally we have that

$$\begin{aligned}&\partial^2_{\underline{x}_1} h_{23} - \partial^2_{\underline{x}_2} h_{13} \\ &= \partial_{\underline{x}_3}\partial^2_{\underline{x}_1}\partial_{\underline{x}_2} g_2 - \partial^2_{\underline{x}_1}\partial^2_{\underline{x}_2} g_3 - \partial_{\underline{x}_3}\partial_{\underline{x}_1}\partial^2_{\underline{x}_2} g_1 + \partial^2_{\underline{x}_2}\partial^2_{\underline{x}_1} g_3 \\ &= \partial_{\underline{x}_3}\partial_{\underline{x}_1} h_{21} = -\partial_{\underline{x}_3}\partial_{\underline{x}_2} h_{12},\end{aligned}$$

which, together with its permutations gives 3 more relations. Now we have to verify that the relations we have found are independent. To this end, we can put the 12 relations in a matrix composed by two blocks obtained by taking the real counterpart of the linear and of the quadratic syzygies, respectively:

$$\begin{bmatrix} U_2 & U_1 & 0 & 0 & 0 & 0 & 0 & 0 \\ 0 & 0 & U_3 & U_1 & 0 & 0 & 0 & 0 \\ 0 & 0 & 0 & 0 & U_3 & U_2 & 0 & 0 \\ U_3 & 0 & U_2 & 0 & 0 & 0 & -U_1 & 0 \\ 0 & U_3 & 0 & 0 & U_1 & 0 & 0 & -U_2 \\ 0 & 0 & 0 & U_2 & 0 & U_1 & U_3 & U_3 \end{bmatrix}$$

and

$$\begin{bmatrix} \{U_1U_3\} & 0 & -U_2U_1 & 0 & 0 & 0 & 0 & U_1^2 \\ 0 & \{U_3U_2\} & 0 & U_1 & U_2 & 0 & -U_2^2 & 0 \\ 0 & 0 & 0 & -U_2U_3 & 0 & \{U_1U_3\} & 0 & U_3^2 \\ 0 & U_3^2 & 0 & -U_2^2 & 0 & U_1U_2 & 0 & 0 \\ U_3^2 & 0 & 0 & U_2U_1 & 0 & -U_1^2 & 0 & 0 \\ 0 & U_3U_1 & U_2^2 & 0 & -U_1^2 & 0 & 0 & 0 \end{bmatrix}.$$

Then we can consider the module generated by the rows of those two matrices. Using CoCoA it is possible to show that the number of generators of this module is 96 which corresponds to the number of rows of the two matrices. This shows that the relations are independent and since they are 12 they are the compatibility conditions of (4.11). □

Note that in the derivation of the above compatibility conditions, once again we made use of no property except the radial algebra relations, and as a result all the syzygies are radial. The syzygies found at the second step give the nonhomogeneous system formed by the 3 equations

$$\partial_{\underline{x}_i} h_{ji} + \partial_{\underline{x}_j} h_{ij} = R_k, \tag{4.14}$$

for $(i,j,k) = (2,1,3)$, $(2,3,1)$ or $(3,1,2)$ together with the 3 equations

$$\partial_{\underline{x}_i} h_{jk} + \partial_{\underline{x}_k} h_{ji} - \partial_{\underline{x}_j} a_k = S_k, \tag{4.15}$$

for $(i,j,k) = (1,2,3)$, $(2,3,1)$ or $(3,1,2)$. Then the 6 equations given by the permutations of $(i,j,k) = (1,2,3)$ in

$$\{\partial_{\underline{x}_i}, \partial_{\underline{x}_j}\} h_{jk} + \partial^2_{\underline{x}_j} a_k - \partial_{\underline{x}_k} \partial_{\underline{x}_j} h_{ji} = T_{ki} \tag{4.16}$$

and, finally, by the 6 permutations of $(i,j,k) = (1,2,3)$ in

$$\partial_{\underline{x}_i} \partial_{\underline{x}_j} h_{kj} + \partial^2_{\underline{x}_k} h_{ji} - \partial^2_{\underline{x}_j} h_{ki} = U_{kj}. \tag{4.17}$$

We also have the following constraint on $T_{ij}$ and $U_{ij}$:

$$T_{23} + T_{32} = \partial_{\underline{x}_1} S_1, \qquad T_{31} + T_{13} = \partial_{\underline{x}_2} S_2, \qquad T_{12} + T_{21} = \partial_{\underline{x}_3} S_3,$$

$$U_{23} + U_{32} = \partial_{\underline{x}_1} R_1, \qquad U_{31} + U_{13} = \partial_{\underline{x}_2} R_2, \qquad U_{12} + U_{21} = \partial_{\underline{x}_3} R_3.$$

These constraints reduce the total number of equations in the system to 12.

**Proposition 4.3.3.** *The* 8 *compatibility conditions of the system (4.14), (4.15), (4.16), (4.17) are given by*

1. *the 3 conditions obtained by the cyclic permutations of* $(1,2,3)$ *in the formula*
$$\partial_{\underline{x}_2} U_{13} - \partial_{\underline{x}_1} U_{32} + \partial^2_{\underline{x}_2} R_2 - \partial^2_{\underline{x}_3} R_3 = 0$$
*of which 2 are independent;*

2. *the 6 permutations of the relation*
$$\partial^2_{\underline{x}_1} S_2 + \partial_{\underline{x}_2} T_{23} - \partial_{\underline{x}_3} \partial_{\underline{x}_1} R_3 - \partial_{\underline{x}_1} U_{12} = 0.$$

*Proof.* As before, we can rewrite the system in a matrix form and check the independence of the 8 relations.

Their derivation again makes use of no property except the radial algebra commutation relations $[\partial_{\underline{x}_i}, \{\partial_{\underline{x}_j}, \partial_{\underline{x}_k}\}] = 0$. From these equations one readily obtains that
$$\partial_{\underline{x}_2} U_{13} = \partial^2_{\underline{x}_2} \partial_{\underline{x}_3} h_{13} + \partial_{\underline{x}_2} \partial^2_{\underline{x}_1} h_{32} - \partial^2_{\underline{x}_3} \partial_{\underline{x}_2} h_{12}$$
and
$$\partial_{\underline{x}_1} U_{32} = \partial^2_{\underline{x}_1} \partial_{\underline{x}_2} h_{32} + \partial^2_{\underline{x}_3} \partial_{\underline{x}_1} h_{21} - \partial^2_{\underline{x}_2} \partial_{\underline{x}_1} h_{31}$$
from which we obtain the first set of relations. The second set of relations follows from
$$\partial^2_{\underline{x}_1} S_2 = \partial^3_{\underline{x}_1} h_{23} + \partial^2_{\underline{x}_1} \partial_{\underline{x}_3} h_{21} - \partial^2_{\underline{x}_1} \partial_{\underline{x}_2} a_2$$
and
$$\begin{aligned}\partial_{\underline{x}_2} T_{23} &= \{\partial_{\underline{x}_3}, \partial_{\underline{x}_1}\} \partial_{\underline{x}_2} h_{12} + \partial^2_{\underline{x}_1} \partial_{\underline{x}_2} a_2 - \partial_{\underline{x}_1} \partial^2_{\underline{x}_2} h_{13} - \partial_{\underline{x}_3} \partial_{\underline{x}_1} R_3 \\ &= -\partial_{\underline{x}_3} \partial_{\underline{x}_1} \partial_{\underline{x}_2} h_{12} - \partial^2_{\underline{x}_1} \partial_{\underline{x}_3} h_{21} - \partial_{\underline{x}_1} U_{12} \\ &= -\partial_{\underline{x}_1} \partial_{\underline{x}_3} \partial_{\underline{x}_2} h_{12} - \partial^3_{\underline{x}_1} h_{23} + \partial_{\underline{x}_1} \partial^2_{\underline{x}_2} h_{13}\end{aligned}$$
which after addition leads to a total cancellation. □

Next we have to consider the nonhomogeneous system consisting of
$$\partial^2_{\underline{x}_i} S_j + \partial_{\underline{x}_j} T_{jk} - \partial_{\underline{x}_k} \partial_{\underline{x}_i} R_k - \partial_{\underline{x}_i} U_{ij} = B_{ij} \tag{4.18}$$
for $(i,j,k) = (1,2,3)$, $(3,1,2)$, $(2,3,1)$ and the 3 relations (in fact only 2 independent relations) obtained by
$$\partial_{\underline{x}_i} U_{kj} - \partial_{\underline{x}_k} U_{ji} + \partial^2_{\underline{x}_j} R_j - \partial^2_{\underline{x}_i} R_i = C_i, \tag{4.19}$$
for $(i,j,k) = (1,2,3)$, $(3,1,2)$, $(2,3,1)$ with the constraint $C_1 + C_2 + C_3 = 0$. We have the following constraints:
$$T_{12} + T_{21} = \partial_{\underline{x}_3} S_3, \quad U_{12} + U_{21} = \partial_{\underline{x}_3} R_3,$$
together with their cyclic permutations. For this system we have the following result.

**Proposition 4.3.4.** *The 3 compatibility conditions of system (4.18) and (4.19) are given by*

$$\{\partial_{\underline{x}_i}, \partial_{\underline{x}_j}\}C_k + \partial_{\underline{x}_j}\partial_{\underline{x}_i}C_j = \partial_{\underline{x}_k}\partial_{\underline{x}_j}B_{ij} + \partial^2_{\underline{x}_j}B_{ik} - \partial_{\underline{x}_k}\partial_{\underline{x}_i}B_{ji} - \partial^2_{\underline{x}_i}B_{jk}$$

*for* $(i,j,k) = (1,2,3),\ (3,1,2),\ (2,3,1)$.

*Proof.* The compatibility relations, in this case, are more difficult to establish. First, we have that

$$\partial_{\underline{x}_3}\partial_{\underline{x}_2}B_{12} = \partial_{\underline{x}_3}\partial_{\underline{x}_2}\partial^2_{\underline{x}_1}S_2 + \partial_{\underline{x}_3}\partial^2_{\underline{x}_2}T_{23} - \partial_{\underline{x}_3}\partial_{\underline{x}_2}\partial_{\underline{x}_3}\partial_{\underline{x}_1}R_3 - \partial_{\underline{x}_3}\partial_{\underline{x}_2}\partial_{\underline{x}_1}U_{12},$$

$$\partial^2_{\underline{x}_2}B_{13} = \partial^2_{\underline{x}_2}\partial^2_{\underline{x}_1}S_3 + \partial^2_{\underline{x}_2}\partial_{\underline{x}_3}T_{32} - \partial^3_{\underline{x}_2}\partial_{\underline{x}_1}R_1 - \partial^2_{\underline{x}_2}\partial_{\underline{x}_1}U_{13},$$

$$-\partial_{\underline{x}_3}\partial_{\underline{x}_1}B_{21} = -\partial_{\underline{x}_3}\partial_{\underline{x}_1}\partial^2_{\underline{x}_2}S_1 - \partial_{\underline{x}_3}\partial^2_{\underline{x}_1}T_{13} + \partial_{\underline{x}_3}\partial_{\underline{x}_1}\partial_{\underline{x}_3}\partial_{\underline{x}_2}R_3 + \partial_{\underline{x}_3}\partial_{\underline{x}_1}\partial_{\underline{x}_2}U_{21},$$

$$-\partial^2_{\underline{x}_1}B_{23} = -\partial^2_{\underline{x}_1}\partial^2_{\underline{x}_2}S_3 - \partial^2_{\underline{x}_1}\partial_{\underline{x}_3}T_{31} + \partial^2_{\underline{x}_1}\partial_{\underline{x}_2}R_1 + \partial^2_{\underline{x}_1}\partial_{\underline{x}_2}U_{23}.$$

Now, using the constraint

$$T_{12} + T_{21} = \partial_{\underline{x}_3}S_3$$

and its cyclic permutations, one can transform within the sum of the above terms, all the $T_{ij}$ into $S_k$ and then verify that all the $S_k$ cancel. Hence the sum of the above terms reduces to

$$\partial_{\underline{x}_3}\partial_{\underline{x}_2}B_{12} + \partial^2_{\underline{x}_2}B_{13} - \partial_{\underline{x}_3}\partial_{\underline{x}_1}B_{21} - \partial^2_{\underline{x}_1}B_{23}$$

$$= -\partial_{\underline{x}_3}\partial_{\underline{x}_2}\{\partial_{\underline{x}_3}, \partial_{\underline{x}_1}\}R_3 + \partial_{\underline{x}_3}\{\partial_{\underline{x}_2}, \partial_{\underline{x}_1}\}U_{21} - \partial^3_{\underline{x}_2}\partial_{\underline{x}_1}R_2 - \partial^2_{\underline{x}_2}\partial_{\underline{x}_1}U_{13}$$

$$+\partial_{\underline{x}_3}\partial_{\underline{x}_1}\partial_{\underline{x}_3}\partial_{\underline{x}_2}R_3 + \partial^3_{\underline{x}_1}\partial_{\underline{x}_2}R_1 - \partial^2_{\underline{x}_1}\partial_{\underline{x}_2}U_{32} + \partial_{\underline{x}_2}\partial^3_{\underline{x}_1}R_1, \tag{4.20}$$

where we have also used the constraints on the terms $U_{ij}$. Finally we have that

$$\{\partial_{\underline{x}_1}, \partial_{\underline{x}_2}\}C_3 = \{\partial_{\underline{x}_1}, \partial_{\underline{x}_2}\}\partial_{\underline{x}_3}U_{21} - \{\partial_{\underline{x}_1}, \partial_{\underline{x}_2}\}\partial_{\underline{x}_2}U_{13} + \partial^3_{\underline{x}_1}\partial_{\underline{x}_2}R_1$$

$$+\partial_{\underline{x}_2}\partial^3_{\underline{x}_1}R_1 - \partial_{\underline{x}_1}\partial_{\underline{x}_2}\partial^2_{\underline{x}_3}R_3 - \partial_{\underline{x}_2}\partial_{\underline{x}_1}\partial^2_{\underline{x}_3}R_3, \tag{4.21}$$

$$\partial_{\underline{x}_2}\partial_{\underline{x}_1}C_2 = \partial_{\underline{x}_2}\partial_{\underline{x}_1}\partial_{\underline{x}_2}U_{13} - \partial_{\underline{x}_2}\partial^2_{\underline{x}_1}U_{32} + \partial_{\underline{x}_2}\partial_{\underline{x}_1}\partial^2_{\underline{x}_3}R_3 - \partial^3_{\underline{x}_2}\partial_{\underline{x}_1}R_2, \tag{4.22}$$

and adding (4.21), (4.22) we get the the right hand side of (4.20). □

**Proposition 4.3.5.** *The compatibility condition of the system*

$$\{\partial_{\underline{x}_2}, \partial_{\underline{x}_3}\}C_1 + \partial_{\underline{x}_3}\partial_{\underline{x}_2}C_3 - \partial_{\underline{x}_1}\partial_{\underline{x}_3}B_{23} - \partial^2_{\underline{x}_3}B_{21} + \partial_{\underline{x}_1}\partial_{\underline{x}_2}B_{32} + \partial^2_{\underline{x}_2}B_{31} = E_1,$$

$$\{\partial_{\underline{x}_3}, \partial_{\underline{x}_1}\}C_2 + \partial_{\underline{x}_1}\partial_{\underline{x}_3}C_1 - \partial_{\underline{x}_2}\partial_{\underline{x}_1}B_{31} - \partial^2_{\underline{x}_1}B_{32} + \partial_{\underline{x}_2}\partial_{\underline{x}_3}B_{13} + \partial^2_{\underline{x}_3}B_{12} = E_2,$$

$$\{\partial_{\underline{x}_1}, \partial_{\underline{x}_2}\}C_3 + \partial_{\underline{x}_2}\partial_{\underline{x}_1}C_2 - \partial_{\underline{x}_3}\partial_{\underline{x}_2}B_{12} - \partial^2_{\underline{x}_2}B_{13} + \partial_{\underline{x}_3}\partial_{\underline{x}_1}B_{21} + \partial^2_{\underline{x}_1}B_{23} = E_3,$$

*is given by*

$$\partial_{\underline{x}_1}E_1 + \partial_{\underline{x}_2}E_2 + \partial_{\underline{x}_3}E_3 = 0.$$

*Proof.* It suffices to consider the three equations in the system together with the constraint $C_1 + C_2 + C_3 = 0$, from which it readily follows that

$$\partial_{\underline{x}_1} E_1 + \partial_{\underline{x}_2} E_2 + \partial_{\underline{x}_3} E_3 = 0. \qquad \square$$

This discussion implies the validity of the following theorem which answers problem 4.3.2.

**Theorem 4.3.2.** *The maps arising in the complex arising from the system*

$$\begin{cases} \partial_{x_1} f = g_1 \\ \partial_{x_2} f = g_2 \\ \partial_{x_3} f = g_3 \end{cases}$$

*are described in the Propositions 4.3.1 to 4.3.5, where the Dirac operators $\partial_{\underline{x}_1}$, $\partial_{\underline{x}_2}$, $\partial_{\underline{x}_3}$ are replaced by the abstract vector derivatives $\partial_{x_1}$, $\partial_{x_2}$, $\partial_{x_3}$.*

*Proof.* The statement follows from the fact that, in our computations to prove Propositions 4.3.1 to 4.3.5, we have used only the radial algebra relations. Therefore, one may define and compute the same complex if one replaces the Clifford–Dirac operators by abstract vector derivatives $\partial_{x_1}$, $\partial_{x_2}$, $\partial_{x_3}$ acting on an abstract radial algebra. $\square$

If one can prove that Conjecture 4.2.1 holds true, one can generalize the discussion carried out for $m = 6$ (and $m = 5$) to the case of a Clifford algebra over $m$ units, with $m \geq 5$. Note that, for the following result, we are not in need of the full validity of the Fischer decomposition, but only of the Fischer decomposition for polynomials in no more than three variables and low values of the degree.

**Proposition 4.3.6.** *If the first syzygies are at most of second order, then in the case of* 3 *Dirac operators in $\mathbb{C}_m$, $m \geq 5$, there can only be radial syzygies and the maps in the Dirac complex are described in the Propositions 4.3.1 to 4.3.5.*

*Proof.* Indeed, we have to study an equation of the form

$$\underline{x}_1 P_1(\underline{x}_2, \underline{x}_3) + \underline{x}_2 P_2(\underline{x}_3, \underline{x}_1) + \underline{x}_3 P_3(\underline{x}_1, \underline{x}_2) = 0$$

for general bilinear polynomials $P_j$ to which one may apply the Fischer decomposition

$$P_1(\underline{x}_2, \underline{x}_3) = M_1(\underline{x}_2, \underline{x}_3) + \underline{x}_2 P_{12}(\underline{x}_3) + \underline{x}_3 P_{13}(\underline{x}_2)$$

with $M_1$ left-monogenic in both variables $\underline{x}_2$, $\underline{x}_3$. Due to Fischer orthogonality it follows that

$$\underline{x}_1 M_1(\underline{x}_2, \underline{x}_3) + \underline{x}_3 M_3(\underline{x}_1, \underline{x}_2) + \underline{x}_2 M_2(\underline{x}_3, \underline{x}_1) = 0$$

which implies

$$M_1(\underline{x}_2, \underline{x}_3) = 0, \quad M_3(\underline{x}_1, \underline{x}_2) = 0, \quad M_2(\underline{x}_3, \underline{x}_1) = 0.$$

Next, we have to consider the equation

$$\underline{x}_1\underline{x}_2 P_{12}(\underline{x}_3) + \underline{x}_1\underline{x}_3 P_{13}(\underline{x}_2) + \underline{x}_3\underline{x}_1 P_{31}(\underline{x}_2)$$

$$+\underline{x}_3\underline{x}_2 P_{32}(\underline{x}_1) + \underline{x}_2\underline{x}_3 P_{23}(\underline{x}_1) + \underline{x}_2\underline{x}_1 P_{21}(\underline{x}_3) = 0,$$

where again one may write, e.g.,

$$P_{12}(\underline{x}_3) = M_{12}(\underline{x}_3) + \underline{x}_3 c_3$$

and in the case where the dimension $m \geq 5$ this ensures that the relation

$$\underline{x}_1\underline{x}_2 M_{12}(\underline{x}_3) + \underline{x}_1\underline{x}_3 M_{13}(\underline{x}_2) + \underline{x}_3\underline{x}_1 M_{31}(\underline{x}_2)$$

$$+\underline{x}_3\underline{x}_2 M_{32}(\underline{x}_1) + \underline{x}_2\underline{x}_3 M_{23}(\underline{x}_1) + \underline{x}_2\underline{x}_1 M_{21}(\underline{x}_3) = 0$$

has just the solution $M_{12} = \ldots = M_{32} = 0$. This shows that $P_{12}(\underline{x}_3) = \underline{x}_3 c_3$ holds, as well as all its permutations. Thus the syzygies must be radial. □

In all the cases in which the complex of three Dirac operators has only radial syzygies, we can consider the explicit description of the resolution of $M_3$, cokernel of $P^t = [U_1, U_2, U_3]$, which is

$$0 \longrightarrow R^{2^m}(-9) \xrightarrow{P} R^{3 \cdot 2^m}(-8) \longrightarrow R^{8 \cdot 2^m}(-6) \longrightarrow R^{6 \cdot 2^m}(-4) \oplus R^{6 \cdot 2^m}(-5)$$

$$\longrightarrow R^{8 \cdot 2^m}(-3) \longrightarrow R^{3 \cdot 2^m}(-1) \xrightarrow{P^t} R^{2^m} \longrightarrow M_3 \longrightarrow 0. \tag{4.23}$$

We have the following result.

**Theorem 4.3.3.** *The complex*

$$0 \longrightarrow R^{2^m}(-9) \xrightarrow{P} R^{3 \cdot 2^m}(-8) \longrightarrow R^{8 \cdot 2^m}(-6) \longrightarrow R^{6 \cdot 2^m}(-4) \oplus R^{6 \cdot 2^m}(-5)$$

$$\longrightarrow R^{8 \cdot 2^m}(-3) \longrightarrow R^{3 \cdot 2^m}(-1) \xrightarrow{P^t} R^{2^m} \longrightarrow 0, \tag{4.24}$$

*dual to resolution (4.23), is exact except the last spot, i.e.,* $\mathrm{Ext}^i(M_3, R) = 0$ *for* $0 \leq i \leq 4$.

*Proof.* Note that (4.24) starts (on the right) with the map $P^t$ and, if we add the cokernel of $P$ we obtain a minimal resolution of $M_3$ provided that none of the maps contains nonzero constants. By Proposition 1.1.4, (4.24) is isomorphic to the Hilbert resolution from which we started. As a consequence, the complex (4.24) is exact and, in particular, the Ext-modules $\mathrm{Ext}^j(M_3, R)$ vanish for $j = 0, \ldots, 4$. □

As a simple byproduct of our explicit description of the maps in the complex we obtain the following generalization of the Martineau–Harvey theorem to the Clifford setting.

**Corollary 4.3.1.** *For any compact set $K \subset \mathbb{R}^{3m}$ we have that $H^5_K(\mathbb{R}^{3m}, \mathcal{R})$ and $H^0(K, \mathcal{R})$ are strong dual to each other.*

*Proof.* It is a consequence of the description of the Dirac complex in the case in which there are only radial syzygies, which shows that last map in the complex is the transpose of the first one, together with Theorem 2.1.11. □

**Remark 4.3.3.** The previous description would allow one to construct also a theory of Clifford hyperfunctions in three variables, that, so far, is developed only in the case of a single variable, see [185].

### 4.3.3 Case of four operators

The case of systems of four Dirac operators in $\mathbb{C}_m$, $m \geq 4$, is more complicated, from a computational point of view. The symbols of such systems can be obtained by considering a matrix $P = [U_1, \dots, U_4]$, where, as usual, $U_r$ is the matrix symbol of the Dirac operator $\partial_{\underline{x}_r}$ in $\mathbb{C}_m$. The only resolutions we are currently able to entirely compute are those in $\mathbb{C}_m$, $m \leq 4$ that obviously correspond to the one predicted in the Cauchy–Fueter case, using suitable reductions. We have 28 quadratic syzygies at the first step, then 70, 84, 56, 20, and finally 3 linear syzygies. By running CoCoA or Macaulay2, it was impossible to compute the complete resolution for $m \geq 5$, because the computations involve algorithms so computationally expensive that they overcome the capabilities of the computer. It was however possible to compute partially the resolutions we are interested in.

In the case of $\mathbb{C}_6$, (see the previous subsection for the description of the matrices used) we obtained

$$R^{960}(-7) \longrightarrow R^{1344}(-6) \longrightarrow R^{160}(-4) \oplus R^{576}(-5) \longrightarrow$$
$$\longrightarrow R^{160}(-3) \oplus R^{40}(-4) \longrightarrow R^{32}(-1) \longrightarrow R^8 \longrightarrow M_4 \longrightarrow 0.$$

The case of $\mathbb{C}_5$ can be treated directly. In both the cases $m = 5, 6$ we have obtained 20 quadratic syzygies and 5 cubic at the first step and this shows that the cases of $\mathbb{C}_5$ and $\mathbb{C}_6$ are still exceptional in the case of 4 operators where by "exceptional" we mean that some relations do not arise from the radial algebra.

The first nonexceptional cases with 4 variables are the cases of $\mathbb{C}_7$ and $\mathbb{C}_8$, that have a similar behavior. We start with the case $m = 8$.

**Example 4.3.2.** The case of $\mathbb{C}_8$ can be dealt using the spinor formalism introduced in Section 2. The Dirac operators

$$\partial_{\underline{x}_r} = \sum_{i=1}^{8} \partial_{x_{ri}}, \quad r = 1, \dots, 4$$

lead to the complex equations

$$D_{X_r} f - i f D_{Y_r} = g_r,$$

where $f$ and $g_r$ assume values in $(\mathbb{C}_{4,+}) = \mathbb{C}_3$ and

$$D_{X_r} = \sum_{i=1}^{4} e_i \partial_{x_{ri}}, \quad D_{Y_r} = \sum_{i=1}^{4} e_i \partial_{y_{ri}}.$$

By taking the real counterpart of the symbol of this system, we get the matrix $P = [U_1, U_2, U_3, U_4]^t$ where

$$U_r^t = \begin{bmatrix} X_r & Y_r \\ -Y_r & X_r \end{bmatrix} \tag{4.25}$$

$$X_r = \begin{bmatrix} x_{r1} & x_{r2} & x_{r3} & x_{r4} & 0 & 0 & 0 & 0 \\ x_{r2} & -x_{r1} & 0 & 0 & x_{r3} & x_{r4} & 0 & 0 \\ x_{r3} & 0 & -x_{r1} & 0 & -x_{r2} & 0 & x_{r4} & 0 \\ x_{r4} & 0 & 0 & -x_{r1} & 0 & -x_{r2} & -x_{r3} & 0 \\ 0 & x_{r3} & -x_{r2} & 0 & x_{r1} & 0 & 0 & x_{r4} \\ 0 & x_{r4} & 0 & -x_{r2} & 0 & x_{r1} & 0 & -x_{r3} \\ 0 & 0 & x_{r4} & -x_{r3} & 0 & 0 & x_{r1} & x_{r2} \\ 0 & 0 & 0 & 0 & x_{r4} & -x_{r3} & x_{r2} & -x_{r1} \end{bmatrix}$$

$$Y_r = \begin{bmatrix} y_{r1} & -y_{r2} & -y_{r3} & -y_{r4} & 0 & 0 & 0 & 0 \\ y_{r2} & y_{r1} & 0 & 0 & -y_{r3} & -y_{r4} & 0 & 0 \\ y_{r3} & 0 & y_{r1} & 0 & y_{r2} & 0 & -y_{r4} & 0 \\ y_{r4} & 0 & 0 & y_{r1} & 0 & y_{r2} & y_{r3} & 0 \\ 0 & y_{r3} & -y_{r2} & 0 & y_{r1} & 0 & 0 & -y_{r4} \\ 0 & y_{r4} & 0 & -y_{r2} & 0 & y_{r1} & 0 & y_{r3} \\ 0 & 0 & y_{r4} & -y_{r3} & 0 & 0 & y_{r1} & -y_{r2} \\ 0 & 0 & 0 & 0 & y_{r4} & -y_{r3} & y_{r2} & y_{r1} \end{bmatrix}.$$

The resolution obtained in this case is

$$R^{1024}(-6) \longrightarrow R^{576}(-5) \oplus R^{320}(-4) \longrightarrow$$
$$\longrightarrow R^{320}(-3) \longrightarrow R^{64}(-1) \longrightarrow R^{16} \longrightarrow M_4 \longrightarrow 0$$

so that we have 20 quadratic syzygies at the first step, 20 linear plus 36 quadratic at the second step, and at least 64 quadratic syzygies at the third step. By direct computations, one finds the same number and degree of the syzygies also in the case $m = 7$.

On the base of the previous results concerning the Fischer decomposition we conjecture that:

**Conjecture 4.3.1.** *In a Clifford algebra over $m$ units, with $m \geq 2k - 1$, the resolution of the Dirac complex in the case of $k$ operators can only have syzygies consisting of radial polynomials. Moreover the complex can be computed in terms of abstract vector derivatives and remains the same for all values $m \geq 2k - 1$.*

# 4.4 Special systems in Clifford analysis

## 4.4.1 Generalized systems

In this section we will study systems of equations associated to some variations of the Dirac operator obtained by replacing $\sum_{i=1}^{m} e_i \partial_{x_i}$ with $\sum_{i=1}^{m} \underline{u}_i \partial_{x_i}$ where $\underline{u}_i$ are arbitrary elements in a Clifford algebra $\mathbb{C}_M$. In this treatment we restrict ourselves to the case in which the coefficients of the operators are Clifford 1-vectors. More generally, one may also consider the case with coefficients that are $k$-vectors, $k > 1$. Note that we do not even require that the number of derivatives in this second operator is equal to $M$, but we allow that it can be any integer $m$.

Let $x_1, \dots, x_m$ be scalar variables and let $e_1, \dots, e_M$ be a basis of a given Clifford algebra $\mathbb{C}_M$. Then we consider the vectors

$$\underline{u}_j^i = \sum_{l=1}^{M} u_{jl}^i e_l, \quad \text{for } i = 1, \dots n, \ j = 1, \dots, m,$$

to get operators of the form

$$D_{\underline{u}}^i = \sum_{j=1}^{m} \underline{u}_j^i \partial_{x_{ij}}. \tag{4.26}$$

Note that, apart from the choice of the parameters $u_{jl}^i$, the inhomogeneous system

$$D_{\underline{u}}^1 f = g_1, \dots, D_{\underline{u}}^n f = g_n$$

where $f, g_j : \mathbb{R}^{nm} \to \mathbb{C}_M$, $j = 1, \dots, n$, depends on three natural numbers, namely,

1. the total number $m$ of scalar variables,
2. the dimension of the Clifford algebra $\mathbb{C}_M$,
3. the number $n$ of operators considered.

We begin our study with a system of $n$ operators of the type (4.26):

$$\begin{cases} D^1 f := (\underline{u}_1^1 \partial_{x_{11}} + \underline{u}_2^1 \partial_{x_{12}} + \ldots + \underline{u}_m^1 \partial_{x_{1m}}) f = g_1 \\ \quad \cdots\cdots\cdots \\ D^n f := (\underline{u}_1^n \partial_{x_{n1}} + \underline{u}_2^n \partial_{x_{n2}} + \cdots + \underline{u}_m^n \partial_{x_{nm}}) f = g_n, \end{cases} \tag{4.27}$$

where $f : \mathbb{R}^{nm} \to \mathbb{C}_M$. Note that system (4.27) is written in terms of $nm$ scalar coordinates

$$x_{11}, \dots, x_{1m}, \dots, x_{n1} \dots, x_{nm}$$

and $M$ Clifford algebra generators $e_1, \dots, e_M$. In what follows we say that a system of $n$-differential operators in a Clifford algebra $\mathbb{C}_M$ is *Dirac-like* if its resolution behaves as the resolution of a system of $n$ Dirac operators in $\mathbb{C}_M$. Specifically, a resolution is said to be Dirac-like if it has the same length, the same degrees of the mappings, and Betti numbers proportional to the corresponding invariants of the resolution of $n$ Dirac operators. Let us denote by $T^i = [u^i_{jk}]$ the matrix of the coefficients $u^i_{jk}$ for every fixed index $i$. We have the following result:

**Theorem 4.4.1.** *The resolution of the system (4.27) is Dirac–like when $M \le m$ and $T^i$ is of maximal rank for every $i = 1, \dots, n$.*

*Proof.* For every fixed index $i$, let us consider the set of the $m$ variables $x_{i1}, \dots, x_{im}$ involved in the $i$-th equation of the system and let us rewrite the operator $D^i$ as follows:

$$D^i = \sum_{j=1}^{m} \underline{u}^i_j \partial_{x_{ij}} = \sum_{j=1}^{m} \sum_{k=1}^{M} e_k u^i_{jk} \partial_{x_{ij}} = \sum_{k=1}^{M} e_k \partial_{y_{ik}}$$

where we have set

$$\partial_{y_{ik}} = \sum_{j=1}^{m} u^i_{jk} \partial_{x_{ij}}.$$

The operators $\partial_{y_{ik}}$ are a set of independent partial derivatives in a linearly transformed space if and only if the $m \times M$ matrix $T^i = [u^i_{jk}]$ is of maximal rank. This assures that the system of operators

$$D^i = \sum_{k=1}^{M} e_k \partial_{y_{ik}}$$

behaves as a Dirac-like system in $M$ dimensions. □

**Remark 4.4.1.** In the case $M > m$ we cannot say whether or not the system is Dirac-like. With the use of CoCoA and for generic choices of the coefficients $u^i_{jl}$, we have explicitly written the resolutions in some particular cases (for the details on the procedure used, see the explicit description in the next subsection). For $n = 3, 4$, $M = 3, 4$, the resolutions are Dirac-like except the trivial case $m = 1$.

Let us now consider the following system.

$$\begin{cases} D^1 f = (\underline{u}^1_1 \partial_{x_1} + \underline{u}^1_2 \partial_{x_2} + \dots + \underline{u}^1_m \partial_{x_m}) f = g_1 \\ \quad \cdots\cdots\cdots\cdots \\ D^n f = (\underline{u}^n_1 \partial_{x_1} + \underline{u}^n_2 \partial_{x_2} + \dots + \underline{u}^n_m \partial_{x_m}) f = g_n \end{cases} \tag{4.28}$$

where, as above,

$$\underline{u}^i_j = \sum_{l=1}^{M} u^i_{jl} e_l$$

for $i = 1, \dots, n$, $j = 1, \dots, m$ and $f, g_i : \mathbb{R}^m \to \mathbb{C}_M$. In this case we have $m$ scalar coordinates $x_1, \dots, x_m$ and $M$ Clifford generators $e_1, \dots, e_M$. The matrix $T^i$ is defined as above. We have the following result.

**Theorem 4.4.2.** *The system (4.28) has a Dirac-like resolution when $m \geq nM$ and $T = [T^1, \ldots, T^n]^t$ is of maximal rank.*

*Proof.* For every $i = 1, \ldots, n$, the operator $D^i$ can be rewritten as follows:

$$D^i = \sum_{j=1}^{m} \underline{u}_j^i \partial_{x_j} = \sum_{j=1}^{m}\sum_{k=1}^{M} e_k u_{jk}^i \partial_{x_j} = \sum_{k=1}^{M} e_k \partial_{y_{ik}}$$

where we have set

$$\partial_{y_{ik}} = \sum_{j=1}^{m} u_{jk}^i \partial_{x_j}.$$

Thus, we have a set of $nM$ new partial derivatives $\partial_{y_{ik}}$ if and only if the matrix $T$ is of maximal rank. □

As in the previous case, when the condition $m \geq nM$ is not satisfied we cannot assure that the complex coming from system (4.28) is Dirac-like. If we fix the integers $n, M$, there are only a finite number of possibilities for the integer $m$ to be checked, so we can decide whether the resolution is Dirac-like or not by using CoCoA. To show that the behavior Dirac-like or non-Dirac-like are both possible, we have treated in detail the case $M = 4$, $n = 3$ so that, necessarily, $m < 12$. In this case, we will give some details of the construction of the complex. By taking the real components with respect to each unit of the equation $D^i f = g_i$, where $f, g_i : \mathbb{R}^m \to \mathbb{C}_4$, we obtain 16 real equations that can be written in the form $V^i(D)f = g_i$, where $f$ is a 16-tuple of real functions. The symbol of the previous system is the $16 \times 16$ polynomial matrix

$$V^i = \begin{bmatrix} A^i & B^i \\ -(B^i)^t & C^i \end{bmatrix} \tag{4.29}$$

where $A^i$, $B^i$, $C^i$ are given by

$$A^i = \begin{bmatrix} 0 & -P_1^i & -P_2^i & -P_3^i & -P_4^i & 0 & 0 & 0 \\ P_1^i & 0 & 0 & 0 & 0 & P_2^i & P_3^i & P_4^i \\ P_2^i & 0 & 0 & 0 & 0 & -P_1^i & 0 & 0 \\ P_3^i & 0 & 0 & 0 & 0 & 0 & -P_1^i & 0 \\ P_4^i & 0 & 0 & 0 & 0 & 0 & 0 & -P_1^i \\ 0 & -P_2^i & P_1^i & 0 & 0 & 0 & 0 & 0 \\ 0 & -P_3^i & 0 & P_1^i & 0 & 0 & 0 & 0 \\ 0 & -P_4^i & 0 & 0 & P_1^i & 0 & 0 & 0 \end{bmatrix},$$

$$B^i = \begin{bmatrix} 0 & 0 & 0 & 0 & 0 & 0 & 0 & 0 \\ 0 & 0 & 0 & 0 & 0 & 0 & 0 & 0 \\ P_3^i & P_4^i & 0 & 0 & 0 & 0 & 0 & 0 \\ -P_2^i & 0 & P_4^i & 0 & 0 & 0 & 0 & 0 \\ 0 & -P_2^i & -P_3^i & 0 & 0 & 0 & 0 & 0 \\ 0 & 0 & 0 & -P_3^i & -P_4^i & 0 & 0 & 0 \\ 0 & 0 & 0 & P_2^i & 0 & -P_4^i & 0 & 0 \\ 0 & 0 & 0 & 0 & P_2^i & P_3^i & 0 & 0 \end{bmatrix},$$

$$C^i = \begin{bmatrix} 0 & 0 & 0 & -P_1^i & 0 & 0 & -P_4^i & 0 \\ 0 & 0 & 0 & 0 & -P_1^i & 0 & P_3^i & 0 \\ 0 & 0 & 0 & 0 & 0 & -P_1^i & -P_2^i & 0 \\ P_1^i & 0 & 0 & 0 & 0 & 0 & 0 & P_4^i \\ 0 & P_1^i & 0 & 0 & 0 & 0 & 0 & -P_3^i \\ 0 & 0 & P_1^i & 0 & 0 & 0 & 0 & P_2^i \\ P_4^i & -P_3^i & P_2^i & 0 & 0 & 0 & 0 & -P_1^i \\ 0 & 0 & 0 & -P_4^i & P_3^i & -P_2^i & P_1^i & 0 \end{bmatrix}$$

and

$$P_r^i = \sum_{l=1}^{m} u_{li}^i x_l$$

for $r = 1, \dots, 4$. The three blocks $V^1$, $V^2$, $V^3$ form the matrix $V$ associated to the system. We have the following proposition.

**Proposition 4.4.1.** *For generic choices of the coefficients, the resolution of the system (4.28) for $M = 4$, $n = 3$, $5 \leq m < 12$ is Dirac-like. For $m = 2, 3, 4$ the resolutions are not Dirac-like.*

*Proof.* With the use of CoCoA, it suffices to consider the $48 \times 16$ matrix $V$ where

$$P_r^i = \sum_{l=1}^{m} u_{li}^i x_l$$

and $m = 2, \dots, 12$. Since we want $u_{jl}^i$ to vary in $\mathbb{R}$, the elements $u_{jl}^i$ are selected randomly in a given range of values by the computer with the use of the function `Rand( )`. The rows of $P$ generate an $R$-module $M$ whose resolution can be calculated by CoCoA with the command `Res(M)`. □

## 4.4.2 Systems using the Witt basis

In this section we will work in the Hermitian setting, i.e., in $\mathbb{C}_m$, $m$ even, with the Witt basis

$$f_1, \dots, f_m, \quad f_1', \dots, f_m'.$$

This leads to differential operators and systems that are invariant under a suitable action of the unitary group $U(m)$ acting as a subgroup of the spin group

Spin$(2m)$, where the space itself is equipped with a Hermitian inner product. This is the reason to call this a "Hermitian setting" (see also [152]).

Equivalently, one can introduce the basis $f_1, \ldots, f_m, f_1^+, \ldots, f_m^+$ where $f_i^+ = -f_i'$. We will make use of the splitting of the basis into

$$f_j = \frac{1}{2}(e_j - ie_{j+m}), \qquad f_j' = \frac{1}{2}(e_j + ie_{j+m}), \qquad j = 1, \ldots, m$$

so that one can write a 1–vector as

$$\underline{x} = \sum_{j=1}^{m} e_j X_j + e_{j+m} Y_j.$$

Using the Witt basis we obtain

$$\begin{aligned} \underline{x} &= \sum_{j=1}^{m} (f_j + f_j')X_j + i(f_j - f_j')Y_j \\ &= \sum_{j=1}^{m} f_j(X_j + iY_j) + f_j'(X_j - iY_j) = Z - Z^+ \end{aligned}$$

where we have set

$$Z = \sum_{j=1}^{m} f_j(X_j + iY_j), \qquad Z^+ = \sum_{j=1}^{m} f_j^+(X_j - iY_j).$$

Let us consider the Dirac operator $\partial_{\underline{x}}$: it can be written as

$$\begin{aligned} \partial_{\underline{x}} &= \sum_{j=1}^{m} e_j \partial_{x_j} = \sum_{j=1}^{m} e_j \partial_{X_j} + e_{j+m} \partial_{Y_j} \\ &= \sum_{j=1}^{m} (f_j - f_j^+)\partial_{X_j} + i(f_j + f_j^+)\partial_{Y_j} \\ &= \sum_{j=1}^{m} f_j(\partial_{X_j} + i\partial_{Y_j}) - \sum_{j=1}^{m} f_j^+(\partial_{X_j} - i\partial_{Y_j}) \end{aligned}$$

which we now rewrite in the form

$$\partial_{\underline{x}} = 2(\partial_{Z^+} - \partial_Z)$$

where

$$\partial_Z = \sum_{j=1}^{m} f_j^+ \partial_{z_j}, \qquad \partial_{Z^+} = \sum_{j=1}^{m} f_j \partial_{\bar{z}_j},$$

and

$$\partial_{z_j} = \frac{1}{2}(\partial_{X_j} - i\partial_{Y_j}), \qquad \partial_{\bar{z}_j} = \frac{1}{2}(\partial_{X_j} + i\partial_{Y_j}),$$

where $\bar{Z}_j$ denotes the complex conjugation. Below, we will study systems involving an even number of operators of the form

$$\partial_{X_j} = \sum_{\ell=1}^{m} f_j \partial_{x_{j\ell}}, \qquad \partial_{Y_j} = \sum_{\ell=1}^{m} f_j' \partial_{y_{j\ell}}, \tag{4.30}$$

where $x_{j\ell}$, $y_{j\ell}$ are real variables. Since those variables are treated as symbols, resolutions involving the operators above are fully equivalent to the corresponding resolutions with the operators $\partial_{Z_j}$, $\partial_{\bar{Z}_j}$.

**The case $\mathbb{R}_{1,1}$.** Let us begin our study with the simplest case, i.e. the case of the Clifford algebra $\mathbb{R}_{1,1}$. The operators

$$\partial_{X_j} = f_1 \partial_{x_{j1}}, \quad \partial_{Y_j} = f_1' \partial_{y_{j1}}$$

acting on $\mathbb{R}_{1,1}$-valued functions, give rise respectively to the matrices

$$U_j(D) = \begin{bmatrix} -\partial_{x_{j1}} & 0 & 0 & 0 \\ 0 & 0 & \partial_{x_{j1}} & 0 \end{bmatrix}$$

$$V_j(D) = \begin{bmatrix} 0 & -\partial_{y_{j1}} & 0 & 0 \\ \partial_{y_{j1}} & 0 & 0 & -\partial_{y_{j1}} \end{bmatrix}$$

acting on $f = [f_0, f_1, f_2, f_3]$. The systems we are considering are represented by a matrix obtained by combining an even number of matrices of the type $U_j$, $V_j$, i.e., systems of the type

$$\begin{bmatrix} U_1(D) \\ V_1(D) \\ \dots \\ U_N(D) \\ V_N(D) \end{bmatrix} f = A_N(D) f = g, \qquad N \geq 1. \tag{4.31}$$

**Proposition 4.4.2.** *The system (4.31) has a minimal resolution of the following type:*

$$0 \longrightarrow R^4(-1) \longrightarrow < A_1^t > \longrightarrow 0, \qquad \text{if} \quad N = 1,$$

$$0 \longrightarrow R^4(-2) \longrightarrow R^8(-1) \longrightarrow < A_2^t > \longrightarrow 0, \qquad \text{if} \quad N = 2,$$

$$0 \longrightarrow R^4(-3) \longrightarrow R^{12}(-2) \longrightarrow R^{12}(-1) \longrightarrow < A_3^t > \longrightarrow 0, \qquad \text{if} \quad N = 3,$$

$$0 \longrightarrow R^4(-4) \longrightarrow R^{16}(-3) \longrightarrow R^{24}(-2)$$
$$\longrightarrow R^{16}(-1) \longrightarrow < A_4^t > \longrightarrow 0, \qquad \text{if} \quad N = 4,$$

*where $< A_N^t >$ denotes the module generated by the columns of the matrices $A_N^t$.*

*Proof.* The proof is based on a computation with CoCoA. □

**Remark 4.4.2.** If we consider systems with a different number of operators of the type $\partial_{X_j}$ and $\partial_{Y_j}$, we again get linear resolutions. For example, the system involving $\partial_{X_1}$, $\partial_{Y_1}$, $\partial_{Y_2}$, $\partial_{Y_3}$ and the systems involving $\partial_{X_1}$, $\partial_{X_2}$, $\partial_{Y_1}$, $\partial_{Y_2}$, $\partial_{Y_3}$, $\partial_{Y_4}$ associated, respectively, to the matrices that we denote by $A'_2(D)$, $A'_3(D)$, have the following minimal resolutions:

$$0 \longrightarrow R^2(-3) \longrightarrow R^6(-2) \longrightarrow R^8(-1) \longrightarrow < {A'_2}^t > \longrightarrow 0,$$

$$0 \longrightarrow R^2(-4) \longrightarrow R^8(-3) \longrightarrow R^{14}(-2) \longrightarrow R^{12}(-1) \longrightarrow < {A'_3}^t > \longrightarrow 0.$$

Let us come back to the case of system (4.31): we can prove that for all $N \geq 1$ the first syzygies are linear. For the notations used in the following proofs, we refer the reader to Chapter 1, Section 1.1.2. We begin by characterizing a reduced Gröbner basis of $\langle A_N^t \rangle$.

**Proposition 4.4.3.** *A reduced Gröbner basis $\mathcal{G}$ of $\langle A_N^t \rangle$ contains the columns of $A_N^t$ and, the vectors $[0,\ 0,\ 0,\ x_i y_k]^t$ for all $i, k = 1, \dots, N$.*

*Proof.* The $S$-polynomial of any two columns of $U_j^t$ or $V_j^t$ are zero, while the $S$-polynomial of the first column of $U_i^t$ and the second column of $V_k^t$ gives rise to the element $[0,\ 0,\ 0,\ x_i y_k]^t$ that cannot be reduced to zero. Then, by the Buchberger algorithm a Gröbner basis of $\langle A_N^t \rangle$ has the properties stated and contains $r = 4N + N^2$ elements. □

We get the following result.

**Theorem 4.4.3.** *The first syzygies of the module $\langle A_N^t \rangle$ are linear for all $N \geq 1$.*

*Proof.* We can compute $\mathrm{Syz}(\langle A_N^t \rangle)$, i.e., the syzygies of $\langle A_N^t \rangle$, if we know the syzygies for a Gröbner basis of $\langle A_N^t \rangle$. Let us denote by $g_1, \dots, g_r \in R^4$ the elements in the Gröbner basis $\mathcal{G}$ found above and by $G$ the matrix $G = [g_1, \dots, g_r]$ whose columns are the elements $g_i$. Then, from the general theory (see again Chapter 1, Section 1.1.2), there exists a $r \times 4N$ matrix $S$ and a $4N \times r$ matrix $T$ such that $A^t = GS$ and $G = A^t T$. If we order the columns in $G$ to have at the places from 1 to $4N$ the columns of $A^t$ with the same order as in $A^t$, then we easily have that

$$A^t = G \begin{bmatrix} I \\ O \end{bmatrix}$$

where $I$ and $O$ are, respectively, the identity matrix and the zero matrix of size $4N$. In an analogous way, we have that $G = A^t[I|L]$ where $L$ is a suitable $4N \times N^2$ matrix that can be determined as follows. $A^t$ is a block matrix of the form $A^t = [U_1^t\ V_1^t \dots U_N^t\ V_N^t]$. The matrix $A^t L$ must have columns of the form $[0,\ 0,\ 0,\ x_i y_k]^t$, i.e., $(A^t L)_{rs} = 0$ if $r \neq 4$ and $(A^t L)_{4s} = x_i y_k$. To get the element $x_i y_k$ at the place $(4, s)$ it suffices to multiply $A^t$ by a matrix $L$ having the $s$-th column such that $L_{4k-1,s} = -x_i$ and $L_{4(i-1)+1,s} = -y_k$ and all the other elements equal zero. Note that $L_{4k-1,s}$ is the element in the $s$-th column at the fourth place of the $k$-th block $[U_k^t\ V_k^t]$ while $L_{4(i-1)+1,s}$ is the first element in the $i$-th block. As $TS = I$, if we know that $s_1, \dots, s_t$ are generators for $\text{syz}(\langle G\rangle)$, then $\text{syz}(\langle A_N^t\rangle) = \langle Ts_1, \dots, Ts_t, r_1, \dots, r_s\rangle \subseteq R^{4N}$ where $r_1, \dots, r_s$ are the columns of the matrix $I - TS = O$. In this case we have $I - TS = O$, so that $\text{syz}(\langle A_N^t\rangle)) = \langle Ts_1, \dots, Ts_t\rangle$.

The $S$-polynomial of two elements $g_i$, $g_j$ in the Gröbner basis is given by

$$S(g_i, g_j) = \frac{X_{ij}}{X_i} g_i - \frac{X_{ij}}{X_j} g_j$$

where $X_i$, $X_j$ are the leading monomials of $g_i$, $g_j$, respectively, and

$$X_{ij} = \text{lcm}(X_i, X_j).$$

We have that $S(g_i, g_j) = \sum_{\ell=1}^{r} h_{ij\ell} g_\ell$ and we can define

$$s_{ij} = \frac{X_{ij}}{X_i} C_i - \frac{X_{ij}}{X_j} C_j - (h_{ij1}, \dots, h_{ijr}) \in R^r$$

where $C_i$ is the $i$-th column of the identity matrix. The set $\{s_{ij}\}$ is, by construction, a generating set for $\text{syz}(\langle G\rangle)$. Now note that for $i, j \le 4N$, $s_{ij}$ is the $S$-polynomial of two columns of $A_N^t$, so either $s_{ij} = 0$ or $s_{ij}$ is a linear combination of elements $g_\ell$ with $\ell \le 4N$. Moreover, $s_{ij}$ has zero elements at the places from $4N + 1$ up to $r$. The $S$-polynomials of $g_i$, $g_j$ with $i, j \ge 4N$ (that are two elements of the type $[0,\ 0,\ 0,\ x_s y_k]^t$) can be easily computed and they are zero. Finally, if $i \le 4N$ and $j > 4N$, then the $S$-polynomials of $g_i$, $g_j$ are either zero or they reduce to zero. Then, by the discussion above, if $s_{ij} \ne 0$ it has linear entries and it has zero elements starting from the place $4N+1$. Then $Ts_{ij} = s_{ij}$ and the first syzygies are linear. □

**The case $\mathbb{R}_{2,2}$.** The Clifford algebra $\mathbb{R}_{m,m}$ has real dimension $2^{2m}$, so when considering the operators (4.30) in the setting of $\mathbb{R}_{m,m}$ with $m \ge 2$, it is more convenient to use a different formalism that allows us to reduce the dimension of the problem. We will consider operators of the type (4.30) acting on functions with values in $\mathbb{R}_{m,m} I = \mathbb{R}_m I$ where $I = f_1' f_1 \cdots f_m' f_m$ is a primitive idempotent. Instead of $\mathbb{R}_m I$ it is then sufficient to consider functions with values in $\mathbb{R}_{m,+} I$ so that the Witt basis elements can be identified with the element of $\text{End}(\mathbb{R}_{m,+})$ given by

$$f_j : a \to \frac{1}{2}(e_j a - a e_j), \qquad f_j' : a \to \frac{1}{2}(e_j a + a e_j).$$

The system of two equations

$$\begin{cases} \sum_{\ell=1}^{m} f_\ell \partial_{x_\ell} f = g \\ \sum_{\ell=1}^{m} f_\ell' \partial_{y_\ell} f = h, \end{cases}$$

where $f$ is $\mathbb{R}_{m,+}$-valued and $g, h$ are $\mathbb{R}_{m,-}$-valued, can be written in the following form

$$\begin{cases} \partial_{\underline{x}} f - f\partial_{\underline{x}} = g \\ \partial_{\underline{y}} f + f\partial_{\underline{y}} = h \end{cases} \tag{4.32}$$

where now

$$\partial_{\underline{x}} = \sum_{\ell=1}^{m} e_\ell \partial_{x_\ell}, \quad \partial_{\underline{y}} = \sum_{\ell=1}^{m} e_\ell \partial_{y_\ell}$$

are Dirac operators and the number of equations is minimal.

Let us consider the case $m = 2$ and more than two equations. Using the formalism described above we can study the system

$$\begin{cases} \partial_{\underline{x}_j} f - f\partial_{\underline{x}_j} = g_j \\ \partial_{\underline{y}_j} f + f\partial_{\underline{y}_j} = h_j \end{cases} \tag{4.33}$$

where

$$\partial_{x_j} = \sum_{\ell=1}^{2} e_\ell \partial_{x_{j\ell}}, \quad \partial_{y_j} = \sum_{\ell=1}^{2} e_\ell \partial_{y_{j\ell}}$$

and $f$ is $\mathbb{R}_{2,+}$-valued, i.e., $f = f_0 + f_{12}e_1e_2$. In matrix form this system can be written as

$$\begin{bmatrix} 0 & -\partial_{x_{j1}} \\ 0 & \partial_{x_{j2}} \\ \partial_{y_{j1}} & 0 \\ \partial_{y_{j2}} & 0 \end{bmatrix} \begin{bmatrix} f_0 \\ f_{12} \end{bmatrix} = \begin{bmatrix} U_j(D) \\ V_j(D) \end{bmatrix} \begin{bmatrix} f_0 \\ f_{12} \end{bmatrix} = \begin{bmatrix} g_{j1} \\ g_{j2} \\ h_{j1} \\ h_{j2} \end{bmatrix}.$$

More generally, if we consider systems of the form

$$\begin{bmatrix} U_1(D) \\ V_1(D) \\ \cdots \\ U_N(D) \\ V_N(D) \end{bmatrix} \begin{bmatrix} f_0 \\ f_{12} \end{bmatrix} = A_N(D) \begin{bmatrix} f_0 \\ f_{12} \end{bmatrix} = \begin{bmatrix} g_{11} \\ g_{12} \\ \cdots \\ h_{N1} \\ h_{N2} \end{bmatrix} \qquad N \geq 1, \tag{4.34}$$

we get, using CoCoA, the following result

**Proposition 4.4.4.** *A minimal free resolution of the system (4.34) is*

$$0 \longrightarrow R^2(-2) \longrightarrow R^4(-1) \longrightarrow < A_1^t > \longrightarrow 0 \qquad \text{if} \quad N = 1,$$

$$0 \longrightarrow R^2(-4) \longrightarrow R^8(-3) \longrightarrow R^{12}(-2)$$
$$\longrightarrow R^8(-1) \longrightarrow < A_2^t > \longrightarrow 0 \qquad \text{if} \quad N = 2,$$

$$0 \longrightarrow R^2(-6) \longrightarrow R^{12}(-5) \longrightarrow R^{30}(-4) \longrightarrow R^{40}(-3) \longrightarrow R^{30}(-2)$$
$$\longrightarrow R^{12}(-1) \longrightarrow < A_3^t > \longrightarrow 0 \qquad \text{if} \quad N = 3.$$

**Remark 4.4.3.** A natural question that arises is whether the corresponding resolution is linear for every value of $N$. Below we will show that this is the case.

**Proposition 4.4.5.** *A reduced Gröbner basis of the R-module* $< A_N^t >$ *is given by the columns of* $A_N^t$.

*Proof.* The columns of $A_N^t$ belong to a Gröbner basis of $< A_N^t >$. The S-polynomial of any two columns of $A_N^t$ is zero. The statement follows. □

**Proposition 4.4.6.** *All the syzygy modules in a minimal resolution of* $A_N$ *are generated by linear polynomials.*

*Proof.* Let us denote by $\mathrm{reg}(\langle A_N \rangle)$ the Castelnuovo–Mumford regularity of $(\langle A_N \rangle)$. It is well known that (see [14]) $\mathrm{reg}(\langle A_N \rangle) \leq \mathrm{reg}(Lt(A_N))$. Since

$$Lt(A_N) = \langle x_i e_2, y_j e_1 \rangle_{1 \leq i,j \leq N}$$

it follows that $\mathrm{reg}(Lt(A_N))$=$\mathrm{reg}(I_N)$ where $I_N$ is the ideal

$$I_N = \langle x_i, y_j \rangle_{1 \leq i,j \leq N}.$$

$I_N$ is a maximal ideal in the ring $R$, its resolution is linear and its regularity index is 1. It follows that $\mathrm{reg}(Lt(A_N)) = 1$ and the statement follows. □

**The case** $\mathbb{R}_{3,3}$. Using the reduction described in the previous subsection, we consider the system (4.32) where, in this case, $\partial_{\underline{x}_j}$ and $\partial_{\underline{y}_j}$ are three-dimensional Dirac operators and $f = f_0 + f_{12}e_1e_2 + f_{13}e_1e_3 + f_{23}e_2e_3$ is $\mathbb{R}_{3,+}$-valued. In matrix form, if we denote $f = [f_0, f_{12}, f_{13}, f_{23}]$ (and analogously $g_j$, $h_j$ will be the data) the system can be written as

$$\begin{bmatrix} 0 & \partial_{x_{j2}} & \partial_{x_{j3}} & 0 \\ 0 & -\partial_{x_{j1}} & 0 & \partial_{x_{j3}} \\ 0 & 0 & -\partial_{x_{j1}} & -\partial_{x_{j2}} \\ \partial_{y_{j1}} & 0 & 0 & 0 \\ \partial_{y_{j2}} & 0 & 0 & 0 \\ \partial_{y_{j3}} & 0 & 0 & 0 \\ 0 & \partial_{y_{j3}} & -\partial_{y_{j2}} & \partial_{y_{j1}} \end{bmatrix} f = \begin{bmatrix} U_j(D) \\ V_j(D) \end{bmatrix} f = \begin{bmatrix} g_j \\ h_j \end{bmatrix}, \quad j = 1, \dots, N.$$

As above we can look at the resolutions of the whole system

$$\begin{bmatrix} U_1(D) \\ V_1(D) \\ \dots \\ U_N(D) \\ V_N(D) \end{bmatrix} f = A_N(D) f = \begin{bmatrix} g_1 \\ h_1 \\ \dots \\ g_N \\ h_N \end{bmatrix} \qquad N \geq 1, \tag{4.35}$$

and with the use of CoCoA we obtain the following result.

**Proposition 4.4.7.** *A minimal resolution of the system (4.35) is of the type*

$$0 \longrightarrow R(-3) \longrightarrow R^4(-2) \longrightarrow R^7(-1) \longrightarrow < A_1^t > \longrightarrow 0, \qquad \text{if} \quad N = 1,$$

$$0 \longrightarrow R(-6) \longrightarrow R^6(-5) \longrightarrow R^{15}(-4) \oplus R(-6)$$

$$\longrightarrow R^{20}(-3) \oplus R^7(-4) \longrightarrow R^{18}(-2) \oplus R^8(-3)$$

$$\longrightarrow R^{14}(-1) \longrightarrow < A_2^t > \longrightarrow 0, \qquad \text{if} \quad N = 2.$$

**Remark 4.4.4.** Note that a Gröbner basis of the module generated by the columns of $A_1^t$ contains only the columns of $A_1^t$ while the Gröbner basis of $A_2^t$ contains the columns of $A_2^t$ and four elements with quadratic entries, three of which appear as S-polynomials of columns in the matrices of type $U_1^t$, $U_2^t$, and one appears as S-polynomials of two columns in $V_1^t$, $V_2^t$.

**Remark 4.4.5.** When we consider systems with $N \geq 3$ like e.g., those associated to $\partial_{X_1}, \partial_{X_2}, \partial_{X_3}, \partial_{Y_1}$, or to $\partial_{X_1}, \partial_{Y_1}, \partial_{Y_2}, \partial_{Y_3}$, the problem becomes even more complicated. We mention here only the resolutions of the last two cases. The first system has the resolution

$$0 \longrightarrow R(-7) \longrightarrow R^{12}(-5) \longrightarrow R(-3) \oplus R^{25}(-4) \longrightarrow$$

$$R^9(-2) \oplus R^{15}(-3) \longrightarrow R^{13}(-1) \longrightarrow < A_2'^t > \longrightarrow 0$$

while the second has resolution

$$0 \longrightarrow R(-9) \longrightarrow R^9(-8) \longrightarrow R^{36}(-7) \longrightarrow R^{84}(-6)$$

$$\longrightarrow R^{126}(-5) \longrightarrow R^{126}(-4) \longrightarrow R^{84}(-3) \oplus R(-5)$$

$$\longrightarrow R^{37}(-2) \oplus R^3(-3) \longrightarrow R^{15}(-1) \longrightarrow < A_2'^t > \longrightarrow 0.$$

Using the usual reduction we consider the system (4.33) where $\partial_{\underline{x}_j}$ and $\partial_{\underline{y}_j}$ are four-dimensional Dirac operators and the function

$$f = f_0 + f_{12}e_1e_2 + f_{13}e_1e_3 + f_{14}e_1e_4 +$$

$$f_{23}e_2e_3 + f_{24}e_2e_4 + f_{34}e_3e_4 + f_{1234}e_1e_2e_3e_4$$

is $\mathbb{R}_{4,+}$-valued. In matrix form, if we denote

$$f = [f_0, f_{12}, f_{13}, f_{14}, f_{23}, f_{24}, f_{34}, f_{1234}],$$

and with $G_j$ the datum, the system can be written as

$$\begin{bmatrix}
0 & \partial_{x_{j2}} & \partial_{x_{j3}} & \partial_{x_{j4}} & 0 & 0 & 0 & 0 \\
0 & -\partial_{x_{j1}} & 0 & 0 & \partial_{x_{j3}} & \partial_{x_{j4}} & 0 & 0 \\
0 & 0 & -\partial_{x_{j1}} & 0 & -\partial_{x_{j2}} & 0 & \partial_{x_{j4}} & 0 \\
0 & 0 & 0 & -\partial_{x_{j1}} & 0 & -\partial_{x_{j2}} & -\partial_{x_{j3}} & 0 \\
0 & 0 & 0 & 0 & 0 & 0 & 0 & -\partial_{x_{j1}} \\
0 & 0 & 0 & 0 & 0 & 0 & 0 & \partial_{x_{j2}} \\
0 & 0 & 0 & 0 & 0 & 0 & 0 & -\partial_{x_{j3}} \\
0 & 0 & 0 & 0 & 0 & 0 & 0 & -\partial_{x_{j4}} \\
\partial_{y_{j1}} & 0 & 0 & 0 & 0 & 0 & 0 & 0 \\
\partial_{y_{j2}} & 0 & 0 & 0 & 0 & 0 & 0 & 0 \\
\partial_{y_{j3}} & 0 & 0 & 0 & 0 & 0 & 0 & 0 \\
\partial_{y_{j4}} & 0 & 0 & 0 & 0 & 0 & 0 & 0 \\
0 & \partial_{y_{j3}} & -\partial_{y_{j2}} & 0 & \partial_{y_{j1}} & 0 & 0 & 0 \\
0 & \partial_{y_{j4}} & 0 & -\partial_{y_{j2}} & 0 & \partial_{y_{j1}} & 0 & 0 \\
0 & 0 & \partial_{y_{j4}} & -\partial_{y_{j3}} & 0 & 0 & \partial_{y_{j1}} & 0 \\
0 & 0 & 0 & 0 & \partial_{y_{j4}} & -\partial_{y_{j3}} & \partial_{y_{j2}} & 0
\end{bmatrix} f$$

$$:= \begin{bmatrix} U_j(D) \\ V_j(D) \end{bmatrix} f = G_j.$$

When considering systems of the type (4.35) where $U_j(D)$ and $V_j(D)$ are as above, the situation may become complicated very soon.

**Proposition 4.4.8.** *The system of type (4.35), where $U_j(D)$ and $V_j(D)$ are as above, has the resolution*

$$0 \longrightarrow R^2(-4) \longrightarrow R^8(-3) \longrightarrow R^{14}(-2)$$
$$\longrightarrow R^{16}(-1) \longrightarrow < D_1 > \longrightarrow 0, \quad \text{for } N = 1;$$
$$0 \longrightarrow R^2(-8) \longrightarrow R^{16}(-7) \longrightarrow R^{56}(-6) \longrightarrow R^{112}(-5)$$
$$\longrightarrow R^{140}(-4) \oplus R^2(-6) \longrightarrow R^{112}(-3) \oplus R^{14}(-4)$$
$$\longrightarrow R^{62}(-2) \oplus R^{16}(-3) \longrightarrow R^{32}(-1) \longrightarrow < D_2 > \longrightarrow 0, \quad \text{for } N = 2.$$

The case $m = 5$ can be studied with the techniques above but the matrices $U_j(D)$, $V_j(D)$ are of size $32 \times 16$ and $32 \times 15$, respectively, so $A_N(D)$ has size $32 \times 31N$. All those matrices are too big to be written here and we write, for simplicity, only the the resolution in the case $N = 1$ which is

$$0 \longrightarrow R(-5) \longrightarrow R^6(-4) \longrightarrow R^{16}(-3) \longrightarrow R^{26}(-2)$$
$$\longrightarrow R^{31}(-1) \longrightarrow \langle A_1^t \rangle \longrightarrow 0.$$

When $N > 1$ the resolutions are very complicated. To give an example, in the case $N = 2$ the first step of the resolutions is

$$R^{93}(-2) \oplus R^{30}(-3) \oplus R^{30}(-4) \oplus R^6(-5) \longrightarrow R^{62}(-1) \longrightarrow \langle A_2^t \rangle \longrightarrow 0,$$

which means that the first syzygies have degrees which go from 1 to 4. Despite this complication, the resolution becomes linear at the 7-th step and ends as

$$0 \longrightarrow R(-10) \longrightarrow R^{10}(-9) \longrightarrow R^{45}(-8).$$

**Proposition 4.4.9.** *Let $A_N(D)f = 0$ be one of the systems of the type (4.35) in $\mathbb{R}_{m,m}$ and let $(\nu, m) \in \{(2,1),(4,2),(6,3),(8,4)\}$. Let $K$ be a compact convex set in $\mathbb{R}^{\nu N}$, $N \geq 1$, and let $f$ be a solution of $A_N(D)f = 0$ on $\mathbb{R}^{\nu N} \setminus K$. Then $f$ extends to a solution of the system on $\mathbb{R}^{\nu N}$ for all $N \geq 1$ if $m = 2, 3, 4$, and for all $N > 1$ if $m = 1$.*

*Proof.* Let us consider the case $m = 1$. The proof follows from Lemma 2.1.1 and Theorem 2.1.15. In fact, the matrix $A_N$ in (4.31) as maximal rank and its $4 \times 4$ minors are relatively prime. If $m > 1$, then the statement holds for $N \geq 1$ and the proof rests on the same arguments. Note that for $N = 1$ we still have nonzero minors of maximal size involving different variables so that their gcd is 1. □

### 4.4.3 Combinatorial systems

All the systems we have treated in the Clifford analysis setting have produced either a Dirac-like resolution or a behavior that does not reflect any known resolution. We will introduce some systems that we call of combinatorial type, and are often associated to a Koszul-like resolution.

The systems of combinatorial type are constructed starting from some incidence structures that correspond to "finite geometries." The "finite geometries" consist of a set of points, also called *tops*, $\{p_1, \dots, p_m\}$, and a collection of lines or *blocks* $\{b_1, \dots, b_n\}$ where every block $b_j$ is a subset of $\{p_1, \dots, p_m\}$. The systems we have in mind consist of operators of the type $\sum_{jk} \pm e_j \partial_{x_k}$ where $x_1, \dots, x_m$ is a set of $m$ scalar coordinates, and $e_1, \dots, e_M$ are generators of the real Clifford algebra $\mathbb{R}_M$, according to the following axioms:

- $(A_1)$ each operator is of the type $\sum_{jk} \pm e_j \partial_{x_k}$;
- $(A_2)$ every partial derivative $\partial_{x_j}$ occurs at most once in a given operator (within a term $\pm e_j \partial_{x_k}$);
- $(A_3)$ every basis element $e_k$ occurs at most once in a given operator;
- $(A_4)$ every term $e_j \partial_{x_k}$ occurs at most once in the whole system, either preceded by plus or minus sign.

In particular, we are interested in the cases in which

- $(A_5)$ the number $M$ of basis elements $e_k$ is minimal.

We will associate to each point $p_k$ the partial derivative $\partial_{x_k}$, so that each block corresponds to a set of partial derivatives. One may then form an operator of the above type by attaching to each $\partial_{x_j}$ a basis element $e_k$ and a signature and taking the sum over all $j$-indices in the block. A system constructed in this way is not unique and differs from other systems associated to the same finite geometry either for the assignment of a unit $e_j$ to a given partial derivative $\partial_{x_k}$ in a block, as well as for a signature to each top of that block. Obviously, it is always possible to write such a system if one chooses the units in a set of $M$ elements with $M$ large enough. The extra axiom $(A_5)$ correspond to the request that $M$ is minimal.

The construction of a system according to axioms $(A_1), \ldots, (A_5)$ can be translated into the classical problem of coloring the edges of a certain bipartite graph. The points of the graph consist of two disjoint sets: the set of tops of the finite geometry and the set of blocks. The lines of the graph connect a point in the set of tops to a point in the set of blocks if that top belongs to the block. The set of elements $\{e_1, \ldots, e_M\}$ may be seen as a set of colors with which we have to color the edges of the graph such that all edges issuing from a given point in the graph have different colors and axiom $(A_5)$ means that the total number of colors is minimal.

There is another way to construct the operators in a system, namely one can consider a finite geometry of incidence structure with tops $\{p_1, \ldots, p_M\}$ and blocks $\{b_1, \ldots, b_n\}$ where to each top $p_k$ we assign a basis element $e_k$. Then for each fixed block we have to choose now a certain "color" $\partial_{x_j}$ to be assigned to each top $e_k$ of the block and a signature. For this purpose, axiom $(A_5)$ is replaced by

- $(A_5')$ the number $m$ of partial derivatives is minimal.

The systems obtained in this second method are called *"super-dual systems"*, in order to distinguish them from the *"dual systems"* obtained by interchanging the words *"point"* and *"line"* in a given finite geometry.

**Definition 4.4.1.** *We say that a resolution is Koszul-like if it has the same length and Betti numbers proportional to those appearing in the standard Koszul complex multiplied by the dimension of the over all Clifford algebra.*

**Remark 4.4.6.** We have obtained a Koszul-like resolution, see [160], for finite geometries like the Fano plane, which is the projective plane over $\mathbb{Z}_2$, and for the Desargues configuration constituted by 10 points and 10 lines such that every line contains 3 points and every point is the intersection of 3 lines.

*Platonic bodies.* We can also consider finite geometries associated to the Platonic bodies and, interestingly, we discovered that only some of the associated operators have Koszul-like resolution. We first consider a tetrahedron. This Platonic body has 4 vertices, 4 faces. Every vertex belongs to 3 faces and every face has 3 vertices. A system associated to it can be written in the following way:

$$\begin{cases} (e_1\partial_{x_1} + e_2\partial_{x_2} + e_3\partial_{x_3})f = g_1 \\ (e_1\partial_{x_2} + e_2\partial_{x_1} + e_3\partial_{x_4})f = g_2 \\ (e_1\partial_{x_3} + e_2\partial_{x_4} + e_3\partial_{x_2})f = g_3 \\ (e_1\partial_{x_4} + e_2\partial_{x_3} + e_3\partial_{x_1})f = g_4. \end{cases} \tag{4.36}$$

Consider now a cube and its 8 vertices and 6 faces. Every vertex belongs to 3 faces and every 3 faces intersect at most in one point. A system describing this solid is

$$\begin{cases} (e_1\partial_{x_1} + e_2\partial_{x_2} + e_3\partial_{x_3} + e_4\partial_{x_4})f = g_1 \\ (e_1\partial_{x_7} + e_2\partial_{x_8} + e_3\partial_{x_5} + e_4\partial_{x_6})f = g_2 \\ (e_1\partial_{x_5} + e_2\partial_{x_6} + e_3\partial_{x_1} + e_4\partial_{x_2})f = g_3 \\ (e_1\partial_{x_8} + e_2\partial_{x_7} + e_3\partial_{x_4} + e_4\partial_{x_3})f = g_4 \\ (e_1\partial_{x_3} + e_2\partial_{x_5} + e_3\partial_{x_7} + e_4\partial_{x_1})f = g_5 \\ (e_1\partial_{x_6} + e_2\partial_{x_4} + e_3\partial_{x_2} + e_4\partial_{x_8})f = g_6. \end{cases} \tag{4.37}$$

The octahedron has 6 vertices and 8 faces. Every face has 3 vertices and every vertex belongs to 4 faces. A system associated to this solid is

$$\begin{cases} (e_1\partial_{x_1} + e_2\partial_{x_2} + e_3\partial_{x_5})f = g_1 \\ (e_1\partial_{x_2} + e_2\partial_{x_1} + e_3\partial_{x_3})f = g_2 \\ (e_1\partial_{x_4} + e_3\partial_{x_1} + e_4\partial_{x_3})f = g_3 \\ (e_2\partial_{x_5} + e_3\partial_{x_4} + e_4\partial_{x_1})f = g_4 \\ (e_1\partial_{x_6} + e_3\partial_{x_2} + e_4\partial_{x_5})f = g_5 \\ (e_1\partial_{x_3} + e_2\partial_{x_6} + e_4\partial_{x_2})f = g_6 \\ (e_2\partial_{x_3} + e_3\partial_{x_6} + e_4\partial_{x_4})f = g_7 \\ (e_1\partial_{x_5} + e_2\partial_{x_4} + e_4\partial_{x_6})f = g_8. \end{cases} \tag{4.38}$$

The dodecahedron has 20 vertices, 12 faces, every face has 5 vertices and every vertex belongs to 3 faces. A system associated to the dodecahedron consists of 12 equations of the form

$$(e_1\partial_{x_i} + e_2\partial_{x_j} + e_3\partial_{x_k} + e_4\partial_{x_\ell} + e_5\partial_{x_n})f = g_\nu \tag{4.39}$$

where $\nu = 1, 2, \dots, 12$, $(i, j, k, \ell, n)$ (in this order) belongs to

$$\{(1,2,3,4,5),\ (5,1,6,14,15),\ (2,12,1,13,14),$$
$$(3,10,2,11,12),\ (4,5,7,6,8),\ (8,3,4,9,10),$$
$$(6,7,15,19,20),\ (16,17,18,20,19),\ (17,11,12,16,13),$$
$$(7,8,19,18,9),\ (9,18,11,10,17),\ (13,14,20,15,16)\}.$$

Finally, we consider the last Platonic body, the icosahedron. It has 12 vertices, 20 faces, every face has 3 vertices and each vertex belongs to 5 faces. A system associated to this solid contains 20 equations of the type (4.39) where $\nu = 1, 2, \ldots, 20$, $(i, j, k, \ell, n)$ (in this order) belongs to

$$\{(1,2,3,0,0),\ (0,1,2,7,0),\ (0,0,1,3,5),$$

$$(4,0,0,5,3),\ (0,6,5,1,0),\ (8,9,0,2,0),$$

$$(6,0,7,0,1),\ (0,7,0,8,2),\ (3,0,9,0,4),$$

$$(0,0,6,11,7),\ (7,8,11,0,0),\ (0,11,0,10,12),$$

$$(12,0,0,6,11),\ (11,0,8,0,10),\ (0,10,0,9,8),$$

$$(9,4,10,0,0),\ (10,0,4,12,0),\ (2,3,0,0,9),$$

$$(0,5,12,4,0),(5,12,0,0,6)\}.$$

The last two cases have been treated in the Clifford algebra $\mathbb{C}_5$ and the matrix associated to each system can be obtained by suitable formal substitutions from the matrix associated to the Dirac operator in five dimensions.

**Proposition 4.4.10.** *The systems associated to the Platonic bodies (4.36), (4.37), (4.38), (4.39) for* $\nu = 20$ *have the following resolutions:*
*tetrahedron:*

$$0 \longrightarrow R^8(-4) \longrightarrow R^{32}(-3)$$
$$\longrightarrow R^{48}(-2) \longrightarrow R^{32}(-1) \longrightarrow R^8 \longrightarrow M \longrightarrow 0,$$

*cube:*

$$0 \longrightarrow R^{32}(-9) \longrightarrow R^{240}(-8) \longrightarrow R^{768}(-7)$$
$$\longrightarrow R^{1344}(-6) \longrightarrow R^{48}(-4) \oplus R^{1344}(-5)$$
$$\longrightarrow R^{160}(-3) \oplus R^{720}(-4) \longrightarrow R^{192}(-2) \oplus R^{160}(-3)$$
$$\longrightarrow R^{96}(-1) \longrightarrow R^8 \longrightarrow M \longrightarrow 0,$$

*octahedron:*

$$0 \longrightarrow R^{16}(-6) \longrightarrow R^{96}(-5) \longrightarrow R^{240}(-4) \longrightarrow R^{320}(-3)$$
$$\longrightarrow R^{240}(-2) \longrightarrow R^{96}(-1) \longrightarrow M \longrightarrow 0,$$

*icosahedron:*

$$0 \longrightarrow R^8(-12) \longrightarrow R^{96}(-11) \longrightarrow R^{528}(-10)$$
$$\longrightarrow R^{1760}(-9) \longrightarrow R^{3960}(-8) \longrightarrow R^{6336}(-7)$$
$$\longrightarrow R^{7392}(-6) \longrightarrow R^{6336}(-5) \longrightarrow R^{3960}(-4) \longrightarrow R^{1760}(-3)$$
$$\longrightarrow R^{528}(-2) \longrightarrow R^{96}(-1) \longrightarrow R^8 \longrightarrow M \longrightarrow 0.$$

**Remark 4.4.7.** The systems associated to the tetrahedron, to the octahedron, to the icosahedron are Koszul-like while the system associated to the cube has a more complicated structure. Note also that, despite the fact that the system (4.39) contains 20 equations, it behaves like a Koszul system in only 12 operators. This fact proves that not all the equations in the system are independent. The case of the dodecahedron is very complicated and exceeds the possibilities of the computer. What we can prove is that the first syzygies are linear.

**Proposition 4.4.11.** *The system (4.36) associated to a tetrahedron can be reduced to a Koszul system of the type*

$$\begin{cases} \partial_{x_1} f = h_1 \\ \partial_{x_2} f = h_2 \\ \partial_{x_3} f = h_3 \\ \partial_{x_4} f = h_4. \end{cases}$$

*Proof.* If we consider the second equation of the system (4.36) multiplied on the left by $e_{12}$, we get that the first pair of equations in the system become

$$\begin{aligned} (e_1\partial_{x_1} + e_2\partial_{x_2} + e_3\partial_{x_3})f &= g_1 \\ (e_2\partial_{x_2} - e_1\partial_{x_1} + e_{123}\partial_{x_4})f &= e_{12}g_2 \end{aligned}$$

so that they can be rewritten in the form

$$\begin{aligned} 2e_1\partial_{x_1} f &= g_1 - e_{12}g_2 - e_3\partial_{x_3} f + e_{123}\partial_{x_4} f \\ 2e_2\partial_{x_2} f &= g_1 + e_{12}g_2 - e_3\partial_{x_3} f - e_{123}\partial_{x_4} f. \end{aligned} \tag{4.40}$$

Now we can multiply the third equation by $-e_{23}$ so that we arrive at the new equation

$$e_2\partial_{x_2} f = -e_{23}g_3 + e_{123}\partial_{x_3} f + e_3\partial_{x_4} f$$

which can be combined with the second one in the system (4.40) to get

$$(e_3 + 2e_{123})\partial_{x_3} f = g_1 + e_{12}g_2 + 2e_{23}g_3 - (2e_3 + e_{123})\partial_{x_4} f.$$

This allows us to write $\partial_{x_3}$ in terms of $\partial_{x_4}$ since the coefficient $e_3 + 2e_{123}$ admits an inverse. By substituting $\partial_{x_1}$ and $\partial_{x_3}$ in the fourth equation of (4.36) we obtain that $\partial_{x_4}$ has coefficient $(2e_1 - e_2)$, which is invertible, so we can get $\partial_{x_4}$ and, by consequence also $\partial_{x_3}$. This concludes the proof. □

**Remark 4.4.8.** We expect Koszul-like resolutions for structures in which the number of blocks (lines or faces) is greater than or equal the number of points. Reductions similar to those ones we have written in the case of the tetrahedron may be written also in the case of the other Koszul-like systems we have obtained even though, in general, the reduction is not obvious.

# 5

# Some First Order Linear Operators in Physics

It is well known that among the main tools to find fields equations are the variational principles. In this introduction we wish to sketch some general underlying ideas. Suppose considering a physical system which requires several fields $\phi_j(x)$, $j = 1, \ldots, n$ to be specified. We can suppose that the field $\phi_j(x)$ is real (if it is complex the procedure can be repeated taking into account both the real and imaginary parts). The index $j$ may label the components of the same field, for example the components of a vector potential, or it can refer to different independent fields. We restrict ourselves to theories which can be derived by means of a variational principle from an action integral involving a Lagrangian density

$$\mathcal{L} = \mathcal{L}(\phi_j, \partial_\alpha \phi_j), \quad \partial_\alpha \phi_j := \frac{\partial \phi_j(x)}{\partial x_\alpha}.$$

We suppose that the function $\mathcal{L}(\phi_j, \partial_\alpha \phi_j)$ depends on the fields and their first derivatives only. Even though it is not the most general setting, it contains such important cases as the Maxwell, Proca, Dirac and many other fields. We define the action integral $\Gamma(\Omega)$ for an arbitrary region $\Omega$ of the four dimensional space-time continuum by

$$\Gamma(\Omega) = \int_\Omega \mathcal{L}(\phi_j(x), \partial_\alpha \phi_j(x)) d^4x$$

where we use the standard notation in physics, namely, $d^4x = dt d^3\mathbf{x}$. We postulate that the equations of motions, i.e., the field equations, are obtained by the following variational principle (compare with the Hamilton principle in mechanics). For any arbitrary region $\Omega$, we consider variations of the fields

$$\phi_j(x) \to \phi_j(x) + \delta\phi_j(x)$$

vanishing on the boundary $\partial\Omega$ of $\Omega$, i.e.,

$$\delta\phi_j(x) = 0, \quad \text{on} \quad \partial\Omega. \tag{5.1}$$

We also require that for any arbitrary region $\Omega$ and any variation $\phi_j(x) \to \phi_j(x) + \delta\phi_j(x)$ satisfying (5.1), the action $\Gamma(\Omega)$ has a stationary value, i.e.,

$$\delta\Gamma(\Omega) = 0.$$

Computing the variation we get

$$\delta\Gamma(\Omega) = \int_\Omega \Big(\frac{\partial\mathcal{L}}{\partial\phi_j}\delta\phi_j + \frac{\partial\mathcal{L}}{\partial\phi_{j;\alpha}}\delta\phi_{j;\alpha}\Big)d^4x = 0,$$

where we have set for simplicity $\phi_{j;\alpha} := \partial_\alpha\phi_j(x)$. Since the variations $\delta\phi_j$ and $\delta\phi_{j;\alpha}$ are not independent, integrating by parts and taking into account the commutation relations

$$\delta\partial_\alpha\phi_j = \partial_\alpha\delta\phi_j,$$

we obtain

$$\delta\Gamma(\Omega) = \int_\Omega \delta\phi_j\Big(\frac{\partial\mathcal{L}}{\partial\phi_j} - \frac{\partial}{\partial x_\alpha}\frac{\partial\mathcal{L}}{\partial\phi_{j;\alpha}}\Big)d^4x + \int_\Omega \frac{\partial}{\partial x_\alpha}\Big(\frac{\partial\mathcal{L}}{\partial\phi_{j;\alpha}}\delta\phi_j\Big)d^4x = 0.$$

The last integral can be transformed in a surface integral over the boundary $\partial\Omega$ using Gauss's divergence theorem in four dimensions and since $\delta\phi_j = 0$ on $\partial\Omega$, this boundary integral vanishes. So, if

$$\delta\Gamma(\Omega) = 0, \quad \forall\Omega, \quad \forall\delta\phi_j,$$

we get the Euler–Lagrange equations

$$\frac{\partial\mathcal{L}}{\partial\phi_j} - \frac{\partial}{\partial x_\alpha}\frac{\partial\mathcal{L}}{\partial\phi_{j;\alpha}} = 0, \qquad j = 1, \dots, n. \tag{5.2}$$

For more details see [82] or [126].

To give an example of the variational approach, we can consider the case of the electromagnetic field. Let us denote by $A = (A_0, A_1, A_2, A_3)$ the four-potential ($A_0$ being the scalar potential of the electric field while $(A_1, A_2, A_3)$ is the vector potential of the magnetic field), and by $j = (j_0, j_1, j_2, j_3)$ the four-current (where $j_0$ is the electric charge density and $(j_1, j_2, j_3)$ is the vector representing the current density). The electromagnetic field is represented by the antisymmetric $4 \times 4$ tensor, see for example [81]:

$$F^{\mu\nu} = \begin{bmatrix} 0 & -E_1 & -E_2 & -E_3 \\ E_1 & 0 & -B_3 & B_2 \\ E_2 & B_3 & 0 & -B_1 \\ E_3 & -B_2 & B_1 & 0 \end{bmatrix}$$

where $(E_1, E_2, E_3)$ and $(B_1, B_2, B_3)$ are the electric field and the magnetic field, respectively. The Lagrangian density can be shown to be (see [81])

$$\mathcal{L} = -1/4 F_{\mu\nu} F^{\mu\nu} - j_\mu A^\mu.$$

Maxwell's equations can be deduced from $\mathcal{L}$, thanks to the Euler-Lagrange equations (5.2). In covariant form we get

$$\begin{cases} F^{\mu\nu} = \partial^\mu A^\nu - \partial^\nu A^\mu \\ \partial_\mu F^{\mu\nu} = j^\nu. \end{cases}$$

Moreover, it is possible to modify the Lagrangian density by adding a mass $m$ in Maxwell's theory in order to study massive vector fields. Examples are particles such as $W^\pm$ and $Z$ which mediate the electro-weak interaction. In the massive case the Lagrangian is

$$\mathcal{L}_m = -1/4 F_{\mu\nu} F^{\mu\nu} + 1/2 m^2 A_\mu A^\mu - j_\mu A^\mu,$$

from which one deduces the Proca equations

$$\partial_\mu F^{\mu\nu} + m^2 A^\nu = j^\nu.$$

We now want to look at the fields equations from another point of view. Some of them can be written as a single equation that expresses a "regularity" condition in the setting of Clifford algebras. The indices that one needs to write the equations in the tensor form are automatically provided by the structure of the Clifford algebra.

For example, Maxwell, Proca, Dirac and other fields equations can be expressed by means of suitable regularity conditions. We point out that the variational principles are of great importance in many fields of Physics and cannot be replaced by our way of reformulating physical laws, but the possibility of seeing these fields as functions in the kernel of suitable operators in a Clifford algebra or in a spinor space is particularly suitable to apply algebraic analysis methods and to study some of the properties of the equations.

## 5.1 Physics and algebra of Maxwell and Proca fields

We begin this section with some preliminaries on Maxwell system. As is well known, the electromagnetic field is described by the system

$$\begin{cases} \nabla \cdot \mathbf{E} = \rho^e \\ \nabla \cdot \mathbf{B} = 0 \\ \nabla \times \mathbf{E} + \partial_t \mathbf{B} = 0 \\ \nabla \times \mathbf{B} - \partial_t \mathbf{E} = \mathbf{j}^e, \end{cases} \tag{5.3}$$

which can be written in a more general way as

$$\begin{cases} \nabla \cdot \mathbf{E} = \rho^e \\ \nabla \cdot \mathbf{B} = \rho^m \\ -\nabla \times \mathbf{E} - \partial_t \mathbf{B} = \mathbf{j}^m \\ \nabla \times \mathbf{B} - \partial_t \mathbf{E} = \mathbf{j}^e, \end{cases} \tag{5.4}$$

where $\mathbf{E}$, $\mathbf{B}$ are the electric and the magnetic fields, $\rho^e$, $\rho^m$ the density of electric and magnetic charges, while $\mathbf{j}^e$, $\mathbf{j}^m$ are the electric and magnetic currents. Even though the existence of magnetic monopoles is still an open problem, we have written the system in two possible ways since, from a mathematical point of view, we will see that we will get two compatibility conditions for the inhomogeneous system and one of them has meaning only if we consider the Maxwell system in the second form (i.e., with both magnetic monopoles and currents). Moreover, one can show that there exists a magnetic charge conservation law in addition to the well known electric charge conservation law.

If we set $\rho^m = 0$ and $\mathbf{j}^m = 0$, standard calculations show that the Maxwell equations imply the conservation of charge given by the equation

$$\partial_t \rho^e + \nabla \cdot \mathbf{j}^e = 0. \tag{5.5}$$

We will see in what follows that this conservation law can be interpreted in term of syzygies and it follows easily from the algebraic structure of the Maxwell equations. Moreover, from the algebraic analysis point of view, the conservation law is a compatibility condition to be satisfied if we want solvability of the inhomogeneous Maxwell equations.

On the other hand, the Proca equations (that were originally introduced in [149]) generalize Maxwell field for massive spin 1 particles. A possible way to write Proca equations, where we assume the absence of magnetic monopoles, is the following:

$$\begin{cases} \nabla \cdot \mathbf{E} = \rho^e + m^2 A^0 \\ \nabla \cdot \mathbf{B} = 0 \\ \nabla \times \mathbf{E} = -\partial_t \mathbf{B} \\ \nabla \times \mathbf{B} = \partial_t \mathbf{E} + m^2 \mathbf{A} + \mathbf{j}^e \\ \mathbf{B} = \nabla \times \mathbf{A} \\ \mathbf{E} = -\nabla A^0 - \partial_t \mathbf{A}. \end{cases} \tag{5.6}$$

In this case, the potentials $A^0$ and $\mathbf{A}$ are not, as in the case of Maxwell equations, a useful tool to deduce more rapidly the electric and the magnetic fields, but they are unknown fields coupled with the fields $\mathbf{E}$ and $\mathbf{B}$. From the system (5.6) we can deduce that

$$\partial_t(m^2 A^0 + \rho^e) + \nabla \cdot (m^2 \mathbf{A} + \mathbf{j}^e) = 0 \tag{5.7}$$

which is the well known conservation law for Proca equations. The difference with respect to the electromagnetic field is that the four-current $(\rho^e, \mathbf{j}^e)$ is not necessarily conserved. This fact depends on the particular nature of the source. If we assume that the source is conservative, then we obtain from (5.7) the Lorentz condition

$$\partial_t A^0 + \nabla \cdot \mathbf{A} = 0 \tag{5.8}$$

for the field $(A^0, \mathbf{A})$.

Let us consider systems (5.3) and (5.4) from an algebraic point of view. We set

$$F = [E_1, E_2, E_3, B_1, B_2, B_3]^t,$$

so we can rewrite the Maxwell equations in matrix form as

$$P_{\mathcal{M}}(D)F = G,$$

where the matrix of differential operators $P_{\mathcal{M}}(D)$ is defined by

$$P_{\mathcal{M}}(D) = \begin{bmatrix} \partial_x & \partial_y & \partial_z & 0 & 0 & 0 \\ 0 & 0 & 0 & \partial_x & \partial_y & \partial_z \\ 0 & \partial_z & -\partial_y & -\partial_t & 0 & 0 \\ -\partial_z & 0 & \partial_x & 0 & -\partial_t & 0 \\ \partial_y & -\partial_x & 0 & 0 & 0 & -\partial_t \\ -\partial_t & 0 & 0 & 0 & -\partial_z & \partial_y \\ 0 & -\partial_t & 0 & \partial_z & 0 & -\partial_x \\ 0 & 0 & -\partial_t & -\partial_y & \partial_x & 0 \end{bmatrix}$$

and the subscript $\mathcal{M}$ reminds us of the Maxwell system.

The vector $G$ is

$$[\rho^e, 0, 0, 0, 0, j_1^e, j_2^e, j_3^e]^t$$

in the case we are looking at system (5.3), while $G$ is

$$[\rho^e, \rho^m, j_1^m, j_2^m, j_3^m, j_1^e, j_2^e, j_3^e]^t$$

in the case of system (5.4).

Using the techniques described in Chapter 2, it is possible to compute the resolution of the map $P_{\mathcal{M}}^t : R^8 \to R^6$, where $R$ is the ring of polynomials $\mathbb{C}[t, x, y, z]$. The result is the following:

$$0 \longrightarrow R^2(-2) \xrightarrow{P_1^t} R^8(-1) \xrightarrow{P_{\mathcal{M}}^t} R^6 \longrightarrow M \longrightarrow 0$$

where $M$ is the cokernel of the map $P^t_{\mathcal{M}}$, i.e. $M = R^6/P^t_{\mathcal{M}}R^8$. This means that, in our case, we have two syzygies of degree 1. By giving the command `SyzOfGens`, CoCoA will display the matrix associated to the first syzygies that turns out to be

$$P_1 = \begin{bmatrix} 0 & t & x & y & z & 0 & 0 & 0 \\ t & 0 & 0 & 0 & 0 & x & y & z \end{bmatrix}$$

and can be written in terms of differential operators as

$$P_1(D) = \begin{bmatrix} 0 & \partial_t & \partial_x & \partial_y & \partial_z & 0 & 0 & 0 \\ \partial_t & 0 & 0 & 0 & 0 & \partial_x & \partial_y & \partial_z \end{bmatrix}. \tag{5.9}$$

We now assume that $S$ is one of the sheaves of (generalized) functions used in Chapter 2, though distributions are the most natural setting, and we summarize the discussion in the following proposition, which is a direct consequence of Theorem 2.1.3:

**Proposition 5.1.1.** *The system*

$$P_{\mathcal{M}}(D)F = G$$

*has a solution* $F \in S^6$ *if and only if* $G \in S^8$ *satisfies* $P_1(D)G = 0$.

**Remark 5.1.1.** This means that in the physical case (5.3) where there are no magnetic monopoles, we get only one condition, that is

$$\partial_t \rho^e + \nabla \cdot \mathbf{j}^e = 0, \tag{5.10}$$

while in the case (5.4), in which there are magnetic monopoles and magnetic currents the solvability conditions are

$$\begin{cases} \partial_t \rho^e + \nabla \cdot \mathbf{j}^e = 0 \\ \partial_t \rho^m + \nabla \cdot \mathbf{j}^m = 0. \end{cases} \tag{5.11}$$

In view of Palamodov's results on the characteristic varieties of the Ext modules, it is important for us to determine the characteristic variety of $\mathrm{Ext}^1(M, R)$. We can use the following general result.

**Theorem 5.1.1.** *Let* $R^s \xrightarrow{A} R^q$ *be a linear transformation, where* $q \geq s$, *with* $A$ *of maximal rank* $s$ *and let* $M = R^q/AR^s$. *Then the characteristic variety of* $\mathrm{Ext}^1(M, R)$, *as a set, is the variety defined by the greatest common divisor* $d$ *of all the* $s \times s$ *minors of* $A$.

*Proof.* First, it is well known that $\mathrm{Ext}^1(M, R)$ is the torsion submodule $N$ of $R^q/AR^s$. Let

$$F_1 \xrightarrow{K} F_0 \longrightarrow N \longrightarrow 0$$

be a free resolution of this module. The characteristic variety of $N$ is the variety defined by the ideal $F_0$ generated by the maximal minors of $K$. This ideal is also known as the 0-th Fitting ideal $\mathcal{G}_0$ of $N$. It is also well known that for some $n$,

$$(A_N)^n \subseteq \mathcal{G}_0 \subseteq A_N,$$

where $A_N \subseteq R$ denotes the annihilator of $N$. Therefore the varieties defined by $A_N$ and by $\mathcal{G}_0$ are the same, as sets. We claim that the only prime ideals in $A_N$ are the ideals generated by prime divisors of $d$. Indeed if $d \notin P$, $P$ a prime ideal, then the map

$$R^s \otimes_R R_P \xrightarrow{A} R^q \otimes_R R_P$$

has maximal rank, where $R_P$ denotes the localization of $R$ at $P$. But in this situation, the greatest common divisor of the $s \times s$ minors is 1 (since $d$ is invertible in $R_P$). Thus, $R_P^q/AR_P^s$ has no torsion elements. So $P$ cannot be in the annihilator of $N$ and hence, as a set, the characteristic variety is defined by $d$. □

In view of the previous theorem we have:

**Proposition 5.1.2.** *The characteristic variety of* $\mathrm{Ext}^1(M,R)$ *is given by the points* $(t,x,y,z)$ *satisfying*

$$(t^2 - x^2 - y^2 - z^2)^2 = 0.$$

*In particular,* $\mathrm{Ext}^1(M,R) \neq 0$ *and* $\mathrm{Ext}^2(M,R) = 0$.

The proof is based on a quick computation with CoCoA and shows that the system is not overdetermined, which means, in other words, that the compact singularities cannot be eliminated.

From a physical point of view we can say that the possible existence of compact singularities may be interpreted as the presence of localized charges which generate the electromagnetic field. The fact that we have a complete description of the characteristic variety of $\mathrm{Ext}^1(M,R)$ has a rather remarkable consequence. Indeed, we have the following result.

**Proposition 5.1.3.** *Let $\Omega$ be an open set in $\mathbb{R}^4$ and let $x$ be a point in $\Omega$. Then every solution of the Maxwell system in $\Omega\backslash\{x\}$, whose components can be extended as distributions to all of $\Omega$, is a distribution solution to the Maxwell system to all of $\Omega$.*

*Proof.* This result follows immediately from Corollary 8.14.4 in [142], which establishes that such an extension of the solutions exists if and only if none of the varieties associated to the module $\mathrm{Ext}^1(M,R)$ is hypoelliptic. We recall (see [142]) that a variety $V \subseteq \mathbb{C}^n$ is said to be hypoelliptic if for any $z \in V$ whose imaginary part is $y$, the inequality $|z| \leq B(1+|y|)^{1/\gamma}$ holds for sufficiently large $B$ and $\gamma > 0$ (note that $1/\gamma = 0$ when $V$ is bounded). In view of the fact that the characteristic variety of $\mathrm{Ext}^1(M,R)$ is the light cone which obviously is not hypoelliptic, since its real part is unbounded, we obtain the statement. □

Let us consider now the Proca system. From the mathematical point of view, we consider the system

$$\begin{cases} \nabla\cdot\mathbf{E} - m^2 A^0 = \rho^e \\ \nabla\cdot\mathbf{B} = \rho^m \\ -\nabla\times\mathbf{E} - \partial_t\mathbf{B} = \mathbf{j}^m \\ \nabla\times\mathbf{B} - \partial_t\mathbf{E} - m^2\mathbf{A} = \mathbf{j}^e \\ \mathbf{B} - \nabla\times\mathbf{A} = \mathbf{B}^{ext} \\ \mathbf{E} + \nabla A^0 + \partial_t\mathbf{A} = \mathbf{E}^{ext} \end{cases} \tag{5.12}$$

where we introduce, besides electric and magnetic sources, also $\mathbf{E}^{ext}$, $\mathbf{B}^{ext}$, which we call external fields, to make the system nonhomogeneous. We want to study now the compatibility conditions, which are, in fact, the conservation laws.

We rewrite system (5.12) in matrix form as

$$P_{\mathcal{P}}(D)F = G$$

where $F$, $G$ are the transpose of

$$[E_1, E_2, E_3, B_1, B_2, B_3, A_0, A_1, A_2, A_3],$$

and

$$[\rho^e, \rho^m, j_1^m, j_2^m, j_3^m, j_1^e, j_2^e, j_3^e, B_1^{ext}, B_2^{ext}, B_3^{ext}, E_1^{ext}, E_2^{ext}, E_3^{ext}],$$

respectively, and the matrix of differential operators $P_{\mathcal{P}}(D)$ is defined by

$$\begin{bmatrix}
\partial_x & \partial_y & \partial_z & 0 & 0 & 0 & -m^2 & 0 & 0 & 0 \\
0 & 0 & 0 & \partial_x & \partial_y & \partial_z & 0 & 0 & 0 & 0 \\
0 & \partial_z & -\partial_y & -\partial_t & 0 & 0 & 0 & 0 & 0 & 0 \\
-\partial_z & 0 & \partial_x & 0 & -\partial_t & 0 & 0 & 0 & 0 & 0 \\
\partial_y & -\partial_x & 0 & 0 & 0 & -\partial_t & 0 & 0 & 0 & 0 \\
-\partial_t & 0 & 0 & 0 & -\partial_z & \partial_y & 0 & -m^2 & 0 & 0 \\
0 & -\partial_t & 0 & \partial_z & 0 & -\partial_x & 0 & 0 & -m^2 & 0 \\
0 & 0 & -\partial_t & -\partial_y & \partial_x & 0 & 0 & 0 & 0 & -m^2 \\
0 & 0 & 0 & 1 & 0 & 0 & 0 & 0 & \partial_z & -\partial_y \\
0 & 0 & 0 & 0 & 1 & 0 & 0 & -\partial_z & 0 & \partial_x \\
0 & 0 & 0 & 0 & 0 & 1 & 0 & \partial_y & -\partial_x & 0 \\
1 & 0 & 0 & 0 & 0 & 0 & \partial_x & \partial_t & 0 & 0 \\
0 & 1 & 0 & 0 & 0 & 0 & \partial_y & 0 & \partial_t & 0 \\
0 & 0 & 1 & 0 & 0 & 0 & \partial_z & 0 & 0 & \partial_t
\end{bmatrix}.$$

The computation of the resolution with CoCoA (see Section 2.4 for the details on how to get a minimal resolution) gives the following result:

$$0 \longrightarrow R(-3) \xrightarrow{P_2^t} R^5(-2) \xrightarrow{P_1^t} R^{14}(-1) \xrightarrow{P_{\mathcal{P}}^t} R^{10} \longrightarrow M \longrightarrow 0$$

where $R = \mathbb{C}[t, x, y, z, m]$. The surprising fact is that in addition to the first syzygies, i.e., compatibility conditions on the data of system (5.12), we also have second syzygies, i.e., the compatibility conditions on the data of the inhomogeneous system associated to the matrix $P_1$. By giving the command `SyzOfGens` it is possible to obtain the matrices $P_1$ and $P_2$:

$$P_1 = \begin{bmatrix} 0 & 1 & 0 & 0 & 0 & 0 & 0 & 0 & -x & -y & -z & 0 & 0 & 0 \\ 0 & 0 & -1 & 0 & 0 & 0 & 0 & 0 & -t & 0 & 0 & 0 & z & -y \\ 0 & 0 & 0 & 0 & -1 & 0 & 0 & 0 & 0 & 0 & -t & y & -x & 0 \\ 0 & 0 & 0 & 1 & 0 & 0 & 0 & 0 & 0 & t & 0 & z & 0 & -x \\ 0 & t & x & x & z & 0 & 0 & 0 & 0 & 0 & 0 & 0 & 0 & 0 \end{bmatrix}$$

and

$$P_2 = \begin{bmatrix} -t & x & z & -y & 1 \end{bmatrix}.$$

From the knowledge of the matrices $P_1$ and $P_2$ we can state the following two theorems:

**Theorem 5.1.2.** *The system*

$$P_{\mathcal{P}}(D)F = G$$

*has a solution $F \in \mathcal{S}^{10}$ if and only if $G \in \mathcal{S}^{14}$ satisfies $P_1(D)G = 0$, i.e.,*

$$\begin{cases} \rho^m - \nabla \cdot \mathbf{E}^{ext} = 0 \\ -\mathbf{j}^m - \partial_t \mathbf{B}^{ext} - \nabla \times \mathbf{E}^{ext} = 0 \\ \partial_t \rho^m + \nabla \cdot \mathbf{j}^m = 0. \end{cases} \tag{5.13}$$

This system represents three conservations laws, but note the important fact that those laws depend only on the presence of magnetic charges and currents. In fact, if one supposes $\mathbf{E}^{ext} = \mathbf{B}^{ext} = 0$, then the first two equations become $\mathbf{j}^m = \rho^m = 0$, and the third equation is trivially an identity. The existence of external fields makes the system nontrivial. If we now consider the nonhomogeneous version of the system (5.13)

$$P_1(D)G = H \tag{5.14}$$

by considering the datum $H = [\ell_1, \mathbf{L}, \ell_2]$, we get the following result.

**Theorem 5.1.3.** *The system (5.14) has a solution $G \in \mathcal{S}^{14}$ if and only if $H \in \mathcal{S}^5$ satisfies*

$$P_2(D)H = 0,$$

*i.e., if and only if the data satisfy the equation*

$$-\partial_t \ell_1 + \nabla \cdot \mathbf{L} + \ell_2 = 0. \tag{5.15}$$

We can interpret $-\mathbf{L}$ as a new magnetic current density, $\ell_1$ the relative magnetic monopoles density, and $\ell_2$ the source of the field. If $\ell_2 = 0$, equation (5.15) represents a conservation law for the magnetic density $\ell_1$, while if $\ell_2 \neq 0$ the differential equation (5.15) gives the evolution of $[\ell_1, \mathbf{L}]$ in terms of the source $\ell_2$ in order for system (5.13) to have a solution.

Note that, in the case of the Maxwell system, we only have first syzygies, and therefore the nonhomogeneous system

$$\begin{cases} \partial_t \rho^e + \nabla \cdot \mathbf{j}^e = \ell \\ \partial_t \rho^m + \nabla \cdot \mathbf{j}^m = \ell' \end{cases}$$

can be solved without conditions on $\ell$ and $\ell'$.

Also, in the case of Proca equations, the system is not overdetermined, in fact the matrix $P_{\mathcal{P}}$ is of maximal rank but the greatest common divisor of its minors of maximal rank is not equal to 1. More precisely, we have:

**Proposition 5.1.4.** *The characteristic variety of* $\mathrm{Ext}^1(M, R)$ *for the system (5.12) is given by the points* $(t, x, y, z)$ *such that*

$$(t^2 - x^2 - y^2 - z^2 - m^2)^3 = 0,$$

*so that* $\mathrm{Ext}^1(M, R) \neq 0$. *Moreover* $\mathrm{Ext}^2(M, R) \neq 0$, *while* $\mathrm{Ext}^3(M, R) = 0$.

*Proof.* All the Ext-modules can be computed using CoCoA. Note that the $m$ represents the mass and can be considered as a fixed real constant. □

**Remark 5.1.2.** The analogue of Proposition 5.1.3 holds also in this case.

Note that the characteristic variety of $\mathrm{Ext}^1(M, R)$ is algebraic since it is defined by the ideal generated by the greatest common divisor of the minors of maximum order of $P_{\mathcal{P}}$. The associated reduced variety is the light cone $t^2 - x^2 - y^2 - z^2 = 0$ for the Maxwell system, and the quadric hypersurface $t^2 - x^2 - y^2 - z^2 - m^2 = 0$ for the Proca system.

According to the conservation law for the Proca equations (5.7), we can consider the Proca system under the assumption that the field $(A_0, \mathbf{A})$ is conserved, i.e., suppose that

$$\partial_t A^0 + \nabla \cdot \mathbf{A} = 0.$$

The nonhomogeneous system (5.12) can be rewritten taking into account this new condition, adding the equation

$$\partial_t A^0 + \nabla \cdot \mathbf{A} = A^{ext},$$

where $A^{ext}$ is a given datum. The system is represented by a $15 \times 10$ matrix and the resolution of the corresponding map is

$$0 \longrightarrow R(-3) \xrightarrow{P_2^t} R^6(-2) \xrightarrow{P_1^t} R^{15}(-1) \xrightarrow{P_{\mathcal{P}}^t} R^{10} \longrightarrow M \longrightarrow 0.$$

The compatibility conditions of this new system are given by (5.13) and by the relation

$$A^{ext} + \partial_t \rho^e + \nabla \cdot \mathbf{j}^e = 0.$$

When we set $A^{ext} = 0$, we get the conservation law of the electric charge.

**Remark 5.1.3.** We finally wish to describe the characteristic varieties of the Maxwell and Proca systems, since they are essential to write the integral representation of their solutions by means of the Ehrenpreis–Palamodov Fundamental Principle. The characteristic variety $V_{\mathcal{M}}$ of the Maxwell system can be obtained as the vanishing set of the ideal generated by the $6 \times 6$ minors of the matrix $P_{\mathcal{M}}$, i.e., the ideal generated by the 10 polynomials of the type

$$(x^2 + y^2 + z^2 - t^2)^2 \chi_i(x, y, z, t), \qquad i = 1, \dots, 10$$

where $\chi_i$ are degree two monomials in the variables $x, y, z, t$. The characteristic variety $V_{\mathcal{P}}$ of the Proca system can be written in a similar way, as the vanishing set of the ideal generated by the $10 \times 10$ minors of the matrix $P_{\mathcal{P}}$ associated to the Proca system, i.e., the ideal generated by the 35 polynomials of the type

$$(x^2 + y^2 + z^2 - t^2 + m^2)^3 \eta_i(x, y, z, t), \qquad i = 1, \dots, 35$$

where $\eta_i$ are monomials of degree less than or equal to three in the variables $x$, $y$, $z$, $t$.

## 5.2 Variations on Maxwell system in the space of biquaternions

From a historical point of view, one can find different treatments of Maxwell equations, according to which Clifford algebra is used. Such equations can be written with complex quaternions ([114]), in the Clifford algebra $\mathbb{R}_3$ of the Euclidean space $\mathbb{R}^3$, in the Clifford algebra $\mathbb{R}_{3,1}$ of the Minkowski space-time $\mathbb{R}^{3,1}$ ([95], [132], [151]), or using spinors. Our point of view is the alternative formulation of electrodynamics based on the study of complex quaternionic functions which are regular in the sense of a suitable operator. Such an operator allows one to deduce Maxwell's equations and to study their behavior in two Minkowski space-times, one with electric charges and the other one with magnetic monopoles. This model, in particular, predicts the lack of wave propagation from one Minkowski space-time to the other, so that the presence of magnetic monopoles in the Minkowski space-time $K^m$ is not contradictory to physical reality in the Minkowski space-time $K^e$. From a mathematical point of view, this approach allows to carry out a (non trivial) analysis of the functions in the kernel of this operator.

In this section we define the algebra $\mathbb{BH}$ of biquaternions, and we discuss its elementary algebraic properties. We point out that $\mathbb{BH}$ is essentially the algebra of complex quaternions, i.e., $\mathbb{H}_\mathbb{C} = \mathbb{H} \otimes_\mathbb{R} \mathbb{C}$, but we adopt the notation $\mathbb{BH}$ because in $\mathbb{BH}$ we introduce a basis making use of the Pauli algebra which provides a more elegant physical setting. To this end, we define the associative complex algebra of biquaternions $\mathbb{BH}$ as the complex algebra generated over the basis $\{e_0, e_1, e_2, e_3\}$ where $e_0 = 1$ and $e_k$, $k = 1, 2, 3$, satisfy the Pauli-type algebra relations

$$e_k^2 = 1, \qquad k = 1, 2, 3, \qquad e_\ell e_j = -e_j e_\ell = i e_k, \qquad i = \sqrt{-1} \in \mathbb{C},$$

where $(\ell, j, k)$ is any cyclic permutation of $(1, 2, 3)$. The relations among the units $e_k$, $k = 0, \ldots, 3$ are the same satisfied by the Pauli matrices

$$\sigma_0 = \begin{bmatrix} 1 & 0 \\ 0 & 1 \end{bmatrix} \qquad \sigma_1 = \begin{bmatrix} 0 & 1 \\ 1 & 0 \end{bmatrix} \tag{5.16}$$

$$\sigma_2 = \begin{bmatrix} 0 & -i \\ i & 0 \end{bmatrix} \qquad \sigma_3 = \begin{bmatrix} 1 & 0 \\ 0 & -1 \end{bmatrix}.$$

According to this definitions, a biquaternion $Z$ is an element

$$Z = e_0 z_0 + e_1 z_1 + e_2 z_2 + e_3 z_3,$$

where $z_\mu$ are complex numbers that are written as

$$z_\mu = x_\mu + i y_\mu, \qquad \mu = 0, 1, 2, 3.$$

The element $Z$ can also be written as

$$Z = x_0 + \mathbf{x} + i y_0 + i \mathbf{y},$$

where

$$\mathbf{x} = e_1 x_1 + e_2 x_2 + e_3 x_3, \qquad \mathbf{y} = e_1 y_1 + e_2 y_2 + e_3 y_3$$

and its hyperconjugate $Z^+$ is defined by

$$Z^+ = x_0 - \mathbf{x} + i y_0 - i \mathbf{y}. \tag{5.17}$$

Finally, we say that a biquaternion $X$ is real if it has the form $X = x_0 + \mathbf{x}$. The subset of the real biquaternions is denoted by $\mathbb{H}$ (with a slight abuse of notation, since we are using different units) and is called a "real biquaternion space" (see [92]). Every biquaternion $Z$ in $\mathbb{BH}$ can be written as $Z = X + iX'$, with $X, X' \in \mathbb{H}$. The square pseudonorm of a biquaternion $Z$ is defined by

$$N(Z) = Z^+ Z = Z Z^+ = z_0^2 - z_1^2 - z_2^2 - z_3^2, \tag{5.18}$$

which is, in general, a complex number. The square pseudonorm becomes real if and only if the components $(x_0, x_1, x_2, x_3)$ and $(y_0, y_1, y_2, y_3)$ of $X$ and $X'$, respectively, are orthogonal with respect to the Minkowski space-time inner product, i.e.,

$$< X, g_{\mu\nu} X' >= 0,$$

where

$$g_{\mu\nu} = \mathrm{diag}(1, -1, -1, -1).$$

When we restrict our attention to a real biquaternion $X$, we have $N(X) = x_0^2 - x_1^2 - x_2^2 - x_3^2 \in \mathbb{R}$ and, as Imaeda points out in [92], the metric space structure of the space of real biquaternions is the structure of the Minkowski space-time. It is important to note that the biquaternion algebra, unlike the real quaternion algebra, is not a division algebra because of the existence of zero divisors: they occur when $N(Z) = 0$ but $Z \neq 0$. Given a biquaternion $Z$ such that $N(Z) \neq 0$ we can define its inverse as

$$Z^{-1} = \frac{Z^+}{N(Z)}.$$

The set $\mathbb{BH}$, considered as a ring, contains the ring $\mathbb{H}$ of quaternions as a subring. A quaternion $q$ is written in this case as

$$q = e_0 x_0 + i e_1 y_1 + i e_2 y_2 + i e_3 y_3,$$

where the units $e_k$, $k = 1, 2, 3$ of $\mathbb{H}$ are related to the units $i_\ell$, $\ell = 1, 2, 3$ by

$$i e_1 = i_1, \quad i e_2 = i_2, \quad i e_3 = i_3. \tag{5.19}$$

**Remark 5.2.1.** The set of units $\{1,\ i_1,\ i_2,\ i_3\}$ forms a basis of $\mathbb{H}$ and satisfies the following multiplication relations (note the sign difference with the usual ones introduced in Chapter 3)

$$i_2 i_1 = -i_1 i_2 = i_3, \quad i_1 i_3 = -i_3 i_1 = i_2, \quad i_3 i_2 = -i_3 i_2 = i_1.$$

We are aware that this notation differs from the standard one. In fact, it is possible to relate the units $e_\mu$ with the units of $\mathbb{H}$ in many ways. However, we decided to keep this particular choice to preserve the Minkowski space time metric structure.

In [92] Imaeda gives a notion of (left) regularity for functions of a real biquaternion variable. In order to do so, Imaeda formally replaces $x_k$ by $-i x_k$ $k = 1, 2, 3$ in the Cauchy–Fueter operator, to obtain the operator

$$\mathcal{D} = \frac{\partial}{\partial x_0} - e_1 \frac{\partial}{\partial x_1} - e_2 \frac{\partial}{\partial x_2} - e_3 \frac{\partial}{\partial x_3}. \tag{5.20}$$

A function $F : \mathbb{H} \longrightarrow \mathbb{BH}$,

$$F(X) = \sum_{\nu=0}^{3} (a_\nu(x_\mu) + i b_\nu(x_\mu)) e_\nu,$$

where $a_\nu(x_\mu)$, $b_\nu(x_\mu)$ are real valued functions, is said to be (left) $\mathcal{D}$-regular if it is of class $C^1$ and satisfies the equation

$$\mathcal{D}F(X) = 0. \tag{5.21}$$

If we set

$$(a_0, \mathbf{a}) = (a_0, a_1, a_2, a_3), \qquad (b_0, \mathbf{b}) = (b_0, b_1, b_2, b_3),$$

the regularity condition in vector notation can be written as

$$\begin{cases} \partial_{x_0} a_0 - \nabla \cdot \mathbf{a} = 0, \\ \partial_{x_0} b_0 - \nabla \cdot \mathbf{b} = 0, \\ \partial_{x_0} \mathbf{a} - \nabla a_0 + \nabla \times \mathbf{b} = 0, \\ \partial_{x_0} \mathbf{b} - \nabla b_0 - \nabla \times \mathbf{a} = 0. \end{cases} \tag{5.22}$$

This system represents Maxwell's equations (5.4) if the vectors $\mathbf{a}$ and $\mathbf{b}$ represent the magnetic field and the electric field respectively; $b_0$ is related to the electric density charge $\rho^e$ and to the electric current density $j^e$ by the relations

$$\rho^e = \partial_{x_0} b_0, \qquad j^e = -\nabla b_0.$$

Moreover, we have to assume that the scalar $a_0$ is a constant to avoid the existence of magnetic monopoles.

If we think of $F$ as a vector with 8 entries, we can write system (5.22) in the following matrix form

$$\begin{bmatrix} \partial_{x_0} & -\partial_{x_1} & -\partial_{x_2} & -\partial_{x_3} & 0 & 0 & 0 & 0 \\ -\partial_{x_1} & \partial_{x_0} & 0 & 0 & 0 & 0 & -\partial_{x_3} & \partial_{x_2} \\ -\partial_{x_2} & 0 & \partial_{x_0} & 0 & 0 & \partial_{x_3} & 0 & -\partial_{x_1} \\ -\partial_{x_3} & 0 & 0 & \partial_{x_0} & 0 & -\partial_{x_2} & \partial_{x_1} & 0 \\ 0 & 0 & 0 & 0 & \partial_{x_0} & -\partial_{x_1} & -\partial_{x_2} & -\partial_{x_3} \\ 0 & 0 & \partial_{x_3} & -\partial_{x_2} & -\partial_{x_1} & \partial_{x_0} & 0 & 0 \\ 0 & -\partial_{x_3} & 0 & \partial_{x_1} & -\partial_{x_2} & 0 & \partial_{x_0} & 0 \\ 0 & \partial_{x_2} & -\partial_{x_1} & 0 & -\partial_{x_3} & 0 & 0 & \partial_{x_0} \end{bmatrix} \begin{bmatrix} a_0 \\ a_1 \\ a_2 \\ a_3 \\ b_0 \\ b_1 \\ b_2 \\ b_3 \end{bmatrix} = 0. \tag{5.23}$$

The matrix associated to (5.23) can be written as

$$V(D) = \begin{bmatrix} A(D) & C(D) \\ -C(D) & A(D) \end{bmatrix}, \tag{5.24}$$

where

$$A(D) = A^t(D) = \begin{bmatrix} \partial_{x_0} & -\partial_{x_1} & -\partial_{x_2} & -\partial_{x_3} \\ -\partial_{x_1} & \partial_{x_0} & 0 & 0 \\ -\partial_{x_2} & 0 & \partial_{x_0} & 0 \\ -\partial_{x_3} & 0 & 0 & \partial_{x_0} \end{bmatrix}$$

and

$$C(D) = -C^t(D) = \begin{bmatrix} 0 & 0 & 0 & 0 \\ 0 & 0 & -\partial_{x_3} & \partial_{x_2} \\ 0 & \partial_{x_3} & 0 & -\partial_{x_1} \\ 0 & -\partial_{x_2} & \partial_{x_1} & 0 \end{bmatrix}.$$

We now want to extend the operator defined by the $8 \times 8$ matrix $V(D)$ to an operator acting not only on functions of a real biquaternion variable, but on functions

$$F : \mathbb{BH} \longrightarrow \mathbb{BH}$$

defined on the whole $\mathbb{BH}$. To this end, we define the operator

$$\mathcal{D}_Z = \frac{\partial}{\partial Z^+} = \frac{\partial}{\partial z_0} - \sum_{k=1}^{3} e_k \frac{\partial}{\partial z_k} \tag{5.25}$$

where

$$\frac{\partial}{\partial z_k} = \frac{\partial}{\partial x_k} - i\frac{\partial}{\partial y_k}, \quad k = 0, 1, 2, 3,$$

and $Z^+$ is defined in (5.17). In view of the above considerations it is clear that $\mathcal{D}_Z$ is a generalization of the Cauchy–Riemann operator $\partial/\partial\bar{z}$ for functions of one complex variable or (as we explained above) of the Cauchy–Fueter operator $\partial/\partial\bar{q}$. Where the more explicit notation $\partial/\partial Z^+$ is needed, we will use it instead of $\mathcal{D}_Z$.

**Definition 5.2.1.** *A function $F : U \subseteq \mathbb{BH} \longrightarrow \mathbb{BH}$ of class $C^1$ on the open set $U$ is said to be left $\mathcal{D}_Z$-regular if*

$$\mathcal{D}_Z F = 0$$

*and right $\mathcal{D}_Z$-regular if*

$$F\mathcal{D}_Z = 0$$

*where $F\mathcal{D}_Z = 0$ is defined by*

$$F\mathcal{D}_Z = \frac{\partial F}{\partial z_0} - \sum_{k=1}^{3} \frac{\partial F}{\partial z_k} e_k = 0.$$

It is immediate to verify that the matrix $U(D)$ associated to the operator $\mathcal{D}_Z$ is

$$U(D) = \begin{bmatrix} P(D) & Q(D) \\ -Q(D) & P(D) \end{bmatrix} \tag{5.26}$$

where

$$P(D) = \begin{bmatrix} \partial_{x_0} & -\partial_{x_1} & -\partial_{x_2} & -\partial_{x_3} \\ -\partial_{x_1} & \partial_{x_0} & \partial_{y_3} & -\partial_{y_2} \\ -\partial_{x_2} & -\partial_{y_3} & \partial_{x_0} & \partial_{y_1} \\ -\partial_{x_3} & \partial_{y_2} & -\partial_{y_1} & \partial_{x_0} \end{bmatrix}$$

and

$$Q(D) = \begin{bmatrix} \partial_{y_0} & -\partial_{y_1} & -\partial_{y_2} & -\partial_{y_3} \\ -\partial_{y_1} & \partial_{y_0} & -\partial_{x_3} & \partial_{x_2} \\ -\partial_{y_2} & \partial_{x_3} & \partial_{y_0} & -\partial_{x_1} \\ -\partial_{y_3} & -\partial_{x_2} & \partial_{x_1} & \partial_{y_0} \end{bmatrix}$$

so the Fourier transform of $U(D)$ is (up to imaginary units) the matrix

$$U = \begin{bmatrix} x_0 & -x_1 & -x_2 & -x_3 & y_0 & -y_1 & -y_2 & -y_3 \\ -x_1 & x_0 & y_3 & -y_2 & -y_1 & y_0 & -x_3 & x_2 \\ -x_2 & -y_3 & x_0 & y_1 & -y_2 & x_3 & y_0 & -x_1 \\ -x_3 & y_2 & -y_1 & x_0 & -y_3 & -x_2 & x_1 & y_0 \\ -y_0 & y_1 & y_2 & y_3 & x_0 & -x_1 & -x_2 & -x_3 \\ y_1 & -y_0 & x_3 & -x_2 & -x_1 & x_0 & y_3 & -y_2 \\ y_2 & -x_3 & -y_0 & x_1 & -x_2 & -y_3 & x_0 & y_1 \\ y_3 & x_2 & -x_1 & -y_0 & -x_3 & y_2 & -y_1 & x_0 \end{bmatrix}.$$

**Remark 5.2.2.** Both Maxwell's equations and the Cauchy–Fueter equations are now particular cases of $\mathcal{D}_Z F = 0$ when we restrict the domain and/or the range of the function $F$. Indeed, if we consider $F : \mathbb{H} \longrightarrow \mathbb{BH}$, the operator $\mathcal{D}_Z$ characterizes $\mathcal{D}$-regular functions in the sense of (5.21), and hence leads to Maxwell's equations, while, if we consider $F : \mathbb{H} \subset \mathbb{BH} \longrightarrow \mathbb{H} \subset \mathbb{BH}$ we obtain the usual quaternionic regular functions and hence the Cauchy–Fueter equations.

**Remark 5.2.3.** Another interesting feature of the operator $\mathcal{D}_Z$ is that it contains also the conditions of regularity for functions of two quaternionic variables. In fact, we can split a biquaternion as the sum of two quaternions $Z = q + iq'$, where

$$q = x_0 + i_1 y_1 + i_2 y_2 + i_3 y_3, \qquad q' = y_0 - i_1 x_1 - i_2 x_2 - i_3 x_3 \ \in \mathbb{H},$$

and we can think of a function $F$ defined on $\mathbb{BH}$ as $F = F(q, q')$. If we consider a function $F : \mathbb{BH} \longrightarrow \mathbb{H}$, where $F = F_0 + i_1 F_1 + i_2 F_2 + i_3 F_3$, and we impose $\mathcal{D}_Z F = 0$, we obtain the system

$$\begin{cases} \dfrac{\partial F}{\partial \bar{q}} = 0 \\[2ex] \dfrac{\partial F}{\partial \bar{q}'} = 0 \end{cases}$$

which corresponds to the Cauchy–Fueter system for functions of two quaternionic variables.

Now we provide a possible physical interpretation of the functions in the kernel of the operator $\mathcal{D}_Z$. If we write in vector notation the regularity condition $\mathcal{D}_Z F = 0$ we get the following system

$$\begin{cases} \partial_{x_0} a_0 + \partial_{y_0} b_0 - \nabla_x \cdot \mathbf{a} - \nabla_y \cdot \mathbf{b} = 0 \\ \partial_{y_0} a_0 - \partial_{x_0} b_0 - \nabla_y \cdot \mathbf{a} + \nabla_x \cdot \mathbf{b} = 0 \\ \partial_{x_0} \mathbf{a} + \partial_{y_0} \mathbf{b} - \nabla_x a_0 - \nabla_y b_0 + \nabla_x \times \mathbf{b} - \nabla_y \times \mathbf{a} = 0 \\ \partial_{x_0} \mathbf{b} - \partial_{y_0} \mathbf{a} - \nabla_x b_0 + \nabla_y a_0 - \nabla_x \times \mathbf{a} - \nabla_y \times \mathbf{b} = 0. \end{cases} \tag{5.27}$$

Let $K^e$ and $K^m$ be two Minkowski space-times with coordinates $(x_0, \mathbf{x})$ and $(y_0, \mathbf{y})$, respectively. The variables $x_0$ and $y_0$ represent the time coordinates, while $\mathbf{x}$ and $\mathbf{y}$ represent the spatial variables. We assume that in $K^e$ there are only electric monopoles and in $K^m$ there are only magnetic monopoles. If this holds, then the terms $a_0$ and $b_0$ depend only on some time and spatial variables, because

- $\partial_{y_0} a_0 := \rho^m(y_0, \mathbf{y})$ is the magnetic monopole density in $K^m$
- $\partial_{x_0} b_0 := \rho^e(x_0, \mathbf{x})$ is the electric monopole density in $K^e$
- $-\nabla_y \, a_0 := \mathbf{j}^m(y_0, \mathbf{y})$ is the magnetic current density in $K^m$
- $-\nabla_x \, b_0 := \mathbf{j}^e(x_0, \mathbf{x})$ is the electric current density in $K^e$.

More precisely we have that the functions

$$a_0 = a_0(y_0, \mathbf{y}), \quad b_0 = b_0(x_0, \mathbf{x}) \tag{5.28}$$

depend only on the variables indicated. This implies that

$$\partial_{x_0} a_0(y_0, \mathbf{y}) = \partial_{y_0} b_0(x_0, \mathbf{x}) = \nabla_x \, a_0(y_0, \mathbf{y}) = \nabla_y \, b_0(x_0, \mathbf{x}) = 0. \tag{5.29}$$

We also require that the spaces $K^e$ and $K^m$ are orthogonal with respect to the Minkowski metric $g_{\mu\nu}$. In other words, we require that

$$< (x_0, \mathbf{x}), g_{\mu\nu} \, (y_0, \mathbf{y}) >= 0. \tag{5.30}$$

Because of the symmetry of the problem, in $K^e$ the fields $\mathbf{a}$ and $\mathbf{b}$ cannot depend on the variables $(y_0, \mathbf{y})$ while in $K^m$ they cannot depend on $(x_0, \mathbf{x})$. If we now replace (5.28) in system (5.27), we obtain two systems of Maxwell's equations, one related to the Minkowski space-time $K^e$ and the other one related to $K^m$. If we set $\mathbf{a} := \mathbf{B}$ and $\mathbf{b} := \mathbf{E}$ we obtain the usual Maxwell's equations in $K^e$ (compare with (5.3)), while in $K^m$ we get

$$\begin{cases} \rho^m - \nabla_y \cdot \mathbf{B} = 0 \\ \nabla_y \cdot \mathbf{E} = 0 \\ \partial_{y_0} \mathbf{E} - \nabla_y \times \mathbf{B} = 0 \\ \partial_{y_0} \mathbf{B} + \mathbf{j}^m + \nabla_y \times \mathbf{E} = 0 \end{cases} \tag{5.31}$$

which are Maxwell's equations for magnetic monopoles only. We also have another natural way to split the system (5.27). Let $K^e$ and $K^m$ be the two Minkowski space-times with the coordinate systems specified above. We consider now the mixed pairs of variables $(x_0, \mathbf{y})$ and $(y_0, \mathbf{x})$. We now suppose that we deal with a particular symmetric problem in which only the coordinates $(x_0, \mathbf{y})$ are considered, while $(y_0, \mathbf{x})$ are neglected. In this way it is easy to derive from the system (5.27) the Cauchy–Fueter equations. We obtain the same result if we consider such a problem in the $(y_0, \mathbf{x})$ coordinates neglecting $(x_0, \mathbf{y})$.

It is well known that, in general, regular functions on a non-division algebra are not harmonic, see [171]. In fact, in our case we have that

$$\frac{\partial}{\partial Z}\frac{\partial}{\partial Z^+} = \frac{\partial}{\partial Z^+}\frac{\partial}{\partial Z} = \frac{\partial^2}{\partial z_0^2} - \sum_{k=1}^{3}\frac{\partial^2}{\partial z_k^2}$$

$$= \Delta_{\mathbf{y}} - \Delta_{\mathbf{x}} + \Big(\frac{\partial}{\partial x_0} + i\frac{\partial}{\partial y_0}\Big)^2 - 2i\sum_{k=1}^{3}\frac{\partial^2}{\partial x_k \partial y_k}, \tag{5.32}$$

which is an ultra-hyperbolic operator.

Note that the operator defined in (5.32) in the case of Maxwell's equations becomes

$$\Delta_{\mathbf{x}} - \frac{\partial^2}{\partial x_0^2} \text{ in } K^e \text{ and } \Delta_{\mathbf{y}} - \frac{\partial^2}{\partial y_0^2} \text{ in } K^m. \tag{5.33}$$

These two D'Alembert operators imply that in $K^e$ and $K^m$, seen as separated spaces, it is possible to have wave propagation phenomena.

The operator defined in (5.32) splits into the following Laplace operators:

$$\Delta_{\mathbf{x}} + \frac{\partial^2}{\partial y_0^2} \text{ and } \Delta_{\mathbf{y}} + \frac{\partial^2}{\partial x_0^2} \text{ in } K^e \oplus K^m \tag{5.34}$$

whose solutions do not permit wave propagation from $K^e$ to $K^m$ and vice-versa.

We can summarize the above considerations as follows: if we consider separately $K^e$ and $K^m$ we obtain propagation phenomena, while if we consider an easy symmetric problem related to some coordinates in $K^e \oplus K^m$ we find that electromagnetic waves cannot propagate.

## 5.3 Properties of $\mathcal{D}_Z$-regular functions

Throughout this section, $R$ will be either the ring

$$R = \mathbb{R}[x_0, x_1, x_2, x_3, y_0, y_1, y_2, y_3]$$

of polynomials in the eight real variables $x_0, \dots, y_3$ or the ring

$$R = \mathbb{R}[x_0, x_1, x_2, x_3]$$

of polynomials in the real variables $x_0, \dots, x_3$. The context will make it clear to the reader which case is being used. First, we use CoCoA to compute the Ext-modules associated to the module $M$ generated by the matrix $U$, i.e.,

$$M = R^8/U^t R^8.$$

**Proposition 5.3.1.** *The module* $\text{Ext}^0(M, R)$ *is zero.*

According to Proposition 2.1.2, this result implies the unique continuation property for the solutions of the system $\mathcal{D}_Z F = 0$. We now show that

$$\text{Ext}^1(M, R) \neq 0,$$

in fact, using Theorem 5.1.1 we obtain the following:

**Proposition 5.3.2.** *The characteristic variety of* $\text{Ext}^1(M, R)$ *is given by the polynomial* $f = |N(Z)^2|^2$.

*Proof.* Since the matrix $U$ is square and its determinant can be computed as

$$f = \left[(N(X) - N(X'))^2 + 4 < X, g_{\mu\nu} X' >^2\right]^2 = |N(Z)^2|^2,$$

we conclude that $\text{Ext}^1(M, R) \neq 0$. □

Note that we can obtain the same conclusion noticing that the syzygy module of $M$ is zero and so $\text{Ext}^1(M, R)$ is isomorphic to $M$. This result, while not surprising (a similar phenomenon occurs for holomorphic functions in one complex variable, as well for regular functions in one quaternionic variable), can be given the analytic interpretation that $\mathcal{D}_Z$-regular functions can have compact singularities.

If we wish to repeat the discussion for $\mathcal{D}$-regular functions, it suffices to consider the matrix $V(D)$. Then, again, it is easy to find that

$$\text{Ext}^1(M, R) \neq 0$$

and that its characteristic variety is defined by the polynomial

$$g = (x_0^2 - x_1^2 - x_2^2 - x_3^2)^4 = N(X)^4$$

i.e., the characteristic variety is, geometrically, the light cone in the dual space. The fact that $\text{Ext}^1(M, R) \neq 0$ means that the system is not overdetermined and, therefore, compact singularities cannot be eliminated. Given the physical meaning of the system $V(D)$, the possible existence of compact singularities may be interpreted as the presence of localized charges which generate the electromagnetic field.

One may be tempted to change this situation by looking for special situations in which the fields **a** and **b** are proportional (plane waves) or even $\mathbf{a} = \mathbf{b}$ (remember that one can interpret **a** as the magnetic field and **b** as the electric field). This suggests the study of a new system

$$\begin{cases} (A + C)\mathbf{a} = 0 \\ (A - C)\mathbf{a} = 0 \end{cases}$$

associated to an $8 \times 4$ matrix whose Fourier transform is

$$\begin{bmatrix} x_0 & -x_1 & -x_2 & -x_3 \\ -x_1 & x_0 & -x_3 & x_2 \\ -x_2 & x_3 & x_0 & -x_1 \\ -x_3 & -x_2 & x_1 & x_0 \\ x_0 & -x_1 & -x_2 & -x_3 \\ -x_1 & x_0 & x_3 & -x_2 \\ -x_2 & -x_3 & x_0 & x_1 \\ -x_3 & x_2 & -x_1 & x_0 \end{bmatrix}.$$

At a first glance we might expect overdeterminacy (because the system is not represented by a square matrix) but, in fact, a quick computation shows that, once again, $\mathrm{Ext}^1(M, R) \neq 0$. Physically this may be interpreted by saying that, in the presence of an electromagnetic field, electric charges must exist. Once again, the obstruction to the vanishing of $\mathrm{Ext}^1(M, R)$ is given by the light cone and the characteristic variety is

$$\{X: \ N(X) = x_0^2 - x_1^2 - x_2^2 - x_3^2 = 0\}.$$

Let us come back to the study of $\mathcal{D}_Z$-regular functions. Not surprisingly, we get a result that parallels Theorem 5.1.3.

**Proposition 5.3.3.** *Let $\Omega$ be an open set in $\mathbb{BH}$ and let $P \in \Omega$. Then every $\mathcal{D}_Z$-regular function on $\Omega \setminus \{P\}$ whose components extend as distributions to all of $\Omega$ is a distribution solution to the system $\mathcal{D}_Z$ on all $\Omega$.*

**Remark 5.3.1.** Note that since the system $\mathcal{D}_Z$ is not elliptic distributions solutions are not necessarily $\mathcal{D}_Z$-regular functions, unlike what happens for the classical Cauchy–Fueter system.

**Remark 5.3.2.** It is important to observe that this phenomenon is quite new, as one-point singularities do occur for regular functions (in the Cauchy–Riemann and Cauchy–Fueter sense) in one variable. A completely analogous result can be formulated for $\mathcal{D}$-regular functions.

We now turn our attention to what happens for functions of several biquaternionic variables. For the sake of simplicity we will restrict our attention to the case of a function $F : (\mathbb{BH})^2 \longrightarrow \mathbb{BH}$ of two biquaternionic variables $Z$ and $W$. In this case, we will say that $F$ is $\mathcal{D}_Z$-regular in the variables $Z$, $W$ if

$$\frac{\partial F}{\partial Z^+} = \frac{\partial F}{\partial W^+} = 0.$$

As usual we can use CoCoA to compute the resolution of the module, where $R$ is the ring of polynomials in 16 real variables, 8 for each biquaternionic variable. Without giving any computational detail, we have:

**Theorem 5.3.1.** *The module $M$ associated to the system*

$$\begin{cases} \dfrac{\partial F}{\partial Z^+} = 0 \\[2ex] \dfrac{\partial F}{\partial W^+} = 0 \end{cases}$$

*admits a resolution of length 4*

$$0 \longrightarrow R^8(-4) \longrightarrow R^{16}(-3) \longrightarrow R^{16}(-1) \longrightarrow R^8 \longrightarrow M \longrightarrow 0.$$

*We also have*

$$\text{Ext}^j(M,R) = 0, \qquad j = 0,1,2,$$

*while* $\text{Ext}^3(M,R) \neq 0$. *Therefore, the property of removability of compact singularities holds.*

Note that this resolution mimics the resolution in the case of two Cauchy–Fueter operators therefore, from an algebraic point of view, the complexification gives rise to an operator $\mathcal{D}_Z$ that cannot be distinguished from the real Cauchy–Fueter operator.

From an analytic point of view the two theories are quite different and we provide here some results that highlight the main differences. The set of left $\mathcal{D}_Z$-regular functions on an open set $U \subseteq \mathbb{BH}$ will be denoted by $\mathcal{R}_l^{\mathcal{D}_Z}(U)$ while the set of right $\mathcal{D}_Z$-regular functions will be denoted by $\mathcal{R}_r^{\mathcal{D}_Z}(U)$. When no confusion arises we will omit the indices $l$ or $r$. The following fact is immediate to verify and it is, in fact, a natural property shared by all functions in Clifford Analysis in the kernel of Dirac operators or their variations:

**Proposition 5.3.4.** *$\mathcal{R}_l^{\mathcal{D}_Z}(U)$ is a right $\mathbb{BH}$–module.*

A significant result in this theory is the following:

**Lemma 5.3.1.** (Poincaré Lemma). *Let $U \subseteq \mathbb{BH}$ be a convex open set, and let $g : U \longrightarrow \mathbb{BH}$ be a $C^\infty$ function. Then there is $f \in C^\infty(U)$ such that*

$$\mathcal{D}_Z f = g \quad \text{on} \quad U.$$

*Proof.* The computation of the syzygy module of the columns of the matrix $\mathcal{D}_Z$ was done in Proposition 5.3.1, and shows that the matrix of the compatibility conditions on $g$ is the null matrix. □

We now introduce some differential forms aimed at the formulation of a Cauchy–type integral formula. We set

$$DZ = dz_1 \wedge dz_2 \wedge dz_3 + e_1 dz_0 \wedge dz_2 \wedge dz_3$$

$$+ e_2 dz_0 \wedge dz_3 \wedge dz_1 + e_3 dz_0 \wedge dz_1 \wedge dz_2$$

and

$$\nu = dz_0 \wedge dz_1 \wedge dz_2 \wedge dz_3.$$

We have the following immediate generalization of the quaternionic case.

**Proposition 5.3.5.** *On* $\mathbb{BH}$ *we have*

$$d(g\ DZ\ f) = dg \wedge DZ\ f - g\ DZ \wedge\ df = \{(g\mathcal{D}_Z)f + g(\mathcal{D}_Z f)\}\,\nu.$$

**Corollary 5.3.1.** *If $f$ is left $\mathcal{D}_Z$-regular and $g$ is right $\mathcal{D}_Z$-regular, we obtain*

$$d(g\ DZ\ f) = 0.$$

*Moreover, a real differentiable function $f$ is $\mathcal{D}_Z$-regular at the point $Z$ if and only if*

$$DZ \wedge\ df = 0$$

*at the point $Z$.*

From this last property and Stokes' Theorem, we can obtain the Cauchy formula for $\mathcal{D}_Z$-regular functions. First, we define a function which is both left and right regular which we will call Cauchy–Fueter kernel for biquaternions:

**Definition 5.3.1.** *We define the Cauchy–Fueter kernel for biquaternions as*

$$G(Z) = \frac{Z^+}{N(Z)^2},$$

*where $N(Z)$ is defined by (5.18).*

**Proposition 5.3.6.** *The Cauchy–Fueter kernel $G(Z)$ is left and right regular on $\mathbb{BH} \setminus \{Z \mid N(Z) = 0\}$, that is $\mathcal{D}_Z G = G\mathcal{D}_Z = 0$.*

*Proof.* We prove that $G(Z)$ is left regular. The other case is similar. Consider the explicit definition

$$G(Z) = \frac{x_0 - \sum_{\ell=1}^3 e_\ell x_\ell + iy_0 - i\sum_{\ell=1}^3 e_\ell y_\ell}{(z_0^2 - z_1^2 - z_2^2 - z_3^2)^2}.$$

Then, an easy computation gives

$$\begin{aligned}
\partial_{x_0} G(Z) &= N(Z)^{-4}(N(Z)^2 - 4z_0 Z^+ N(Z)), \\
\partial_{x_\ell} G(Z) &= N(Z)^{-4}(4z_\ell Z^+ N(Z) - e_\ell N(Z)^2), \qquad \ell = 1,2,3, \\
\partial_{y_0} G(Z) &= N(Z)^{-4}(iN(Z)^2 - 4iz_0 Z^+ N(Z)), \\
\partial_{y_\ell} G(Z) &= N(Z)^{-4}(4iz_\ell Z^+ N(Z) - ie_\ell N(Z)^2), \qquad \ell = 1,2,3,
\end{aligned}$$

so that applying operator $\mathcal{D}_Z$ to $G(Z)$ and replacing the derivatives calculated above we get

$$\mathcal{D}_Z G(Z) = \frac{8N(Z)^2 - 4Z^+(2z_0 + 2z_1 + 2z_2 + 2z_3)N(Z)}{N(Z)^4}$$

$$= \frac{8N(Z)^2 - 8Z^+ZN(Z)^2}{N(Z)^4} = 0.$$

This completes the proof. □

**Theorem 5.3.2.** (Cauchy formula I). *Let $U \subset \mathbb{BH}$ be an open set, and let $\Sigma$ be a compact 3-chain, boundary of a 4-chain $S$ in $U$. Then, if $g$ is right $\mathcal{D}_Z$-regular and $f$ is left $\mathcal{D}_Z$-regular, the identity*

$$\int_\Sigma g \; DZ \; f = 0$$

*holds.*

*Proof.* It is a direct consequence of the Stokes' Theorem. In fact, from the equality

$$\int_\Sigma g \; DZ \; f = \int_S d(g \; DZ \; f)$$

and from the relation

$$d(g \; DZ \; f) = \{(g\mathcal{D}_Z)f + g(\mathcal{D}_Z f)\}\nu$$

which follows from standard computations, we get the statement. □

It is known that, in general, the Cauchy formula holds without limitations in any real Clifford algebra. However, in complexified Clifford algebras that are not division algebras, the Cauchy kernel at a point $P$ is not necessarily defined, and in our case, it is defined only outside the translated light cone, i.e., in $\mathbb{BH} \setminus \{N(Z - Z_P) = 0\}$. Moreover, as the authors point out in [35] (where some integral formulas in hypercomplex analysis are given), if a contour of integration $\Sigma$ is homologically trivial in an open set $\Omega$ contained in a complexified Clifford algebra, it is not necessarily true that $\Sigma$ is homologically equivalent in $\Omega \setminus \{N(Z - Z_P) = 0\}$ to a sphere around $P$, so a suitable hypothesis is needed.

**Definition 5.3.2.** *We define the light cone with vertex in $P$ to be the set* $\mathbb{C}N_P = \{Z \in \mathbb{BH} \; : \; N(Z - Z_P) = 0\}$.

**Remark 5.3.3.** Identifying $\mathbb{BH}$ with the space of $2 \times 2$ matrices with complex coefficients and the biquaternion $Z = e_0 z_0 + \ldots + z_3$ with the matrix

$$\begin{bmatrix} z_0 + z_3 & z_1 - iz_2 \\ z_1 + iz_2 & z_0 - z_3 \end{bmatrix},$$

we see that $N(Z)$ is its determinant and therefore one can identify $\mathbb{BH} \setminus \mathbb{C}N_O$ with $GL(2, \mathbb{C})$. This has the homotopy type of the maximal compact subgroup $U(2)$, and the nontrivial cohomology of this space is $\mathbb{Z}$ only in dimensions 0, 1, 3, 4. Following [35] it is possible to compute explicitly a generator for the group $H_3(\mathbb{BH} \setminus \mathbb{C}N_P, \mathbb{Z})$, and so to characterize the cycles that can be used to write the Cauchy formula.

**Theorem 5.3.3.** *A generator for the group $H_3(\mathbb{BH}\backslash\mathbb{C}N_P, \mathbb{Z}) \cong \mathbb{Z}$ is the element*

$$S^3 = \{Z_P + Z \;:\; Z = x_0 + iy_1 + jy_2 + ky_3 \in \mathbb{H},\ x_0^2 + \sum_{i=0}^{3} y_i^2 = 1\}.$$

*Proof.* Without loss of generality, we can assume that $P$ is the zero biquaternion $O$. Obviously, there exists an inclusion $i : S^3 \longrightarrow \mathbb{BH}\backslash\mathbb{C}N_O$, so it suffices to show that the morphism

$$i^* : H_3(S^3, \mathbb{Z}) \longrightarrow H_3(\mathbb{BH}\backslash\mathbb{C}N_O, \mathbb{Z})$$

induced by $i$ is an isomorphism. Let us define the hypersurfaces $S^1$ and $S^7$ by

$$S^1 = \{z = x_0 + iy_0 \in \mathbb{C} \;:\; x_0^2 + y_0^2 = 1\},$$

$$S^7 = \{Z \in \mathbb{BH} \;:\; \sum_{i=0}^{3}(x_i^2 + y_i^2) = 1\}.$$

Let $E = S^7 \cap (\mathbb{BH}\backslash\mathbb{C}N_O)$ and let $p : E \longrightarrow S^1$ be the fibration

$$Z \in E \longrightarrow \frac{N(Z)}{|N(Z)|},$$

where $|N(Z)|$ denotes the absolute value of the complex number $N(Z)$ computed in $\mathbb{C}$. Let us now consider $F_0 = p^{-1}(1)$. We have that

$$F_0 = \Big\{Z \in \mathbb{BH} \;:\; \sum_{i=0}^{3}(x_i^2 + y_i^2) = 1,\ x_0y_0 - \sum_{i=1}^{3} x_iy_i = 0,$$

$$x_0^2 - y_0^2 - \sum_{i=1}^{3}(x_i^2 - y_i^2) > 0\Big\}.$$

We obtain the following inclusions:

$$S^3 \overset{i''}{\hookrightarrow} F_0 \overset{i'}{\hookrightarrow} E \overset{i}{\hookrightarrow} \mathbb{BH}\backslash\mathbb{C}N_O.$$

The proof that $i'$ and $i$ induce isomorphisms at the homology level can be done exactly as in [35]. In fact, the projection along the rays from the origin onto $S^7$ shows that $E$ is a deformation retract of $\mathbb{BH}\backslash\mathbb{C}N_O$ so that $i'^*$ is an isomorphism. The Wang sequence (see [135]) assures us that $i^*$ is an isomorphism. Finally, $S^3$ is a deformation retract of $F_0$. In fact, let us define

$$\theta_t : F_0 \to F_0, \qquad t \in [0, 1]$$

as

$$\theta_t(Z) = \rho x_0 + ity_0 + e_1(tx_1 + i\rho y_1) + e_2(tx_2 + i\rho y_2) + e_3(tx_3 + i\rho y_3),$$

where

$$\rho = \sqrt{\frac{1 - t^2\beta^2}{\alpha^2}}, \qquad \sqrt{x_0^2 + \sum_{i=1}^{3} y_i^2}, \qquad \sqrt{y_0^2 + \sum_{i=1}^{3} x_i^2}.$$

For $t = 1$ we have $\rho = 1$ and $\theta_1$ is the identity, while for $t = 0$, we have $\rho = 1/\alpha$ and $\theta_0 : \ F_0 \to S^1$, so $\theta_t$ is a retraction and also $i''^*$ is an isomorphism. □

As we observed before, it is necessary to require a suitable assumption on $\Omega$ to guarantee that, when the contour of integration $\Sigma$ is near the cone $\mathbb{C}N_P$, we have the possibility to follow rays on $\mathbb{C}N_P$ to approach the point $P$. This is done in the following:

**Definition 5.3.3.** *A domain $\Omega \subset \mathbb{BH}$ is said to be null-convex if for all $Z$, $Z' \in \Omega$ such that $N(Z - Z') = 0$, the whole segment $ZZ'$ belongs to $\Omega$. Note that if $\Omega$ is null convex, then a cycle $\Sigma$ can be deformed to a sphere around a point $P$ near the cone $\mathbb{C}N_P$.*

**Theorem 5.3.4.** (Cauchy formula II). *Let $\Omega \subset \mathbb{BH}$ be a null-convex domain and let $f \in \mathcal{R}^{\mathcal{D}_Z}(\Omega)$. If $P \in \Omega$, then*

$$f(P) Ind_\Sigma(P) = \frac{1}{2\pi^2} \int_\Sigma G(Z - P)\ DZ\ f(Z)$$

*where $\Sigma \subset \Omega$ is any cycle homologous to the 3-sphere $S^3$.*

*Proof.* We prove the theorem in the case $P = O$. It is obvious that the theorem holds for $\Sigma = S^3$. In fact, it suffices to repeat the arguments given in [206] since $S^3$ is a sphere in $\mathbb{H}$. In the general case, it suffices to use the fact that $\Sigma \sim nS^3$. □

To have a theory of $\mathcal{D}_Z$-regular functions that parallels that of regular functions of a quaternionic variable, it is also necessary to write Taylor series expansions for $\mathcal{D}_Z$-regular functions. We observe that the operator $\mathcal{D}_Z$ can be decomposed into two operators as

$$\mathcal{D}_Z = \mathcal{D}_{x_0,\mathbf{x}} + i\mathcal{D}_{y_0,\mathbf{y}},$$

where

$$\mathcal{D}_{x_0,\mathbf{x}} := \frac{\partial}{\partial x_0} - \sum_{j=1}^{3} e_j \frac{\partial}{\partial x_j}$$

so it coincides with $\mathcal{D}$, and

$$\mathcal{D}_{y_0,\mathbf{y}} := \frac{\partial}{\partial y_0} - \sum_{j=1}^{3} e_j \frac{\partial}{\partial y_j}.$$

Let us set

$$p_\nu(Z) = \frac{1}{n!} \sum_{1 \le k_1, \dots, k_n \le 3} (z_{k_1} + e_{k_1} z_0) \dots (z_{k_n} + e_{k_n} z_0),$$

where $\nu = (n_1, n_2, n_3)$ is such that $n_\ell$ is the number of indices equal to $k_\ell$ and $n_1 + n_2 + n_3 = n$; the sum is taken over all different orderings of $n_1$ elements equal to 1, $n_2$ 2's, $n_3$ 3's. Let us denote by $\sigma_n$ the set of $[n_1, n_2, n_3]$ such that $n_1 + n_2 + n_3 = n$. If $F$ is a $\mathcal{D}$-regular function, then

$$F(Z) = \sum_{n=0}^{+\infty} \sum_{\nu \in \sigma_n} a_\nu p_\nu(Z), \qquad a_\nu \in \mathbb{BH}. \tag{5.35}$$

A comparison with the results in Section 3.1 shows that the polynomials $p_\nu(Z)$ are not only $\mathcal{D}$-regular (i.e., $\mathcal{D}_{x_0,\mathbf{x}}$-regular, with our notation) but also $\mathcal{D}_{y_0,\mathbf{y}}$-regular because the $x$ variables and $y$ variables play a symmetric role. So expansion (5.35) holds not only for $\mathcal{D}$-regular functions (with respect to the variables $x$ or $y$) but also for $\mathcal{D}_Z$-regular functions.

**Remark 5.3.4.** Expansion (5.35) implies that a $\mathcal{D}_Z$-regular function is a real infinitely differentiable function.

Now we show how it is possible to generalize the notion of hyperfunctions to the setting of $\mathcal{D}$-regular functions. The set of (left) $\mathcal{D}$-regular functions on an open set $U \subseteq \mathbb{H}$ will be denoted by $\mathcal{R}^{\mathcal{D}}$. Let us start with the following obvious fact whose proof we leave to the reader.

**Proposition 5.3.7.** *Let $U$ be any open set in $\mathbb{H}$. The assignment $U \longrightarrow \mathcal{R}^{\mathcal{D}}(U)$ is a sheaf of right $\mathbb{BH}$-modules.*

Now we prove two basic facts necessary to be able to define hyperfunctions. The first result is the following.

**Proposition 5.3.8.** *Let $\Omega$ be a relatively compact set in an open set $U \subset \mathbb{H}$. Then $H^0_\Omega(U, \mathcal{R}^{\mathcal{D}}) = 0$.*

*Proof.* We have shown that $\mathrm{Ext}^0(M, R) = 0$. This fact (see [13]) is equivalent to the vanishing of $H^0_\Omega(U, \mathcal{R}^{\mathcal{D}}) = 0$. □

Another basic result that we need is the cohomological version of the Mittag–Leffler theorem (whose proof is standard).

**Theorem 5.3.5.** *Let $U \subset \mathbb{H}$ be an open set. Then*

$$H^1(U, \mathcal{R}^{\mathcal{D}}) = 0.$$

This last result does not hold if we consider $\mathcal{D}$-regular or $\mathcal{D}_Z$-regular functions of several variables, because Hartogs' phenomenon holds. Let $\Omega$ be a relatively compact set in an open set $U$ contained in $\mathbb{H}$. We have the long exact sequence

$$0 \longrightarrow H^0_\Omega(U, \mathcal{R}^{\mathcal{D}}) \longrightarrow H^0(U, \mathcal{R}^{\mathcal{D}}) \longrightarrow H^0(U \backslash \Omega, \mathcal{R}^{\mathcal{D}})$$

$$\longrightarrow H^1_\Omega(U,\mathcal{R}^\mathcal{D}) \longrightarrow H^1(U,\mathcal{R}^\mathcal{D}) \longrightarrow \dots.$$

We know that $H^0_\Omega(U,\mathcal{R}^\mathcal{D}) = 0$ and $H^1(U,\mathcal{R}^\mathcal{D}) = 0$. Thus, we obtain the isomorphism

$$H^1_\Omega(U,\mathcal{R}^\mathcal{D}) \cong \frac{H^0(U\backslash\Omega,\mathcal{R}^\mathcal{D})}{H^0(U,\mathcal{R}^\mathcal{D})}.$$

According to the notation given in (3.13) we can give the following definition:

**Definition 5.3.4.** *Let $\Omega$ be an open set in $\widetilde{\mathbb{H}}$ and let $U$ be an open set in $\mathbb{H}$ such that $\Omega$ is relatively closed in $U$. Then the right module defined by*

$$\mathcal{F}(\Omega) \cong H^1_\Omega(U,\mathcal{R}^\mathcal{D}) \cong \frac{H^0(U\backslash\Omega,\mathcal{R}^\mathcal{D})}{H^0(U,\mathcal{R}^\mathcal{D})}$$

*is called the module of (left) $\mathbb{H}$-hyperfunctions.*

**Remark 5.3.5.** The above proposition and theorem imply that the definition is well-posed and does not depend on the open set $U$.

**Theorem 5.3.6.** *The correspondence*

$$\Omega \longrightarrow \mathcal{F}(\Omega),$$

*for any open set $\Omega \subset \widetilde{\mathbb{H}}$ defines a flabby sheaf on $\widetilde{\mathbb{H}}$.*

## 5.4 The Dirac equation and the linearization problem

The nonrelativistic wave mechanics of particles with spin 1/2 rests on the assumption that the two-components wave function obeys the Schrödinger equation

$$i\partial_t\Psi(t,x) = -\frac{1}{2m}\Delta\Psi(t,x) + U\Psi(t,x)$$

which can be obtained replacing the quantum mechanical operators $i\partial_t$, $-i\nabla_x$, associated to energy and momentum in the classical mechanical conservation principle

$$E = \frac{1}{2m}P^2 + U,$$

where $E$ is the total energy, $P^2/(2m)$ is the kinetic energy, and $U$ is the potential energy. Obviously, the operator associated to the potential energy $U$ is the multiplication.

This equation splits into two equations for the components $\Psi^+$ and $\Psi^-$ which are independent if the potential $U$ does not involve the spin. To deduce the relativistic equation for the two components wave function is not so simple. Since we are looking for a relativistic wave equation and in the theory of relativity space and time are treated in a unified way, we require that the equation must be of the first order with respect to both space and time derivatives. Moreover, in order that the superposition principle remains valid, we require it to be linear. Finally, we also ask that the components $\Psi^+$ and $\Psi^-$ are coupled even in the absence of external spin-dependence forces, in such a way that, even for a free particle, an interdependence of the translation and spin degrees of freedom will be established. The equation must be invariant under space and time translations, i.e., it may not contain the spatial variables $x$ or the time variable $t$ explicitly, and it must be invariant under space rotations. These conditions are required in order that, for a free particle, the four-momentum and the angular momentum be constants of motion. Let us introduce the vector $\sigma$ which is proportional to the magnetic moment associated to the spin angular momentum.

The rotation invariance can be satisfied only if the operator $\sigma$ occurs in the scalar product $\sigma \cdot \mathbf{v}$ where $\mathbf{v}$ is a vector. Moreover, to take into account the space derivatives we introduce the term $\sigma \cdot \nabla_x$. From the above considerations we look for an equation of type

$$[\partial_t + \lambda_1 \sigma \cdot \nabla_x + \lambda_2]\Psi = 0 \tag{5.36}$$

and we try to determine the constants $\lambda_1$ and $\lambda_2$. We recall that the identity

$$(\sigma \cdot \nabla_x) \cdot (\sigma \cdot \nabla_x) = \Delta + i\varepsilon_{rst}\sigma_t \nabla_{x_r} \nabla_{x_s} = \Delta$$

holds thanks to the symmetry of $\nabla_{x_r}\nabla_{x_s}$ and the antisymmetry of the tensor $\varepsilon_{rst}$ defined below in (5.37). We now apply the operator

$$-\partial_t + \lambda_1 \sigma \cdot \nabla_x - \lambda_2$$

to (5.36) to get the wave equation

$$\lambda_1^2 \Delta\Psi(t,x) - \partial_{tt}^2\Psi(t,x) - 2\partial_t\Psi(t,x) - \lambda_2^2\Psi(t,x) = 0.$$

We observe that the spin has disappeared and we require that the equation has plane waves solutions of type

$$\Psi(t,x) = e^{i\langle (k_0,\mathbf{k}),(x_0,\mathbf{x})\rangle}$$

so that, since $(k_0,\mathbf{k}) = (k_0,k_1,k_2,k_3)$ and $(x_0,\mathbf{x}) = (x_0,x_1,x_2,x_3)$, we get

$$k_0^2 + 2\lambda_2 k_0 - \lambda_1^2|\mathbf{k}|^2 - \lambda_2^2 = 0.$$

The solutions of this equation are

$$k_0 = -i\lambda_2 \pm \lambda_1|\mathbf{k}|$$

which have physical meaning only in the case $\lambda_1 = 1$ and $\lambda_2 = 0$, which yields

$$k_0 = \pm|\mathbf{k}|.$$

In fact, this is the energy momentum relation for a particle of zero rest mass, and the wave equation for $\lambda_1 = 1$ and $\lambda_2 = 0$, becomes the Klein–Gordon equation for massless particles. This leads us to the Weyl equation, which is relativistic, for particles of spin 1/2 but does not contain the mass.
To deduce the Dirac equation, i.e., to take also the mass into account, we cannot assume $\lambda_1$ and $\lambda_2$ to be real numbers. We shall see that they must be matrices.

Let us start from the relativistic energy conservation law without the potential term

$$E^2 = p^2 + m^2,$$

i.e.,

$$E = \sqrt{p^2 + m^2}.$$

Replacing formally the energy and momentum operators, we obtain the equation

$$i\partial_t \Psi(t,x) = \sqrt{m^2 - \Delta}\ \Psi(t,x).$$

Even though the problem of the square root of the operator $m^2 - \Delta$ can be solved, the main trouble of this equation is that it is not covariant under the homogeneous Lorentz group. To get the covariant form of the equation, we look for a linear equation of the previous type

$$E\psi(t,x) = (\alpha \cdot \mathbf{p} + \beta m)\psi(t,x)$$

that is, we want to linearize the square root

$$\sqrt{p^2 + m^2} = \alpha \cdot \mathbf{p} + \beta m,$$

assuming now that $\alpha$ and $\beta$ are Hermitian matrices. The relativistic energy momentum can be obtained naturally taking the square

$$E^2\Psi(t,x) = (\alpha \cdot \mathbf{p} + \beta m)^2 \Psi(t,x) = (p^2 + m^2)\Psi(t,x).$$

We now require that this relation holds for all $\Psi$, obtaining

$$(\sum_{j=1}^{3} \alpha_j p_j + \beta m)^2 = (\sum_{j=1}^{3} \alpha_j p_j)^2 + \beta^2 m^2 + m \sum_{j=1}^{3} \alpha_j p_j \beta + m\beta \sum_{j=1}^{3} \alpha_j p_j.$$

By setting

$$\{A, B\} := AB + BA$$

we have

$$m \sum_{j=1}^{3} \alpha_j p_j \beta + m\beta \sum_{j=1}^{3} \alpha_j p_j = m \sum_{j=1}^{3} \{\beta, \alpha_j\} p_j + \frac{1}{2}\{\alpha_j, \alpha_k\}_{k \neq j} p_j p_k.$$

To get the conservation principle, we have to impose the conditions

$$\beta^2 = (\alpha_j)^2 = 1,$$

$$\{\beta, \alpha_j\} = \{\alpha_j, \alpha_k\}_{k \neq j} = 0.$$

The matrices $\alpha_j$, $\beta$ can be determined using the following two propositions whose proofs are left to the reader.

**Proposition 5.4.1.** *The matrices $\beta$ and $\alpha_j$, $j = 1, 2, 3$ have null trace and eigenvalues $\pm 1$.*

**Proposition 5.4.2.** *The size of the matrices $\beta$ and $\alpha_j$, $j = 1, 2, 3$ is even and strictly greater then 2.*

In the case where we choose the size to be 4 we can take

$$\beta = \begin{bmatrix} \sigma_0 & 0 \\ 0 & -\sigma_0 \end{bmatrix},$$

$$\alpha_j = \begin{bmatrix} 0 & \sigma_j \\ \sigma_j & 0 \end{bmatrix}, \qquad j = 1, 2, 3$$

where 0, $\sigma_j$ are $2 \times 2$ matrices and the $\sigma_j$ are the Pauli matrices. We finally obtain the Dirac equation

$$i\partial_t \Psi(t, x) = -i \sum_{j=1}^{3} \alpha_j \partial_{x_j} \Psi(t, x) + m\beta \Psi(t, x).$$

## 5.5 Octonionic Dirac equation

In this section we investigate how to generalize the Dirac equation using the nonassociative algebra of octonions. Even though this is not the first attempt to use octonions in this framework, see [94] and [144], our formulation is based on a new nonassociative product of matrices, which allows us to write an $8 \times 8$ matrix which represents the Dirac equation in a novel way. In [94] Joshi obtains an octonionic wave equation whose properties are studied splitting an octonion in a pair of quaternions and then studying the corresponding matrix representation, while in the interesting paper [144] the linearization is done by using the bimodule structure of quaternions. From a mathematical point of view, however, the consideration of quadratic equations in a bimodule cannot be properly formalized, since multiplication is not allowed. In this section, we will give an appropriate octonionic setting in which to study the Dirac equation.

To construct the algebra of octonions one can, for example, use the Cayley–Dickson doubling process: complex numbers can be obtained as pairs of real numbers with componentwise addition and a suitable product. Similarly, quaternions can be considered as pairs of complex numbers with the sum defined componentwise and a suitable product involving conjugation. In a similar way, octonions can be defined as pairs of quaternions. We then have

$$\mathbb{C} \cong \mathbb{R} \oplus \mathbb{R}i$$

$$\mathbb{H} \cong \mathbb{C} \oplus \mathbb{C}j$$

$$\mathbb{O} \cong \mathbb{H} \oplus \mathbb{H}\ell$$

for $\ell$ new imaginary unit such that $\ell^2 = -1$ which anticommutes with $i$, $j$, $k$, units of $\mathbb{H}$. It is well known that if one repeats this procedure, it is not possible to obtain other division algebras (see [172]). So the only real division algebras are the real and complex fields, and the algebras of quaternions and of octonions. The multiplication table of the units of octonions can summarized by the Fano plane as in Figure 5.1.

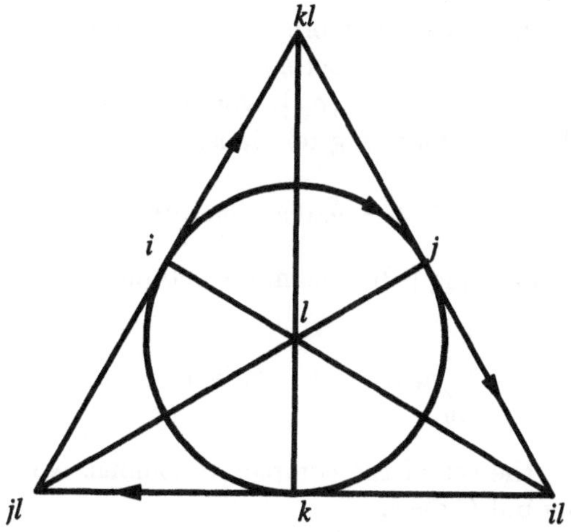


**Figure 5.1.**

In the literature the symbol $\mathbb{O}$ denotes the nonassociative real algebra of octonions with respect to the basis

$$\{e_0, e_1, \ldots, e_7\}$$

whose elements satisfy the relations

$$e_r e_s = -\delta_{rs} e_0 + \varepsilon_{rst} e_t,$$

where $\delta_{rs}$ is the Kronecker delta, $\varepsilon_{rst}$ are totally antisymmetric in $r, s, t$ and

$$\varepsilon_{rst} = +1 \quad \text{for} \quad rst = 123, 145, 176, 572, 347, 365, 246. \tag{5.37}$$

We will denote an octonion by

$$X = \sum_{r=0}^{7} e_r x_r, \quad x_r \in \mathbb{R},$$

and we define its conjugate by setting $e_r^+ = -e_r$, $r = 1, \ldots, 7$ and $e_0^+ = e_0$ therefore

$$X^+ = e_0 x_0 - \sum_{r=1}^{7} e_r x_r.$$

The square norm of an octonion $X$ is given by

$$|X|^2 = XX^+ = X^+X = \sum_{r=0}^{7} x_r^2.$$

For our purposes, as we did for the biquaternion space $\mathbb{BH}$, it is useful to introduce a different basis that is more suitable when dealing with octonions with complex coefficients (complex octonions).

Let us consider the set $\{a_0, a_1, \ldots, a_n\}$, where $a_0 = e_0$, $a_r = ie_r$, $r = 1, \ldots, 7$, $i = \sqrt{-1} \in \mathbb{C}$. We have the following relations:

$$a_r a_s = \delta_{rs} a_0 + i\varepsilon_{rst} a_t.$$

We will also define, following [93], a complex octonion $Z$ as

$$Z = \sum_{r=0}^{7} a_r z_r, \qquad z_r = x_r + iy_r \in \mathbb{C}$$

and $\mathbb{CO}$ will denote the set of the complex octonions. Note that, as a vector space $\mathbb{CO}$ is nothing but $\mathbb{O} \otimes_{\mathbb{R}} \mathbb{C}$.

Since complex octonions, as well as real octonions, form a nonassociative algebra it is not possible to provide a matrix representation with the usual matrix multiplication. As shown in [94], [172], it is possible to consider $\mathbb{CO}$ as a bimodule over a suitable set of matrices; on the other hand, we need to consider a linearization of the relativistic energy–momentum quadratic equation, and bimodules do not lend themselves to the study of quadratic equations. We therefore need a matrix algebra representation of $\mathbb{CO}$. Using the Pauli matrices in (5.16), where $i = \sqrt{-1} \in \mathbb{C}$ and the following matrices

$$\sigma_4 = \begin{bmatrix} 1 & 0 \\ 0 & i \end{bmatrix} \qquad \sigma_5 = \begin{bmatrix} 1 & 0 \\ 0 & -i \end{bmatrix}$$

$$\sigma_6 = \begin{bmatrix} 0 & 1 \\ i & 0 \end{bmatrix} \qquad \sigma_7 = \begin{bmatrix} 0 & 1 \\ -i & 0 \end{bmatrix}$$

we can define the eight matrices

$$A_0 = \begin{bmatrix} \sigma_0 & 0 & 0 & 0 \\ 0 & \sigma_0 & 0 & 0 \\ 0 & 0 & \sigma_0 & 0 \\ 0 & 0 & 0 & \sigma_0 \end{bmatrix} \qquad A_1 = \begin{bmatrix} \sigma_1 & 0 & 0 & 0 \\ 0 & \sigma_2 & 0 & 0 \\ 0 & 0 & \sigma_2 & 0 \\ 0 & 0 & 0 & -\sigma_2 \end{bmatrix}$$

$$A_2 = \begin{bmatrix} 0 & \sigma_4 & 0 & 0 \\ \sigma_5 & 0 & 0 & 0 \\ 0 & 0 & 0 & -i\sigma_0 \\ 0 & 0 & i\sigma_0 & 0 \end{bmatrix} \qquad A_3 = \begin{bmatrix} 0 & \sigma_7 & 0 & 0 \\ i\sigma_7 & 0 & 0 & 0 \\ 0 & 0 & 0 & \sigma_2 \\ 0 & 0 & \sigma_2 & 0 \end{bmatrix}$$

$$A_4 = \begin{bmatrix} 0 & 0 & \sigma_4 & 0 \\ 0 & 0 & 0 & i\sigma_0 \\ \sigma_5 & 0 & 0 & 0 \\ 0 & -i\sigma_0 & 0 & 0 \end{bmatrix} \qquad A_5 = \begin{bmatrix} 0 & 0 & \sigma_7 & 0 \\ 0 & 0 & 0 & -\sigma_2 \\ i\sigma_7 & 0 & 0 & 0 \\ 0 & -\sigma_2 & 0 & 0 \end{bmatrix}$$

$$A_6 = \begin{bmatrix} 0 & 0 & 0 & \sigma_5 \\ 0 & 0 & -i\sigma_3 & 0 \\ 0 & i\sigma_3 & 0 & 0 \\ \sigma_4 & 0 & 0 & 0 \end{bmatrix} \qquad A_7 = \begin{bmatrix} 0 & 0 & 0 & \sigma_6 \\ 0 & 0 & -i\sigma_1 & 0 \\ 0 & i\sigma_1 & 0 & 0 \\ -i\sigma_6 & 0 & 0 & 0 \end{bmatrix}.$$

**Remark 5.5.1.** The columns of each matrix $A_j$ are the complex octonions $a_ja_0, a_ja_1, \ldots, a_ja_7$, viewed as vectors in $\mathbb{C}^8$. This means that $A_j$ corresponds to the matrix of the linear transformation of the complex vector space $\mathbb{CO}$ defined by the left multiplication by $a_j$. In complete analogy, one could construct matrices $E_j$ whose columns are the octonions $e_je_0, e_je_1, \ldots, e_je_7$ viewed as vectors in $\mathbb{R}^8$. The matrices $A_j$, $j = 1, \ldots, 7$ are Hermitian, anticommuting and traceless.

If we denote by O the complex vector space of the $8 \times 8$ matrices generated by $A_0, A_1, \ldots, A_7$, it is immediate to prove the following:

**Proposition 5.5.1.** *The identification $a_j \to A_j$ is a vector spaces isomorphism between $\mathbb{CO}$ and O.*

We now consider the multiplication structure of O. We note that we have the following antisymmetric equations:

$$A_1A_2 = iH_{12}A_3, \qquad A_1A_4 = iH_{14}A_5, \qquad A_1A_7 = iH_{17}A_6,$$

$$A_5A_7 = iH_{57}A_2, \qquad A_3A_4 = iH_{34}A_7,$$

$$A_3A_6 = iH_{36}A_5, \qquad A_2A_4 = iH_{24}A_6,$$

where

$$H_{12} = \begin{bmatrix} \sigma_0 & 0 & 0 & 0 \\ 0 & \sigma_0 & 0 & 0 \\ 0 & 0 & -\sigma_0 & 0 \\ 0 & 0 & 0 & -\sigma_0 \end{bmatrix} \qquad H_{14} = \begin{bmatrix} \sigma_0 & 0 & 0 & 0 \\ 0 & -\sigma_0 & 0 & 0 \\ 0 & 0 & \sigma_0 & 0 \\ 0 & 0 & 0 & -\sigma_0 \end{bmatrix}$$

$$H_{17} = \begin{bmatrix} \sigma_0 & 0 & 0 & 0 \\ 0 & -\sigma_0 & 0 & 0 \\ 0 & 0 & -\sigma_0 & 0 \\ 0 & 0 & 0 & \sigma_0 \end{bmatrix} \qquad H_{57} = \begin{bmatrix} \sigma_3 & 0 & 0 & 0 \\ 0 & \sigma_3 & 0 & 0 \\ 0 & 0 & -\sigma_3 & 0 \\ 0 & 0 & 0 & -\sigma_3 \end{bmatrix}$$

$$H_{34} = \begin{bmatrix} \sigma_3 & 0 & 0 & 0 \\ 0 & -\sigma_3 & 0 & 0 \\ 0 & 0 & \sigma_3 & 0 \\ 0 & 0 & 0 & -\sigma_3 \end{bmatrix} \qquad H_{36} = \begin{bmatrix} \sigma_3 & 0 & 0 & 0 \\ 0 & -\sigma_3 & 0 & 0 \\ 0 & 0 & -\sigma_3 & 0 \\ 0 & 0 & 0 & \sigma_3 \end{bmatrix}$$

$$H_{24} = \begin{bmatrix} \sigma_3 & 0 & 0 & 0 \\ 0 & \sigma_3 & 0 & 0 \\ 0 & 0 & \sigma_3 & 0 \\ 0 & 0 & 0 & \sigma_3 \end{bmatrix};$$

for other values of $r$, $s$ we have $A_r A_s = i\varepsilon_{rst} H_{rs} A_t$. The description above allow us to define an antisymmetric binary operation in $\mathbb{CO}$ which is extended linearly from the relations

$$A_1 * A_2 = A_1 A_2 i H_{12} = A_3, \qquad A_1 * A_4 = A_1 A_4 i H_{14} = A_5,$$

$$A_1 * A_7 = A_1 A_7 i H_{17} = A_6, \qquad A_5 * A_7 = A_5 A_7 i H_{57} = A_2,$$

$$A_3 * A_4 = A_3 A_4 i H_{34} = A_7, \qquad A_3 * A_6 = A_3 A_6 i H_{36} = A_5,$$

$$A_2 * A_4 = A_2 A_4 i H_{24} = A_6, \qquad A_j * A_j = A_j A_j = -A_0, \;\; j = 0, 1, \dots, 7.$$

This operation endows $\mathbb{CO}$ with an algebra structure albeit nonassociative, so we have a representation of octonions as matrices with respect to the $*$ operation. Note that a similar structure can be introduced also for the algebra of octonions by replacing, in the relations above, $A_r$ by $E_r$ and $iH_{rs}$ by $H_{rs}$. This representation is different from the one used in [94] where the octonions are characterized as a bimodule using the general construction of Schafer [172]. In [94], the octonion $e_\alpha$ is represented by a pair of matrices $(L_\alpha, R_\alpha)$ which corresponds to a pair of linear transformations on the space of the octonions, viewed as a real vector space; the matrices $L_\alpha$ describe the left multiplication by $e_\alpha$ (our matrices $E_r$) while the matrices $R_\alpha$ describe the right multiplication by $e_\alpha$. From the discussion above, it is immediate to prove the following proposition.

**Proposition 5.5.2.** *The algebra* $(\mathbb{O}, +, *)$ *is isomorphic to the algebra* $\mathbb{C}\mathbb{O}$.

**Proposition 5.5.3.** *The set*

$$\mathcal{H} = \{I, H_{12}, H_{14}, H_{17}, H_{57}, H_{34}, H_{36}, H_{24}\}$$

*endowed with the usual product forms a group isomorphic to the group* $\mathbb{Z}_2 \oplus \mathbb{Z}_2 \oplus \mathbb{Z}_2$.

Following the ideas of Dirac, we will use the basis $A_\nu$, $\nu = 0, \dots, 7$ of complex octonions to linearize the relativistic energy conservation law

$$E^2 = p^2 + m^2. \tag{5.38}$$

Let $p_\nu$ and $m_\mu$, $\nu = 1, \dots, 3$, $\mu = 4, \dots, 7$ be real numbers. It is easy to verify that if we set

$$P := \sum_{\nu=1}^{3} A_\nu p_\nu, \qquad M := \sum_{\mu=4}^{7} A_\mu m_\mu,$$

we get the equality

$$(P + M)^2 := (P + M) * (P + M) = P * P + M * M = E^2.$$

By the relations above we obtain the linear equation

$$E = P + M. \tag{5.39}$$

Then we define the following quantum mechanical operators

$$\hat{E} = iA_0\partial_0, \qquad \hat{P} = -i\sum_{\nu=1}^{3} A_\nu\partial_\nu, \qquad \hat{M} = \sum_{\mu=4}^{7} A_\mu m_\mu, \qquad m_\mu \in \mathbb{R}, \tag{5.40}$$

where $\hat{E}$, $\hat{P}$, $\hat{M}$ are the operators related to energy, momentum and mass, respectively. In the octonionic case the mass operator has four components, the first is the usual mass associated to the field, while the other components represent other physical observables, see [94]. Inserting (5.40) in (5.39) we get the octonionic Dirac equation

$$\hat{E}\psi = (\hat{P} + \hat{M})\psi,$$

that is,

$$iA_0\partial_0\psi = \Big[ -i\sum_{\nu=1}^{3} A_\nu\partial_\nu + \sum_{\mu=4}^{7} A_\mu m_\mu \Big]\psi \tag{5.41}$$

where $\psi = (\psi_0, \dots, \psi_7)^t$ and $\psi_j$, $j = 0, \dots, 7$ are complex functions of time and space variables.

**Remark 5.5.2.** From the octonionic Dirac equation (5.41) it is possible to obtain the the Klein–Gordon equation. Indeed, let us define the operator

$$\mathcal{Q} = iA_0\partial_0 + i\sum_{\nu=1}^{3} A_\nu\partial_\nu - \sum_{\mu=4}^{7} A_\mu m_\mu \tag{5.42}$$

and its conjugate

$$\mathcal{Q}^+ = iA_0\partial_0 - i\sum_{\nu=1}^{3} A_\nu\partial_\nu + \sum_{\mu=4}^{7} A_\mu m_\mu$$

obtained by the relations $A_0^+ = A_0$ and $A_\alpha^+ = -A_\alpha$ for $\alpha = 1, \dots, 7$. Writing the Dirac equation as

$$\mathcal{Q}\psi = 0 \tag{5.43}$$

and applying the operator $\mathcal{Q}^+$ to (5.43) we get

$$\mathcal{Q}^+\mathcal{Q}\psi = [-\partial_0^2 + \sum_{\nu=1}^{3} \partial_\nu^2 - \sum_{\mu=4}^{7} m_\mu^2] I\psi = 0 \tag{5.44}$$

which is the Klein–Gordon equation.

**Remark 5.5.3.** In [94] the author formally linearized (5.38) and obtained the equation

$$[i\sum_{\nu=0}^{3} L_\nu\partial_\nu - \sum_{\mu=4}^{7} L_\mu m_\mu]\psi = 0. \tag{5.45}$$

Even though (5.45) is similar to the Dirac equation it does not allow us to deduce the Klein–Gordon equation; in fact, unfortunately, the algebraic properties of the matrices $L_\alpha$ do not allow us to obtain a hyperbolic operator. To verify this, define the operator

$$\Gamma = L_0\partial_0 + \sum_{\nu=1}^{3} L_\nu\partial_\nu + i\sum_{\mu=4}^{7} L_\mu m_\mu$$

and its conjugate

$$\Gamma^{+} = L_0\partial_0 - \sum_{\nu=1}^{3} L_\nu\partial_\nu - i\sum_{\mu=4}^{7} L_\mu m_\mu.$$

Then, equation (5.45) can be written as

$$\Gamma\psi = 0,$$

and applying $\Gamma^{+}$ we obtain

$$\Gamma^{+}\Gamma\psi = [\partial_0^2 + \sum_{\nu=1}^{3} \partial_\nu^2 - \sum_{\mu=4}^{7} m_\mu^2]I\psi = 0.$$

We observe that it is not possible to obtain the Klein–Gordon equation, in fact $\Gamma^{+}\Gamma\psi = 0$ is an elliptic equation while the Klein–Gordon equation is not.

**Remark 5.5.4.** It is worth noticing that since the octonionic Dirac equation (5.41) has the form of the classical Dirac equation, it is possible to deduce most of the classical properties of the Dirac equation.

For example, we now prove that the wave function can be interpreted as a probability density. We point out that this result is essentially due to the fact that our wave function is a column vector so that the nonassociative product with octonions does not affect the standard proof.

**Proposition 5.5.4.** *The norm of the solutions of the octonionic Dirac equation does not depend on time.*

*Proof.* Rewrite the equation $Q\psi = 0$ as

$$iA_0\partial_0\psi = -i\sum_{\nu=1}^{3} A_\nu\partial_\nu\psi + \sum_{\mu=1}^{4} A_\mu m_\mu\psi := H\psi.$$

If $\psi^\dagger$ denotes the adjoint of the transpose of $\psi$, we obtain the following equations:

$$\psi^\dagger(iA_0\partial_0\psi) = \psi^\dagger(H\psi) \quad \text{and} \quad (iA_0\partial_0\psi)^\dagger\psi = (H\psi)^\dagger\psi.$$

We now have two matrix equations that can be subtracted to get

$$iA_0\partial_0(\psi^\dagger\psi) = -i\sum_{\nu=1}^{3} \psi^\dagger(A_\nu\partial_\nu\psi) - i\sum_{\nu=1}^{3}(A_\nu\partial_\nu\psi^\dagger)\psi.$$

This implies that

$$\partial_0(\psi^\dagger A_0\psi) + \sum_{\nu=1}^{3} \partial_\nu(\psi^\dagger A_\nu\psi) = 0.$$

As in the case of the classical Dirac equation, we have that the eight-current $\psi^\dagger\psi$ is conserved (i.e., does not depend on time). Now, if we integrate this equation and use, as in the usual Dirac equation, periodic boundary conditions to eliminate the spatial part, we find that the following quantity is a constant of motion

$$\int_{\mathbb{R}^3} \psi^\dagger \psi d^3 r. \qquad \square$$

**Remark 5.5.5.** We may add electromagnetic interactions to the Dirac octonionic equation using the well-known minimal substitution [82] on the four-momentum $(A_0\partial_0, -iA_1\partial_1, -iA_2\partial_2, -iA_3\partial_3)$. Let $(A^0, A^j)$ for $j = 1, 2, 3$ be the electromagnetic potentials and let $e$ be the electric charge. Then the most natural way to obtain the minimal substitution is to replace

$$A_0\partial_0 \quad \text{by} \quad A_0(\partial_0 - eA^0)$$

and

$$-\sum_{\nu=1}^{3} A_\nu\partial_\nu \quad \text{by} \quad -\sum_{\nu=1}^{3}(A_\nu\partial_\nu - eA^\nu)$$

in the Dirac equation. We obtain the equation

$$[A_0\partial_0 + i\sum_{\nu=1}^{3} A_\nu\partial_\nu - \sum_{\mu=4}^{7} A_\mu m_\mu + eA_0A^0 + i\sum_{\nu=1}^{3} A_\nu A^\nu]\psi = 0.$$

We can also give such an equation a Hamiltonian form by defining the operators

$$H := i\sum_{\nu=1}^{3} A_\nu\partial_\nu - \sum_{\mu=4}^{7} A_\mu m_\mu$$

and

$$H^{el} := -eA_0A^0 - i\sum_{\nu=1}^{3} A_\nu A^\nu$$

where $H^{el}$ is the Hamiltonian related to the electromagnetic field. In a more compact way, we can rewrite the octonionic Dirac equation in an electromagnetic field as

$$A_0\partial_0\psi = (H + H^{el})\psi. \tag{5.46}$$

In the case of radial potential $V(r)$ such as the Coulomb or the Yukawa potentials, the equation (5.46) can be written in the form

$$A_0\partial_0\psi = (H + A_0V(r))\psi.$$

Finally, we discuss some algebraic properties of the octonionic Dirac equation. The point of view which we will take is that such equation is satisfied by functions $f : \mathbb{O} \to \mathbb{C}\mathbb{O}$ and can therefore be interpreted as a system of differential equations for (vector-valued) functions:

$$f : \mathbb{R}^8 \to \mathbb{C}^8$$

where the link between $f = (f_0, f_1, \dots, f_7)$ and $f$ is obviously given by

$$f = f_0 a_0 + f_1 a_1 + \dots + f_7 a_7.$$

The system we are considering can be written as

$$A(D) f = 0, \tag{5.47}$$

where the matrix $A(D)$ is defined by

$$\begin{bmatrix} i\partial_{x_0} & i\partial_{x_1} & i\partial_{x_2} & i\partial_{x_3} & -m_4 & -m_5 & -m_6 & -m_7 \\ i\partial_{x_1} & i\partial_{x_0} & \partial_{x_3} & -\partial_{x_2} & im_5 & -im_4 & -im_7 & im_6 \\ i\partial_{x_2} & -\partial_{x_3} & i\partial_{x_0} & \partial_{x_1} & im_6 & im_7 & -im_4 & -im_5 \\ i\partial_{x_3} & \partial_{x_2} & -\partial_{x_1} & i\partial_{x_0} & im_7 & -im_6 & im_5 & -im_4 \\ -m_4 & -im_5 & -im_6 & -im_7 & i\partial_{x_0} & \partial_{x_1} & \partial_{x_2} & \partial_{x_3} \\ -m_5 & im_4 & -im_7 & im_6 & -\partial_{x_1} & i\partial_{x_0} & -\partial_{x_3} & \partial_{x_2} \\ -m_6 & im_7 & im_4 & -im_5 & -\partial_{x_2} & \partial_{x_3} & i\partial_{x_0} & -\partial_{x_1} \\ -m_7 & -im_6 & im_5 & im_4 & -\partial_{x_3} & -\partial_{x_2} & \partial_{x_1} & i\partial_{x_0} \end{bmatrix}. \tag{5.48}$$

Let us consider the ring $R = \mathbb{C}[x_0, x_1, x_2, x_3, m_4, m_5, m_6, m_7]$ of polynomials of eight real variables $x_0, \dots, m_7$. The Fourier transform of the matrix defined in (5.48), denoted by $A$, is given by

$$\begin{bmatrix} -x_0 & -x_1 & -x_2 & -x_3 & -m_4 & -m_5 & -m_6 & -m_7 \\ -x_1 & -x_0 & ix_3 & -ix_2 & im_5 & -im_4 & -im_7 & im_6 \\ -x_2 & -ix_3 & -x_0 & ix_1 & im_6 & im_7 & -im_4 & -im_5 \\ -x_3 & ix_2 & -ix_1 & -x_0 & im_7 & -im_6 & im_5 & -im_4 \\ -m_4 & -im_5 & -im_6 & -im_7 & -x_0 & ix_1 & ix_2 & ix_3 \\ -m_5 & im_4 & -im_7 & im_6 & -ix_1 & -x_0 & -ix_3 & ix_2 \\ -m_6 & im_7 & im_4 & -im_5 & -ix_2 & ix_3 & -x_0 & -ix_1 \\ -m_7 & -im_6 & im_5 & im_4 & -ix_3 & -ix_2 & ix_1 & -x_0 \end{bmatrix} \tag{5.49}$$

and the module $M$ associated to the Dirac equation is the quotient $R^8/A^t R^8$. Using CoCoA it is possible to compute the syzygy module of the columns of the matrix $A^t$. A quick computation shows that the first syzygies are trivial so that the inhomogeneous system

$$A(D)\psi = \varphi$$

can be solved for any $\varphi \in \mathcal{S}$ where $\mathcal{S}$ is one of the usual sheaves of generalized functions. Moreover $\mathrm{Ext}^0(M, \mathcal{S}) = 0$, which implies the unique continuation property for the solution of the system $\mathcal{Q}\psi = 0$. We have that $\mathrm{Ext}^1(M, \mathcal{S}) \neq 0$. In fact, a CoCoA computation shows that the determinant of the matrix $A$ can be computed as

$$\mathrm{Det}(A) = (x_1^2 + x_2^2 + x_3^2 - x_0^2 + m_4^2 + m_5^2 + m_6^2 + m_7^2)^4$$

and the result follows by Lemma 2.1.1. This last fact implies that the $\mathcal{S}$-solutions of the octonionic Dirac equation can have compact singularities.

When we do not take into account external fields the octonionic Dirac equation becomes a linear system with constant coefficients and it is possible to apply the Fundamental Principle. The characteristic variety of the system (5.48) is the 7-dimensional variety of $\mathbb{R}^8$ given by

$$x_1^2 + x_2^2 + x_3^2 - x_0^2 + m_4^2 + m_5^2 + m_6^2 + m_7^2 = 0. \tag{5.50}$$

The knowledge of the characteristic variety of the system (5.48) allows us to employ the Ehrenpreis–Palamodov Fundamental Principle to give a complete description of its solutions.

**Remark 5.5.6.** From a theoretical point of view it is possible to introduce another operator obtained substituting in $\mathcal{Q}$ defined in (5.42) the real derivatives with complex ones. Let us complexify the variables $x_\nu$ setting $z_\nu = x_\nu + iy_\nu$. Among the many operators that one can introduce, we consider the following:

$$iA_0\frac{\partial}{\partial \bar{z}_0} + i\sum_{\nu=1}^{3} A_\nu \frac{\partial}{\partial z_\nu} - \sum_{\mu=4}^{7} A_\mu m_\mu, \tag{5.51}$$

where

$$\frac{\partial}{\partial z_\nu} = \frac{\partial}{\partial x_\nu} - i\frac{\partial}{\partial y_\nu}, \qquad \frac{\partial}{\partial \bar{z}_\nu} = \frac{\partial}{\partial x_\nu} + i\frac{\partial}{\partial y_\nu}.$$

In the massless case, the operator (5.51) can be written as

$$\mathcal{M} = iA_0\frac{\partial}{\partial \bar{z}_0} + i\sum_{\nu=1}^{3} A_\nu \frac{\partial}{\partial \bar{z}_\nu}.$$

The Maxwell system is contained, as a particular case, in the equation $\mathcal{M}f = 0$. In fact, a function $f$ depending only on the variables $y_\nu$ satisfies $\mathcal{M}f = 0$ if and only if it satisfies the system, i.e.,

$$(A_0\frac{\partial}{\partial y_0} - \sum_{\nu=1}^{3} A_\nu \frac{\partial}{\partial y_\nu})f = 0. \tag{5.52}$$

Observing that the matrices $\{A_0, A_1, A_2, A_3\}$ form an associative algebra isomorphic to the algebra of quaternions, and comparing with systems (5.3) and (5.27), we deduce that system (5.52) is the Maxwell system. Obviously, if a function $f$ depends only on the $x_\nu$ variables, it satisfies $\mathcal{M}f = 0$ if and only if it satisfies the Dirac massless equation.

# 6

# Open Problems and Avenues for Further Research

The work we have done so far and which we have described in the last five chapters shows the power of algebraic analysis as well as the importance of the accessibility of suitable computational techniques. We have demonstrated how these ideas can greatly contribute to the development of a function theory for solutions of suitable systems of differential operators. However, many delicate questions remain open. In this short chapter we will analyze several lines of research which are strictly connected with the positive results obtained so far.

## 6.1 The Cauchy–Fueter system

**Problem 6.1.1.** In Chapter 3 we constructed minimal resolutions for the Cauchy–Fueter system in several variables. In his famous paper [13], Baston utilizes the Bernstein–Gelfand–Gelfand sequences to construct an (a priori different) resolution for the same system. A recent work of Somberg [183] indicates that our results can be interpreted as the quaternionic complexes of invariant differential operators discussed in [13]; similar remarks are presented by Soucek [199]. The first problem which we want to pose consists in establishing explicitly the equivalence between our resolution and the one found by Baston.

**Problem 6.1.2.** In Chapter 3, we have provided significant explicit information on the nature of the resolution of the Cauchy–Fueter system. In particular, we know its length, its Betti numbers and, in some cases, the explicit form of the maps which appear in the resolution. We propose that it would be very interesting to provide an explicit expression (possibly in quaternionic term) of all the maps that appear in our sequences.

**Problem 6.1.3.** In the theory of several complex variables the exactness of the Koszul complex has many consequences, one of which is the Martineau–Harvey duality theorem. For the case of regular functions of several quaternionic variables, we have a general form of such a duality theorem, but the specific form of the dualizing sheaf is known only for two quaternionic variables. We propose discovering the explicit form of such a sheaf in order to give concrete meaning to the duality theorem.

**Problem 6.1.4.** Another important consequence of our theory, and an original motivation for our research is the construction of a quaternionic based hyperfunction theory. This was done in [72], [138] and in [165] for the case of one quaternionic variable. We propose that it would be important to construct such a theory for several quaternionic variables. Let us observe that the fundamental facts to construct a theory of hyperfunctions are:

1. the flabby dimension of the sheaf $\mathcal{O}$ equal to $n$;
2. the pure $n$-codimensionality of $\mathbb{R}^n$ with respect to the sheaf $\mathcal{O}$.

A direct consequence of those two facts is that the assignment

$$U \to H^n_{\mathbb{R}^n}(U, \mathcal{O})$$

defines a flabby sheaf: the sheaf of hyperfunctions on $\mathbb{R}^n$.

In the quaternionic case we know the flabby dimension of the sheaf $\mathcal{R}_l$ to be equal to $2n-1$ if $n$ is the number of quaternionic variables considered. However, for $n > 1$, we still do not know which is the quaternionic analogue of $\mathbb{R}^n$. So we ask which is the purely $(2n-1)$-codimensional variety with respect to sheaf $\mathcal{R}_l$.

## 6.2 The Dirac system

**Problem 6.2.1.** In Chapter 4 we have shown that in a large number of cases, the complex derived from the Dirac system is actually simpler and better behaved than the one derived by the Cauchy–Fueter, at least when the dimension of the algebra is high enough. Despite its apparent simplicity, however, there are some crucial informations on this complex which are still unknown. The first problem we propose is the study of the resolution in the case of $n$-Dirac operators. In particular we want to compute its length, the degrees of its maps, and the associated characteristic variety.

**Problem 6.2.2.** A strictly related question consists in proving the validity of the Fischer decomposition for polynomials in several Clifford variables. As we have shown in Chapter 4, if we can prove the Fischer decomposition, we obtain also that the Dirac complex depends only on the radial algebra relations thus providing the description of all the maps in it.

**Problem 6.2.3.** It is reasonable to expect that the function theory for the solutions of the Dirac system could be constructed similarly to what we have done for the Cauchy–Fueter system. The case of one variable is quite clear, while an analysis in the case of several variables has still to be investigated. As an additional problem, we wish to obtain a theory of hyperfunctions in several Clifford variables following the lines sketched in Problem 6.1.4.

**Problem 6.2.4.** In our paper [164] and in Section 4, we have shown that the behavior of the Dirac complex depends on the dimension of the Clifford algebra with respect to the number of operators considered. It will be useful to look at those complexes from the point of view of invariant operators theory to explain this feature and to predict in which cases one can expect that the radial algebra relations produce all the maps in the resolution.

**Problem 6.2.5.** The resolution, in the case of $n$-Dirac operators, has still to be constructed. We expect that at least in the radial case, one can explicitly write all the maps in the resolution. A first insight into this problem has been given in [163] in which we use to theory of the so-called megaforms to recover the complex in the case of two and three operators. Since the application of this theory is quite complicate, it will be useful to understand how to apply this method in a more efficient way.

## 6.3 Miscellanea

**Problem 6.3.1.** A fundamental tool in this book is the sequence (2.6), i.e.,

$$0 \longrightarrow R^{r_m} \xrightarrow{P^t_{m-1}} R^{r_{m-1}} \longrightarrow \dots \xrightarrow{P^t_1} R^{r_1} \xrightarrow{P^t} R^{r_0} \longrightarrow M \longrightarrow 0$$

which, in general, cannot be easily computed. We ask if it is possible to use the special structure of the matrix $P^t$ to improve the existing algorithms.

**Problem 6.3.2.** Another interesting issue related to the sequence (2.6) is the homogeneity of the entries of $P^t$ which, in many instances, have different degrees. It is well known from Sato's fundamental theorem [169] that the sheaf of microfunctions solutions to linear partial differential equations is trivial outside the variety defined by the minors of the homogeneous components of highest degree, $P^t_{\max}$ of $P^t$. The free resolution of $M' = R^{r_0}/\langle P^t_{\max}\rangle$ may be very different from the resolution of $M$ and that of the module $M'' = R^{r_0}/\langle P^t_h\rangle$ where $P^t_h$ is the homogeneized matrix. So we propose studying the relationship between the resolutions of $M$, $M'$ and $M''$ and their analytic relevance.

**Problem 6.3.3.** Suppose we have resolution (2.6). Its dual, as we explain in Chapter 2, is a complex. Set $M := R^{r_1}/\langle P\rangle$ and consider its dual

$$M^* = \mathrm{Hom}(M, R) \cong \mathrm{Syz}(P^t) = \langle P_1^t\rangle.$$

The resolution of $R^{r_0}/\langle P^t\rangle$ can be used to resolve $M^*$:

$$0 \longrightarrow R^{r_m} \xrightarrow{P_{m-1}^t} R^{r_{m-1}} \longrightarrow \dots \xrightarrow{P_2^t} R^{r_2} \xrightarrow{P_1^t} M^* \longrightarrow 0.$$

Applying the functor $\mathrm{Hom}(-, R)$ we get the following complex

$$0 \longrightarrow M^{**} \xrightarrow{P_1} R^{r_2} \xrightarrow{P_2} R^{r_3} \longrightarrow \dots \xrightarrow{P_{m-1}} R^{r_m} \longrightarrow 0 \qquad (6.1)$$

whose maps are given by the transposed of the matrices in resolution (2.6). If $M = M^{**}$, then we can extend complex (6.1) to get the complex dual of (2.6) with the first two homology modules equal to zero (the kernel of the canonical map $\phi: M \to M^{**}$ is the first homology module in the dual of (2.6) while the cokernel is its second homology module). Note, moreover, that there is a map $M \to M^{**}$ and this map is one-to-one when $M$ is torsion free. Determine the class of matrices for which $M = M^{**}$ (an example is the matrix associated to the Cauchy–Fueter system).

**Problem 6.3.4.** The Fundamental Principle discussed in Chapter 2 is an extremely powerful tool to represent the solutions of a system of linear partial differential equation with constant coefficients. It becomes effective once we know the primary decomposition of the variety associated to the module and the noetherian operators associated to its primary components. The difficulty in computing the primary decomposition of the variety is well known, so one can concentrate, at the beginning, to the case of primary varieties. The construction of the noetherian operators associated to the module is given by Palamodov in [142], but the procedure is not explicit enough to be made into an algorithm. We would like to find an algorithm which, at least under suitable hypothesis, allows the construction of the noetherian operators associated to a given module.

**Problem 6.3.5.** We have studied systems arising from combinatorial structures (colored bipartite graphs). In our description, the points correspond to coordinate derivatives $\partial_{x_j}$, the lines to operators of Dirac type satisfying axioms $(A_1), \dots, (A_4)$ and the colors are Clifford basis elements $\pm e_j$. When considering a combinatorial structure of this type, it seems natural to consider the minimality constraint $(A_5)$ on the number of colors, thus reducing the number of possible systems associated to a given geometry. Two such systems will be called equivalent if they can be identified by performing a permutation on the set of points, lines or colors and possibly changing also the signs. An interesting question is to study the equivalence class containing a given system.

One may consider the following group actions on the set $\{x_1, \dots, x_m\}$ of variables and on the set $\{e_1, \dots, e_M\}$ of colors:

1. ($G_1$) to perform a permutation of the variables together with possible changes of sign, i.e.,
$$x_k \to (-1)^{s_k} x_{\pi(k)};$$
2. ($G_2$) to perform a permutation of the colors together with possible changes of sign, i.e.,
$$e_j \to (-1)^{s_j} e_{\pi(j)}.$$

One can also permute the lines, but clearly this action does not alter the system. Two natural problems associated with the consideration of this group are:
**Question 1**. Any action of a group element of type $G_1$ or $G_2$ changes a system into an equivalent system, with the same properties. How many essentially inequivalent system does a given finite geometry admit (with or without assuming $A_5$)?
**Question 2**. What is the subgroup of pairs $(g_1, g_2) \in G_1 \times G_2$ leaving a given system invariant?

**Problem 6.3.6.** Another interesting problem is to study the invariance properties of combinatorial systems under coordinate transformations, in particular under linear transformations. In the case of a single Dirac operator $\sum e_j \partial_{x_j}$ acting on functions $f(x)$, $x = \sum e_j x_j \in \mathbb{R}^m$, we consider the action of the group Spin$(m)$ given by

$$s \in \text{Spin}(m) \to L(s) \ : \ f(\bar{x}) \to sf(sxs)$$

and one can easily verify the relation

$$\partial_x L(s) = L(s)\partial_x$$

expressing the Spin-invariance of the Dirac operator $\partial_x$ (see also [55], [78]). The coordinate transformation which corresponds to this action of Spin$(m)$ is the rotation
$$h(s) \in SO(m) \ : \ x \to sxs.$$

A natural question is whether a similar group invariance exists for combinatorial systems. Although a detailed study of this interesting problem is still to be done, it seems that the minimality condition ($A_5$) is rather restrictive for this and we have examples of combinatorial systems, not satisfying ($A_5$), which are Spin$(m)$-invariant.

**Problem 6.3.7.** Finally, looking at the systems arising from combinatorial structures, we have found in several cases that the resolution is proportional to a Koszul complex. We conjecture that this happens in the case of all affine and projective geometries over $\mathbb{Z}_p$, $p$ prime, and also in case of self–dual designs and for structures in which the number of blocks (lines or faces) is greater than or equal to the number of points.

# Bibliography

[1] R. A. Adams, *Sobolev Spaces*, Pure and Appl. Math. Academic Press, Inc., Vol. 65, New York, London, 1975.

[2] W. W. Adams, C. A. Berenstein, P. Loustaunau, I. Sabadini, D. C. Struppa, Regular functions of several quaternionic variables and the Cauchy–Fueter complex, *J. Geom. Anal.*, **9** (1999), 1–15.

[3] W. W. Adams, C. A. Berenstein, P. Loustaunau, I. Sabadini, D. C. Struppa, On compact singularities for regular functions of one quaternionic variable, *Compl. Var.*, **31** (1996), 259–270.

[4] W. W. Adams, P. Loustaunau, *An Introduction to Gröbner Bases*, American Mathematical Society, Graduate Studies in Mathematics, Vol. 3, 1994.

[5] W. W. Adams, P. Loustaunau, Analysis of the module determining the properties of regular functions of several quaternionic variables, *Pacific J.*, **196** (2001), 1–15.

[6] W. W. Adams, P. Loustaunau, V. P. Palamodov, D. C. Struppa, Hartogs' phenomenon for polyregular functions and projective dimension of related modules over a polynomial ring, *Ann. Inst. Fourier*, **47** (1997), 623–640.

[7] L. A. Aizenberg, S. A. Dautov, *Differential Forms Orthogonal to Holomorphic Functions or Forms, and Their Properties*, Translations of Mathematical Monographs, Vol. 56, A.M.S., Providence, R.I., 1983.

[8] A. Andreotti, M. Nacinovich, Complexes of Partial Differential Operators, *Ann. Scuola Norm. Sup. di Pisa*, **3** (1976), 553–621.

[9] J. Aniansson, *Some integral representation in real and complex analysis. Peano–Sard kernels and the Fischer kernel*, Ph.D dissertation Stockholm, 1999.

[10] M. F. Atiyah, *Geometry of Yang–Mills Fields*, Lezioni Fermiane, Pisa, 1979.

[11] M. F. Atiyah, I. G. MacDonald, *Introduction to Commutative Algebra*, Addison-Wesley, Reading MA, 1969.

[12] R. Baer, Abelian groups that are direct summands of every containing abelian group, *Bull. Amer. Math. Soc.*, **46** (1940), 800–806.

[13] R. J. Baston, Quaternionic Complexes, *J. Geom. Phys.*, **8** (1992), 29–52.

[14] D. Bayer, M. Stillman, A Criterion for Detecting $m$-regularity, *Invent. Math.*, **87** (1987), 1–11.

[15] D. Bayer, M. Stillman, On the complexity of computing syzygies, *J. Symb. Comp.*, **6** (1988), 135–147.

[16] D. Bayer, M. Stillman, Computation of Hilbert Functions, *J. Symb. Comp.*, **14** (1992), 31–50.

[17] C. A. Berenstein, R. Gay, *Complex Analysis and Special Topics in Harmonic Analysis*, Springer, New York, 1995.

[18] C. A. Berenstein, T. Kawai, D. C. Struppa, Interpolating varieties and the Fabry–Ehrenpreis–Kawai gap theorem, *Adv. Math.*, **122** (1996), 280–310.

[19] C. A. Berenstein, T. Kawai, D. C. Struppa, *A fundamental principle for hyperfunction solutions of systems of infinite order differential equations with constant coefficients*, unpublished, 1996.

[20] C. A. Berenstein, B. A. Taylor, Interpolation problems in $\mathbb{C}^n$ with applications to harmonic analysis, *J. Anal. Math.*, **38** (1980), 188–254.

[21] C. A. Berenstein, I. Sabadini, D. C. Struppa, Boundary values of regular functions of quaternionic variables, *Pitman Res. Notes in Math. Ser.*, Vol. 347 (1996), 220–232.

[22] C. A. Berenstein, D. C. Struppa, 1–inverses for polynomial matrices of non–constant rank, *Systems and Control Letters*, **6** (1986), 309–314.

[23] C. A. Berenstein, D. C. Struppa, *Sheaves of holomorphic fuctions with growth conditions*, in *D-Modules and Microlocal Geometry*, Lisbon 1990, (1993), 63–74.

[24] G. Bernardes, F. Sommen, Monogenic functions of higher spin by Cauchy–Kowalevska extension of real–analytic functions, *Compl. Var.*, **39** (1999), 305–325.

[25] S. Bochner, Analytic and meromorphic continuation by means of Green's formula, *Ann. Math.*, **44** (1943), 652–673.

[26] S. Bochner, W. T. Martin, *Several Complex Variables*, Princeton, 1948.

[27] N. N. Bogoliubov, D. V. Shirkov, *Introduction to the Theory of Quantized Fields*, Wiley-Interscience Publ., J. Wiley & Sons 1980.

[28] W. M. Boothby, *Introduction to Differentiable Manifolds and Riemannian Geometry*, Second Edition, Pure and Appl. Math., Vol. 120, Academic Press Inc., Orlando, 1986.

[29] F. Brackx, R. Delanghe, F. Sommen, *Clifford Analysis*, Pitman Res. Notes in Math. Ser., Vol. 76, 1982.

[30] G. E. Bredon, *Sheaf Theory*, McGraw–Hill, New York, 1967.

[31] B. Buchberger, *Introduction to Gröbner bases*, Gröbner bases and Applications (Linz, 1998), London Math. Soc. Lecture Note Ser. **251**, Cambridge Univ. Press, 1998.

[32] D. Buchsbaum, A generalized Koszul complex, I, *Trans. Am. Math. Soc.*, **111** (1964), 183–196.

[33] J. Bures, F. Sommen, V. Souček, P. Van Lancker, Rarita–Schwinger type operators in Clifford analysis, *J. Funct. Anal.*, **185** (2001), 425–455.

[34] J. Bures, F. Sommen, V. Souček, P. Van Lancker, Symmetric analogue of Rarita–Schwinger equations, *Ann. Glob. Anal. Geom.*, **21** (2002), 215–240.

[35] J. Bures, V. Souček, Generalized hypercomplex analysis and its integral formulas, *Compl. Var.*, **5** (1985), 53–70.

[36] E. Cartan, *Nombre complexes*, in Œuvres Complètes, Partie II, Gauthier-Villars, Paris, 1953, 107–246.

[37] H. Cartan, S. Eilenberg, *Homological Algebra*, Princeton University Press, Princeton, NJ, 1956.

[38] C. C. Chea, *Microfunctions for sheaves of holomorphic functions with growth conditions*, Ph.D dissertation, University of Maryland, 1994.

[39] C. Chevalley, *The Algebraic Theory of Spinors*, Columbia Univ. Press, New York, 1954.

[40] C. Chevalley, *The Algebraic Theory of Spinors and Clifford Algebras*, Springer, Berlin, 1997.

[41] W. K. Clifford, Application of Grassmann's extensive algebra, *Amer. J. Math.*, **1** (1878), 350–358.

[42] W. K. Clifford, *On the classification of geometric algebras*, in Mathematical papers by William Kingdon Clifford, R. Tucker ed., Macmillan, London, 1882; reprinted by Chelsea, New York 1968.

[43] F. Colombo, A. Damiano, I. Sabadini, D. C. Struppa, *Koszul-type complexes for commuting matrices of differential operators*, preprint 2003.

[44] F. Colombo, P. Loustaunau, I. Sabadini, D. C. Struppa, Regular functions of biquaternionic variables and Maxwell's Equations, *J. Geom. Phys.*, **26** (1998), 183–201.

[45] F. Colombo, I. Sabadini, D. C. Struppa, Dirac equation in the octonionic algebra, *Contemp. Math.*, **251** (2000), 117–134.

[46] F. Colombo, I. Sabadini, F. Sommen, D. C. Struppa, Syzygies and conservation laws, *Found. Phys. Lett.*, **15** (2002), 507–522.

[47] F. Colombo, I. Sabadini, D. C. Struppa, An introduction to Computational Algebraic Analysis, *Milan J. Math.*, **71** (2003), 1–36.

[48] D. Constales, *The relative position of $L^2$-domains in complex and Clifford analysis*, Ph.D dissertation, Gent, 1990.

[49] D. Cox, J. Little, D. O'Shea, *Ideals, Varieties, and Algorithms*, UTM, Springer-Verlag, New York, 1996.

[50] D. Cox, J. Little, D. O'Shea, *Using Algebraic Geometry*, GTM Vol. 185, Springer-Verlag, New York, 1998.

[51] A. Crumeyrolle, *Orthogonal and Symplectic Clifford Algebras*, Kluwer, 1990.

[52] A. Damiano, *Applicazioni dell'Algebra Computazionale allo studio di alcuni Operatori Differenziali*, Tesi di Laurea, Genova, 2001.

[53] J.W. De Roever, *Complex Fourier Transformation and Analytic Functionals with Unbounded Carriers*, Mathematical Centre Tracts, Vol. 89, Amsterdam, 1978.

[54] R. Delanghe, Clifford analysis: history and perspective, *Comp. Meth. Funct. Theor.*, **1** (2001), 107–153.

[55] R. Delanghe, F. Sommen, V. Souček, *Clifford Algebra and Spinor-Valued Functions*, Math. and its Appl. Vol. 53, Kluwer Acad. Publ. Dordrecht, 1992.

[56] R. Delanghe, V. Souček, On the structure of spinor-valued differential forms, *Compl. Var.*, **18** (1992), 223–236.

[57] J. Dieudonné, L. Schwartz, La dualité dans les espaces F et LF, *Ann. Inst. Fourier Grenoble*, **1** (1949), 61–101.

[58] P.A.M. Dirac, *Principles of Quantum Mechanics*, Oxford, 1930.

[59] M.G. Eastwood, R. Penrose, R.O. Wells, Cohomology and massless fields, *Comm. Math. Phys.*, **78** (1980/81), 305–351.

[60] L. Ehrenpreis, Solutions of some problems of division I, *Am. J. Math.*, **76** (1954), 883–903.

[61] L. Ehrenpreis, Solutions of some problems of division II, *Am. J. Math.*, **77** (1955), 286–292.

[62] L. Ehrenpreis, Solutions of some problems of division III, *Am. J. Math.*, **78** (1956), 685–715.

[63] L. Ehrenpreis, Sheaves and Differential Equations, *Proc. Am. Math. Soc.*, **7** (1956), 1131–1138.

[64] L. Ehrenpreis, The Fundamental Principle for Linear Constant Coefficients Partial Differential Equations, in *Proc. Intern. Symp. Linear Spaces*, Jerusalem, 1960, 161–174.

[65] L. Ehrenpreis, Solutions of some problems of division IV, *Am. J. Math.*, **82** (1960), 522–588.

[66] L. Ehrenpreis, A New proof and an Extension of Hartogs' Theorem, *Bull. Am. Math. Soc.*, **67** (1961), 507–509.

[67] L. Ehrenpreis, Solutions of some problems of division V, *Am. J. Math.*, **84** (1962), 324–348.

[68] L. Ehrenpreis, *Fourier Analysis in Several Complex Variables*, Wiley Interscience, New York, 1970.

[69] L. Ehrenpreis, *Personal Communication*, 1989.

[70] D. Eisenbud, *Commutative Algebra with a View Toward Algebraic Geometry*, GTM Vol. 150, Springer-Verlag, New York, 1994.

[71] L. Euler, De integratione aequationum differentialium altiorum graduum, *Misc. Berol.*, **7**, 193–242.

[72] A. Fabiano, G. Gentili, D. C. Struppa, Sheaves of quaternionic hyperfunctions and microfunctions, *Compl. Var.*, **24** (1994), 161–184.

[73] E. Fischer, Über algebraische Modulsysteme und lineare homogene partielle Differentialgleichungen mit konstanten Koeffizienten, *J. Reine Angew. Math.*, **140** (1911), 48–81.

[74] O. Forster, *Lectures on Riemann Surfaces*, GTM Vol. 81, Springer-Verlag, New York, 1981.

[75] R. Fueter, Über Hartogs'schen Satz, *Comm. Math. Helv.*, **12** (1939), 75–80.

[76] R. Fueter, Über einen Hartogs'schen Satz in der Theorie der analytischen Funkionen von $n$ Komplexen Variables, *Comm. Math. Helv.*, **14** (1942), 394–400.

[77] R. Fueter, Die Funktionentheorie der Dirac'schen Differentialgleichungen, *Comm. Math. Helv.*, **16** (1944), 19–28.

[78] J. Gilbert, M. Murray, *Clifford Algebras and Dirac Operators in Harmonic Analysis*, Cambridge Univ. Press, Cambridge, UK, 1990.

[79] R. Godement, *Topologie Algébrique et Théorie des Faisceaux*, Hermann, Paris, 1958.

[80] H. Grauert, On Levi's problem and the embedding of real analytic manifolds, *Ann. Math.*, **68** (1958), 460–472.

[81] W. Greiner, J. Reinhardt, *Field Quantization*, Springer-Verlag, Berlin, Heidelberg, 1996.

[82] F. Gross, *Relativistic Quantum Mechanics and Field Theory*, J. Wiley, 1993.

[83] K. Guerlebeck, W. Sprößig, *Clifford Analysis and Elliptic Boundary Value Problems*, Math. Appl. Vol. 321, Kluwer, Acad. Publ., Dordrecht, 1995.

[84] S. Hansen, Localizable analytically uniform spaces and the fundamental principle, *Trans. Am. Math. Soc.*, **264** (1981), 235–250.

[85] F. Hartogs, Einige Folgerungen aus der Cauchyschen Integralformel bei Funktionen meherer Veränderlichen, *Sitzungber. Köngl. Bayer. Akad. Wissen*, **36** (1906), 223–241.

[86] D. Hestenes, G. Sobczyk, *Clifford Algebra to Geometric Calculus. A Unified Language for Mathematics and Physics*, Fundamental Theories of physics, D. Reidel Publ., Dordrecht, 1984.

[87] D. Hilbert *Hilbert's Invariant Theory Papers*, English transl. by M. Ackermann. Lie Groups: History, Frontiers and Applications. Vol. VIII, Math. Sci. Press, Boston, MA, 1978.

[88] L. Hörmander, Differentiability properties of solutions of systems of differential equations, *Ark. Mat.*, **3** (1958), 527–535.

[89] L. Hörmander, *Linear Partial Differential Operators*, Springer-Verlag, Berlin, 1963.

[90] L. Hörmander, *An Introduction to Complex Analysis in Several Variables*, Van Nostrand, Princeton, 1966.

[91] L. Hörmander, *The analysis of Linear Partial Differential Operators, I*, Springer-Verlag, Berlin, 1983.

[92] K. Imaeda, A new formulation of classical electrodynamic, *Nuovo Cimento*, **32B** (1976), 138–162.

[93] K. Imaeda, H. Tachibana, M. Imaeda, S. Ohta, Solutions of the octonion wave equation and the theory of functions of an octonion variable, *Nuovo Cimento*, **100B** (1987), 53–71.

[94] G.C. Joshi, *Octonionic unification and C, P and T symmetries*, *Lett. Nuovo Cimento*, **44** (1985), 449–454.

[95] G. Juvet, A. Schindlof, Sur les nombres hypercomplexes de Clifford et leurs applications à l'analyse vectorielle ordinaire, à l'électromagnetisme de Minkowski et à la théorie de Dirac, *Bull. Soc. Neuchat. Sci. Nat.*, **57** (1932), 127–147.

[96] A. Kaneko, Representation of hyperfunctions by measures and some of its applications, *J. Fac. Sci. Univ. Tokyo Sect. IA*, **19** (1972), 321–357.

[97] A. Kaneko, *Hyperfunctions and pseudo-differential equations*, Springer LNM, Vol. 287 (1973), 122–134.

[98] A. Kaneko, *Introduction to Hyperfunctions*, Mathematics and its Applications, Kluwer, 1988.

[99] M. Kashiwara, Algebraic Study of Systems of Partial Differential Equations, Mem. Soc. Math. de France, Vol. 63, 1995.

[100] M. Kashiwara, T. Kawai, T. Kimura, *Foundations of Algebraic Analysis*, Princeton, 1986.

[101] M. Kashiwara, P. Schapira, *Sheaves on Manifolds*, Springer-Verlag, 1990.

[102] G. Kato, D. C. Struppa, *Fundamentals of Algebraic Microlocal Analysis*, Marcel Dekker Inc., New York, 1999.

[103] T. Kawai, On the theory of Fourier hyperfunctions and its applications to partial differential equations with constant coefficients, *J. Fac. Sci. Univ. Tokyo*, **17** (1970), 467–517.

[104] T. Kawai, Removable singularities of solutions of systems of linear differential equations, *Bull. A.M.S.*, **81** (1975), 461–463.

[105] T. Kawai, D. C. Struppa, An existence theorem for holomorphic solutions of infinite order differential equations with constant coefficients, *Int. J. Math.*, **1** (1990), 63–82.

[106] H. Komatsu, Resolution by hyperfunctions of sheaves of solutions of differential equations with constant coefficients, *Math. Ann.*, **176** (1968), 77–86.

[107] H. Komatsu, *Introduction to Sato's hyperfunctions*, Sūrikaisekikenkyūsho Kōkyūroku **188**, Kyoto Univ., Kyoto 1973.

[108] H. Komatsu, *Relative cohomology of sheaves of solutions of differential equations*, Springer LNM, Vol. 287 (1973), 192–261.

[109] G. Köthe, *Die Randverteilungen analytischer Funktionen*, Math. Zeit., **57** (1952), 13–33.

[110] S. Krantz, *Theory of Several Complex Variables*, Wadsworth and Brooks/ Cole Advanced Books and Software, Belmont, California, 1992.

[111] V.V. Kravchenko, *Applied Quaternionic Analysis*, Research and Exposition in Mathematics, Vol. 28, Heldermann Verlag, 2003

[112] V.V. Kravchenko, M. Shapiro, *Integral Representation for Spatial Models of Mathematical Physics*, Pitman Res. Notes in Math. Series, Vol. 351 (1996).

[113] M. Kreuzer, L. Robbiano, *Computational Commutative Algebra I*, Springer-Verlag, Berlin, Heidelberg, 2000.

[114] C. Lanczos, *Collected published papers with commentaries*, Vol. 1–4, North Carolina State University, 1998.

[115] R. Larsen, *Functional Analysis: an Introduction*, Marcel Dekker, Inc., New York, 1973.

[116] F.W. Lawvere, S.H. Schanuel, *Conceptual Mathematics. A First Introduction to Categories*, reprint of 1991, Cambridge Univ. Press, Cambridge, 1997.

[117] B. Y. Levin, *Distribution of Zeroes of Entire Functions*, Providence, R.I. AMS, 1964.

[118] P. Lounesto, *Clifford Algebras and Spinors*, London, Math. Soc. Lecture Notes Series, Vol. 286, Cambridge Univ. Press, Cambridge 2001.

[119] J. Lützen, *The Prehistory of the Theory of Distributions*, Studies in the History of Mathematics and Physical Sciences, Vol. 7, Springer-Verlag, New York, Berlin, 1982.

[120] S. Mac Lane, *Categories for the Working Mathematician*, Second Edition, GTM, Springer-Verlag, New York, 1998.

[121] F. Macaulay, Some properties of Enumeration in the Theory of Modular Systems, *Adv. Math.*, **46** (1927), 531–555.

[122] B. Malgrange, Existence et approximation des solutions des équations aux dérivées partielles et des équations de convolution, *Ann. Inst. Fourier (Grenoble)*, **6** (1956), 271-355.

[123] B. Malgrange, *Sur les systèmes differéntiels à coefficients constants*, Séminaire Leray, Collège de France, Exposés 8 et 8a, 1961–62.

[124] B. Malgrange, *Systèmes differéntiels à coefficients constants*, Séminaire Bourbaki, Vol. 246, 1962–63.

[125] H. Malonek, *The concept of hypercomplex differentiability and related differential forms*, Pitman Res. Notes, Vol. 256, Longman, 1991, 193–202.

[126] F. Mandl, G. Shaw, *Quantum Field Theory*, J. Wiley & Sons, New York, Toronto, 1984.

[127] A. Martineau, *Les hyperfonctions de M. Sato*, Sém. Bourbaki, **6** Exp. 214, 1960, 127–139, Soc. Math. France, Paris 1995, 1–13.

[128] A. Martineau, Sur les fonctionnelles analytiques et la transformation de Fourier–Borel, *J. An. Math.*, **11** (1963), 1–164.

[129] A. Martineau, *Distributions et valeurs au bord des fonctions holomorphes*, Proc. Intern. Sum. Course on the Theory of Distributions, Lisbon, 1964, 195–326.

[130] E. Martinelli, Alcuni teoremi integrali per le funzioni analitiche di più variabili complesse, *Mem. R. Accad. Italia*, **9** (1938), 269–283.

[131] E. Martinelli, Sopra una dimostrazione di R. Fueter per un teorema di Hartogs, *Com. Math. Helv.*, **15** (1942-43), 340–349.

[132] A. Mercier, *Expression des èquations de l'électromagnetisme au moyen des nombres de Clifford*, thesis, Univ. de Genève, 1935.

[133] A. Meril, D. C. Struppa, Equivalence of Cauchy problems for entire and exponential type functions, *Bull. London Math. Soc.*, **17** (1985), 469–473.

[134] A. Meril, D. C. Struppa, Syzygies of modules and applications to propagation of regularity phenomena, *Publ. Mat.*, **34** (1990), 349–377.

[135] J. W. Milnor, *Singular Points of Complex Hypersurfaces*, Princeton University Press, 1968.

[136] G. Moisil, N. Theodorescu, Functions holomorphes dans l'espace, *Mathematica (Cluj)*, **5** (1931), 142–159.

[137] M. Morimoto, *An Introduction to Sato's Hyperfunctions*, Trans. Math. Monographs, Vol. 129 A.M.S., Providence, Rhode Island, 1993.

[138] D. Napoletani, D. C. Struppa, *On a large class of supports for quaternionic hyperfunctions in one variable*, Pitman Res. Notes Math. Series, Vol. 394 (1999), 170–175.

[139] D. J. Newman, H. S. Shapiro, Certain Hilbert spaces of entire functions, *Bull. Amer. Math. Soc.*, **72** (1966), 971–977.

[140] D. J. Newman, H. S. Shapiro, *Fischer spaces of entire functions*, Proc. Symp. Pure Math., Vol. 11 AMS 1968, 360–369.

[141] T. Oshima, A proof of Ehrenpreis' fundamental theorem in hyperfunctions, *Proc. Japan Acad.*, **50** (1974), 16–18.

[142] V. P. Palamodov, *Linear Differential Operators with Constant Coefficients*, Springer-Verlag, Berlin, 1970.

[143] V. P. Palamodov, Holomorphic synthesis of monogenic functions of several quaternionic variables, *J. Anal. Math.*, **78** (1999), 177–204.

[144] R. Penney, Octonions and isospin, *Nuovo Cimento*, **3** (1971), 95–113.

[145] D. Pertici, Funzioni regolari di piú variabili quaternioniche, *Ann. Mat. Pura e Appl. Serie IV*, CLI, (1988) 39–65.

[146] H. Poincaré, *Leçons de Mécanique Céleste*, II, Paris, 1905.

[147] H. Poincaré, Les fonctions analytiques de deux variables et la représentation conforme, *Rend. Circ. Mat. Palermo*, **23** (1907), 185–220.

[148] I. R. Porteous, *Clifford Algebras and the Classical Groups*, Cambridge Studies in Advanced Mathematics, Vol. 50, Cambridge, 1995.

[149] A. Proca, Sur la théorie ondulatoire des électrons positifs et négatifs, *Le J. de Phys. et le Rad.*, **7** (1936), 347–353.

[150] R. M. Range, Extension phenomena in multidimensional complex analysis: correction of the historical record, *Math. Intel.*, **24** (2002), 4–12.

[151] M. Riesz, *Clifford numbers and spinors*, in E. F. Bolinder, P. Lounesto eds., Kluwer, Dordrecht, The Netherlands, 1993.

[152] R. Rocha-Chavez, M. Shapiro, F. Sommen, *Integral Theorems for Functions and Differential Forms in* $\mathbb{C}^m$, Research Notes in Math., Vol. 428, Chapman & Hall/CRC, 2001.

[153] W. Rudin, *Functional Analysis*, New York, 1973.

[154] J. Ryan, Duality in complex Clifford analysis, *J. Funct. Anal.*, **61** (1985), 117–135.

[155] J. Ryan, Applications of complex Clifford analysis to the study of solutions to generalized Dirac and Klein–Gordon equations with holomorphic potentials, *J. Diff. Eq.*, **67** (1987), 295–329.

[156] J. Ryan, Hypercomplex algebras, hypercomplex analysis and conformal invariance, *Comp. Math.*, **61** (1987), 61–80.

[157] I. Sabadini, *Verso una teoria delle iperfunzioni quaternioniche*, PhD dissertation, Milan, 1995.

[158] I. Sabadini, M. Shapiro, D. C. Struppa, Algebraic analysis of the Moisil–Theodorescu system, *Compl. Var.*, **40** (2000), 333–357.

[159] I. Sabadini, F. Sommen, Special first order systems and their resolutions, *Z. Anal. Anw.*, **21** (2002), 27–55.

[160] I. Sabadini, F. Sommen, *Combinatorics and Clifford analysis*, in Clifford Algebras and its Applications, NATO Sci. Series II, Kluwer, Vol. 25 (2001), 267–282.

[161] I. Sabadini, F. Sommen, Hermitian Clifford analysis and resolutions, *Math. Met. Appl. Sci.*, **25** (2002), 1395–1413.

[162] I. Sabadini, F. Sommen, *Clifford analysis on the space of vectors, bivectors and* $\ell$*-vectors*, Trends in Mathematics, Advances in Analysis and Geometry, Birkhäuser, 2004, 161–185.

[163] I. Sabadini, F. Sommen, D. C. Struppa, The Dirac complex on abstract vector variables: megaforms, *Exp. Math.*, **12** (2003), 351–364.

[164] I. Sabadini, F. Sommen, D. C. Struppa, P. Van Lancker, Complexes of Dirac operators in Clifford algebras, *Math. Zeit.*, **239** (2002), 215–240.

[165] I. Sabadini, D. C. Struppa, Topologies on quaternionic hyperfunctions and duality theorems, *Compl. Var.*, **30** (1996), 19–34.

[166] I. Sabadini, D. C. Struppa, Some open problems on the analysis of the Cauchy–Fueter system in several variables, Exact WKB analysis and Fourier analysis in the complex domain (Japanese) (Kyoto, 1997). *Sūrikaisekikenkyūsho Kōkyūroku*, **1001** (1997), 1–21.

[167] M. Saito, B. Sturmfels, N. Takayama, *Gröbner Deformation of Hypergeometric Differential Equations*, Springer-Verlag, 2000.

[168] M. Sato, Theory of hyperfunctions, *Sūgaku*, **10** (1958), 1–27.

[169] M. Sato, T. Kawai, M. Kashiwara, *Microfunctions and Pseudo-Differential Equations*, Springer LNM, Vol. 287 (1973), 265–529.

[170] M. Sato, Theory of hyperfunctions, *J. Fac. Sci. Univ. Tokyo, Sect. I.*, **8** (1958), 139–193 and 387–436.

[171] M. Sce, P. Dentoni, Funzioni regolari nell'algebra di Cayley, *Rend. Sem. Mat. Univ. Padova*, **50** (1973), 251–267.

[172] R. D. Schafer, An introduction to nonassociative algebras, *Trans. Am. Math. Soc.*, **72**, 1952.

[173] B. Schuler, Zur Theorie der regulären Funktionen einer Quaternionen-variablen, *Comm. Math. Helv.*, **10** (1937), 327–342.

[174] L. Schwartz, Théorie générale des fonctions de moyenne périodiques, *Ann. Math., II Ser.*, **48** (1947), 857–929.

[175] L. Schwartz, *Théorie des Distributions I, II*, Hermann, Paris, 1950–51.

[176] B. Segre, Sull'estensione delle formule integrali di Cauchy e sui residui degli integrali $n$-upli, nella teoria delle funzioni di $n$ variabili complesse, *Atti I Convegno U.M.I.*, 1937.

[177] J. P. Serre, Un théorème de dualité, *Comm. Math. Helv.*, **29** (1955), 9–26.

[178] H. S. Shapiro, An algebraic theorem of E. Fischer and the holomorphic Goursat problem, *Bull. London Math. Soc.*, **21** (1989), 513–537.

[179] M. Shapiro, *Some remarks on generalizations of the one-dimensional complex analysis: hypercomplex approach*, in Functional Analytic Methods in Complex Analysis and Applications to Partial Differential Equations, World Scientific, 1995, pp. 379–401.

[180] M. Shapiro, Structure of quaternionic modules and some properties of involutive operators, *J. Nat. Geom.*, **1** (1992), 9–37.

[181] M. Shapiro, *On the conjugate harmonic functions of M. Riesz– E. Stein–G. Weiss*, in Topics in Complex Analysis, Differential Geometry and Math. Physics, S. Dimiev and K. Sekigawa (eds.), World Scientific, 1995, pp. 8–32.

[182] M. Shapiro, N. Vasilevski, Quaternionic $\psi$–hyperholomorphic functions, singular integral operators and boundary value problems, I and II, *Compl. Var.*, **27**, (1995), 17–46 and 69–76.

[183] P. Somberg, *Quaternionic complexes in Clifford analysis,* NATO Sci. Series II, Kluwer, Vol. 25 (2001), 293–301.

[184] P. Somberg, V. Souček, *Personal Communication,* (2000).

[185] F. Sommen, Hyperfunctions with values in a Clifford algebra, *Simon Stevin,* **57** (1983), 225–254.

[186] F. Sommen, Monogenic differential calculus, *Trans. Am. Math. Soc.*, **326** (1991), 613–632.

[187] F. Sommen, Clifford tensor calculus, Proc. XXIIth Conference on Diff. Geom. Meth. Theor. Phys., Ixtapa 1993, *Adv. Appl. Cliff. Alg.*, **4** (1994), 423–436.

[188] F. Sommen, An algebra of abstract vector variables, *Port. Math.*, **54** (1997), 287–310.

[189] F. Sommen, Clifford analysis in two and several vector variables, *Appl. Anal.*, **73** (1999), 225–253.

[190] F. Sommen, *An extension of Clifford analysis towards super–symmetry,* Clifford Algebras and their Applications in Mathematical Physics, (Ixtapa 1999), Progr. Phys., Birkhäuser, **19** (2000), 199-224.

[191] F. Sommen, Clifford analysis on super–space, Proc. Cetraro Conference 1998, *Adv. Appl. Cliff. Alg.*, **11** (2001), 291–304.

[192] F. Sommen, *Clifford analysis on the level of abstract vector variables,* in Clifford Algebras and its Applications, NATO Sci. Series II, Kluwer, Vol. 25 (2001), 303–322.

[193] F. Sommen, G. Bernardes, Multivariable monogenic functions of higher spin, *J. Nat. Geom.*, **18** (2000), 101–114.

[194] F. Sommen, B. Jancewicz, Explicit solutions of the inhomogeneous Dirac equation, *J. Anal. Math.*, **71** (1997), 59–74.

[195] F. Sommen, V. Souček, Monogenic differential forms, *Compl. Var.*, **19** (1992), 81–90.

[196] F. Sommen, N. Van Acker, Invariant differential operators on polynomial-valued functions, Clifford Algebras and their Applications to Mathematical Physics (Deinze 1993), *Fund. Theor. Phys.*, Kluwer, Vol. 55 (1993), 203–212.

[197] L. Sorgsepp, J. Lohmus, About nonassociativity in Physics and Cayley–Graves' octonions, *Hadr. J.*, **2** (1979), 1388–1459.

[198] V. Souček, Complex quaternionic analysis applied to spin $\frac{1}{2}$ massless fields, *Compl. Var.*, **1** (1983), 327–346.

[199] V. Souček, *Clifford analysis as a study of invariant operators*, NATO Sci. Series II, Kluwer, Vol. 25 (2001), 323–339.

[200] R. Stanley, *Combinatorics and Commutative Algebra*, Second Edition, Progress in Math. Vol. 41, Birkhäuser, 1996.

[201] E. M. Stein, G. Weiss, Generalization of the Cauchy–Riemann equations and representation of the rotation group, *Amer. J. Math.*, **90** (1968), 163–196.

[202] D.C. Struppa, *The Fundamental Principle for Systems of Convolution Equations*, Mem. AMS Vol. 273, 1983.

[203] D. C. Struppa, The first eighty years of Hartogs' theorem, *Sem. Geom. Dip. Mat. Bologna*, (1987), 127–209.

[204] D. C. Struppa, *An extension of Fantappiè's theory of analytic functionals*, in Geometry and Complex Variables, S. Coen ed., Marcel Dekker Inc., New York, Basel, Hong Kong, 1991, 329–356.

[205] D. C. Struppa, *Gröbner bases in partial differential equations*, Gröbner bases and Applications (Linz, 1998), London Math. Soc. Lecture Note Ser., Cambridge Univ. Press, **251** (1998), 235–245.

[206] A. Sudbery, Quaternionic analysis, *Math. Proc. Camb. Phil. Soc.*, **85** (1979), 199–225.

[207] P. Van Lancker, F. Sommen, D. Constales, Models for irreducible representations of Spin($m$), Proc. Conf. Cetraro 1998, *Adv. Appl. Cliff. Alg.*, **11** (2001), 271–289.

[208] V. S. Vladimirov, *Methods of the Theory of Functions of Several Complex Variables*, MIT Press, 1966.

[209] V. S. Vladimirov, *Le Distribuzioni nella Fisica Matematica*, Edizioni Mir, 1981.

[210] K. Yosida, *Functional Analysis*, Springer-Verlag, Berlin, Heidelberg, New York, 1980.

[211] O. Zariski, P. Samuel, *Commutative Algebra*, Vol. I, II, Graduate Texts in Math. Vols. 28, 29, Springer-Verlag, 1975.

# Index